Toward a Philosophy of Error in Science

DOUGLAS ALLCHIN

OXFORD
UNIVERSITY PRESS

Oxford University Press is a department of the University of Oxford.
It furthers the University's objective of excellence in research, scholarship,
and education by publishing worldwide. Oxford is a registered trade mark of
Oxford University Press in the UK and in certain other countries.

Published in the United States of America by Oxford University Press
198 Madison Avenue, New York, NY 10016, United States of America.

CIP data is on file at the Library of Congress.

ISBN 9780197827673

DOI: 10.1093/9780197827703.001.0001

Printed by Marquis Book Printing, Canada

The manufacturer's authorized representative in the EU for product safety is
Oxford University Press España S.A. of Parque Empresarial San Fernando de Henares,
Avenida de Castilla, 2 – 28830 Madrid (www.oup.es/en or product.safety@oup.com).
OUP España S.A. also acts as importer into Spain of products made by the manufacturer.

Chapter Summaries

Preface

Given the philosophical heritage of characterizing science in terms of its reliable knowledge, focusing on its errors may seem counterproductive, or even perverse. But scientists regularly discover errors. And learn from them. While many philosophers (as well as historians, sociologists, and scientists themselves) have disparately commented on error, we need a comprehensive and systematic approach—a philosophy of error—to organize our understanding and guide scientists in practice.

1. Error itself

Errors permeate the history of science. What can these cases tell us about science as a process? First, what precisely do we mean by error? Defining error proves to be surprisingly difficult, because the identification of an error straddles different times, contexts, and bodies of evidence. I focus on the change in justification—from a claim once considered justified to its later status as unjustified. This cryptic shift will guide our analysis, and invite us to think more fully about what is meant by scientific justification. An important dimension of interpreting error is not the content of the claim itself, but the epistemic posture toward or commitment to the justification of the claim in question. Error is definite, even if a negative claim perhaps. Error thus differs from uncertainty or vague disclaimers of "tentativeness."

2. Observational errors

Experimentalists frequently refer to sources of error, the factors ranging from dirty glassware and contaminated samples to confounding variables and mistaken assumptions that can produce misleading results or faulty conclusions. Here, I generalize this concept to the whole process of science, applied at three levels: observational, conceptual, and social (discoursive).

At each level, a survey of historical cases yields general error types—the many potential pitfalls that might ultimately guide more reliable scientific practice. They support assembling an error inventory as a reference guide. Observational errors arise in laboratory experiments, field studies, or other forms of measurement or collection of data. They include material errors, instrument errors, human perceptual deficits, observer bias, observer effects (including artifacts), and various forms of misframing—sampling bias, small sample size, incomplete sampling, proxy variables and heuristic gaps, and confounders.

3. Conceptual errors

Conceptual errors arise in reasoning about observations at many levels, from processing data and analyzing them statistically to conventional logic (and its fallacies) and interpretive practices (some shaped by heuristics and inherent cognitive biases). They include overgeneralizations, faulty assumptions, theory-laden judgment and confirmation bias, heuristic gaps, cognitive lapses, unaddressed alternatives, cryptic alternatives, and various forms of cultural bias—such as those based on religion, gender, race, class, or politics.

4. Social-level errors

Errors also occur at the social level, in the customary discourse among scientists. That is, there are institutional, even if informal, mechanisms of "quality control" in the scientific community. They are designed to regulate accurate and full reporting, expertise, theoretical bias, and deception. Error types include communal confirmation bias, communal cultural bias, ineffective peer review/publication, credibility bias and fraud, and conflict of interest. Consensus arising from reciprocal criticism is thus an important benchmark.

5. From incongruence to error

How, indeed, are errors remedied? Errors are errors in part because the flaws in their justification are hidden. Exposing them, and ascertaining their status as errors, requires *work*—epistemic work, not the mere passage of time. The first step is an awareness that something is awry. Researchers may encounter incongruences in the following cases: (1) in their conflicting observations (discordances), (2) in the match between theory and observations (anomalies), or (3) in alternative theoretical interpretations (ambiguities, or disagreement). The next step is to isolate the error by tracing it to any of the error types. Prospective errors are confirmed through testing, especially through controlled experiments. The growth of knowledge poses an epistemic puzzle: how can new information upend a verdict once deemed to be justified by more limited evidence? How are learning and *unlearning* related?

6. Deeper evidence

But how does the relevant additional evidence arise? Deeper evidence may emerge through (at least) modified replications, increased sample size, more diverse sampling, wider scope, the increased resolving power of instruments or methods, new technology, filling of heuristic gaps, new conceptual perspectives, or happenstance. These strategies complement the inventory of error types and underscore the pervasive work of troubleshooting and error correction in science.

7. The conundrum of bias

Can errors ever be eliminated from science? Individual perspectives (or "biases") may generate blind spots and errors. But they also seem to underlie important insights and discoveries. Bias can be fruitful, as well as misleading. The cost of innovation is the risk of failure. Science can harness diverse standpoints through a social system of checks and balances. Reciprocal criticism helps expose adverse bias and filter out various forms of individual error, while allowing new and marginal ideas opportunity.

8. Managing error

A deeper awareness of error in science may underscore that the practices of science are focused as much on regulating sources of error as on "seeking the truth." We benefit from negative knowledge. Indeed, most of the familiar abstract methodological norms of science have concrete historical roots, based on particular encounters with certain error types. Errors may seem to threaten the notion of scientific progress, but by focusing on the escalating standards of proof, we may acknowledge growth in the quality of knowledge, even if errors lead us to abandon some concepts or theories. We may articulate field-specific error repertoires and identify corresponding error signatures which may assist in mitigating future errors and in diagnosing problems in research. The familiar concept of checklists may be adapted in envisioning a series of checkpoints, occasions where reviewing sources of error seems especially appropriate. A deeper awareness of error types may also lead us to a view that confirming evidence alone is insufficient. To deepen reliability, we need error probes to actively search for possible loose ends and qualifications. "Nothing's concluded until error is excluded." All these projects may foster further study in the Philosophy of Error and in developing Error Analytics, a set of philosophically informed practices for managing errors in everyday scientific practice.

Contents

List of Figures

Preface

Conventionally, philosophers have championed science by characterizing its distinctive methodologies and disciplined protocols for establishing reliable knowledge. Yet ironically, the history of science is littered with errors. For many, perhaps, these are simply embarrassing failures, to ruefully acknowledge and cast into the shadows. In this book, by contrast, I *celebrate* these errors. Indeed, I show how, counter-intuitively perhaps, they are integral to the process—and progress—of science. The cost of new knowledge is the risk of error.

Further, I detail how scientists respond to and identify these errors. From an extended analysis, I also suggest how, looking ahead, from a more pragmatic perspective, scientists may fruitfully manage the errors that they inevitably encounter.

While many philosophers (as well as historians and sociologists—and scientists themselves) have disparately commented on error, we need a comprehensive and systematic approach—a philosophy of error in science—to organize our understanding and to guide scientists in practice.

This book, therefore, endeavors to engage in a thoughtful reflection about error in science.

Indeed, a concerted study of error in science yields some unexpected conclusions (see Chapter 8). For example, "negative" knowledge has a "positive" role. Understanding particular errors contributes to deepening the precision and accuracy of knowledge—even when some earlier concepts are abandoned as "wrong." Accordingly, we might reconceptualize knowledge—not as "true" *versus* "false," but embracing them both as forms of knowledge. That is, we should contrast true-or-false (the known) to uncertainty (the unknown).

Errors lead to improving methodologies at multiple levels. Thus, standards of proof escalate. The quality of knowledge improves. Looking ahead, we can make scientific practice even more effective through more systematic attention to error. We can nurture a habit of deepening existing knowledge by deliberately probing for possible unresolved sources of errors.

Accordingly, this book will be valuable to anyone concerned with how science works (and thus how, on occasions, it doesn't work). While of primary concern for philosophers of science, it will also be relevant to many historians, sociologists, ethnographers, and others in Science Studies. It will be of interest to many scientists—those who reflect on the nature of their practice. Thus, it may inform science graduate students and other aspiring scientists, science teachers, and science administrators. And science journalists. And even judges and legislators. Indeed, anyone who thinks carefully about the epistemic foundations of science and how science informs public policy or personal decision making. It offers an intimate glimpse into the nature of science, essential for the task of assessing the reliability of scientific claims.

The book is a timely contribution addressing several significant issues. First, it provides perspective on the "reproducibility crisis" and the documented failures to replicate many apparently landmark experiments, leading to doomsday warnings that (according to one prominent critic) "most scientific findings are false." The apparently embarrassing "failures" are a prominent topic in *Nature* or *Science* magazines and in informal conversation (and gossip) among practicing scientists, as well as philosophers. Equally important are concerns about the scandalous rise of misconduct and "questionable research practices"—cold fusion, organic semiconductors, room temperature superconductivity, and other such outlandish claims. A fruitful response to the current moment of perceived crisis, in my view, is to adopt a more holistic approach: to examine all the types of error in science, how they occur, and how they are remedied.

Another enduring concern is the status of the "pessimistic induction." Philosophers have mused on the history of science for decades, and observing that past theories seem to eventually fall by the wayside, some conclude that current theories must inevitably be wrong, as well. A dire prospect, if true. I take a more pragmatic view and articulate, in particular, the importance of negative knowledge and its role in the growth of knowledge. Error can embody progress.

I also address the persistent related ambiguity between viewing science as "tentative" and yet reliable. Here, I provide a sequel to and reconciliation of positivist traditions and radical Kuhnian perspectives. We may resolve this tension by articulating both how (and when) scientists have indeed erred and how (and when) they have found and remedied their errors. An empirically based science *of* science, perhaps?

Finally, many philosophers and science advocates of many stripes admit that science (at the margin) is not perfect and that individual scientists, at least, do sometimes make errors. But they tend to add hastily that science is *self-correcting*. No cause for alarm. No need to inquire further. But just *how* science is self-correcting is less clear. Two mechanisms are typically cited: peer review and replication. Yet, when considered historically, these processes prove quite weak. We need to reappraise and reflect on the entrenched rhetoric of a "self-correcting" mythos. We need, instead, to describe how errors are, ultimately, found and fixed, allowing us to understand more clearly the various contingencies that shape the trustworthiness of science.

The book is shaped, in part, by my own background participating in several research projects—on tree gap succession in the forests by the Chesapeake Bay, on sexual selection of flowers in the Rocky Mountains, on species turnover in a topical rainforest in Panama, and on modeling the dynamics of "information-center" foraging among honeybees and other animals. It is further informed by my work as a historian of science, a philosopher of science, and science teacher. And by many, many informal conversations with working scientists who encounter error in their everyday practice.

For many philosophers, the central aim of science is to produce reliable knowledge. Given such a perspective, how could error be anything but an annoyance and a distraction along the way? I hope to show, by contrast, how error is integral to learning and to increasing the reliability of knowledge, and thus to the very aims of science. Surprisingly, perhaps, to err is science.

1

Error Itself

The philosophical conundrum of error • error naturalized • what counts as error • epistemic posture • error and uncertainty • taking stock • a guide to the book

The philosophical conundrum of error

Traditionally, science is regarded as the epitome of fact. Philosophers, for their part, have endeavored to articulate the reasons for its special status as a way of knowing, whether it is characterized in terms of verifiable truth,[1] reliable knowledge,[2] theoretical explanation,[3] progress,[4] problem-solving,[5] technological performance,[6] or other. The scientific enterprise is idealized.[7] It is demarcated.[8] Its practices are probed for their unique epistemic qualities.[9]

Yet sometimes scientists get entangled in the "wrong" answers, imagining them to be fully justified. The history of science is filled with errors. There are abundant familiar examples—Ptolemaic astronomy, diluvial geology, humoral medicine, chemical affinities, species immutability, vitalistic physiology, racist anthropology, sexist primatology, and eugenics, to name just a few (see Figure 1.1). Thus, if we want to understand science and its trustworthiness, then we surely need to explore why and how science errs.

For many philosophers, perhaps, the historical errors are simply embarrassing failures, to ruefully acknowledge and then cast into the shadows. Yet few would deny that science is a human enterprise. And to err is human (so they say). So no one should be surprised that scientists might err. To err is science?

Paradoxically, perhaps, errors provoke a unique response in science. When errors emerge, investigators typically attend to them. They delve into the misleading or problematic results and the reasons(s) for them. As a result, they ultimately develop more reliable knowledge. Namely, finding errors and remedying them is *also* science. However, to date, these remedial efforts have not received the philosophical attention they deserve. This

Toward a Philosophy of Error in Science. Douglas Allchin, Oxford University Press. © Douglas Allchin (2026).
DOI: 10.1093/9780197827703.003.0001

acidifying principle
chemical affinities
animal magnetism
ausonium
black bile
caloric
cold fusion
coronium
draeptomania
electrical fluid
fifth force
founder cells in slime molds
hesperium
high-energy intermediates of oxidative phosphorylation
hysteria
J-phenomenon
magneto-optical effect
mesosomes
metagons
miasmas
moons of Venus
N-rays
nebulium
oops-Leon particle
pangenesis
phlogiston
polar wandering
polywater
vitamin P
planet Vulcan

Figure 1.1 A sampling of nonexistent entities and phenomena once part of science. How did these once seem justified, and how did deeper evidence later change that assessment?

dimension of scientific work has yet to be fully articulated and characterized epistemically. We are in need of an informative Philosophy of Error in science.[10]

•

Science is work. Challenging work. Stereotypically, scientists follow a widely touted "Scientific Method," which supposedly leads unerringly to a triumphant discovery. However, as often as not, the results are unexpected or ambiguous. No easy "aha!" moment. What happens then?

Different experiments may yield apparently contradictory results. Measurements may not align with theoretically predicted values. Evidence may be ambiguous and susceptible to multiple interpretations. Various incongruences arise. Discordances. Anomalies. Incompatible explanations. Something is "wrong." (Something.) There is an error lurking somewhere. This book is about those occasions.

For me, this is when the nitty-gritty work of science *really* begins. How can one reconcile the conflicting findings? What further observations will help resolve the uncertainties? It requires creatively designed investigation. Collecting more relevant data. Imaginatively reconfiguring all the evidence. Back to the field, back to the lab. To a surprising degree, perhaps, scientists may spend more time troubleshooting problematic results than simply testing the hypotheses they originally set out to confirm or disconfirm. But what comes from that meandering trajectory is genuine discovery—learning something no one suspected before. Yes, scientists *learn* from error.

In the introduction to the English edition the *Report of the Royal Commission of 1784 on Mesmer's System of Animal Magnetism*, the anonymous translator (likely William Godwin) commented on the susceptibility to the practices of Anton Mesmer:

> Perhaps the history of the errors of mankind, all things considered, is more valuable and interesting than that of their discoveries.[11]

Here, he posed a provocative challenge, of sorts, to philosophers of science and others concerned with how knowledge develops. Do we learn more about the process of science from its achievements or its mistakes? Which are more informative in interpreting how science works: discoveries or errors? At the very least, he considered the errors more *interesting*!

Appropriately viewed, errors—far from being blemishes—are tools for learning. Accordingly, in this book, I *celebrate* these errors. Indeed, I show how they are integral to the process—and progress—of science. The cost of new knowledge is the risk of error.

Error naturalized

The first step on this adventure is to *naturalize* error. That is, we need to (1) screen off idealizations about the nature of science, and (2) decouple methodological norms from the scientific conclusions themselves, imagining that the former provide sufficient justification for the later.

For most philosophers, science is idealized. It consists of the construction of knowledge, not its errors or mistakes along the way. Error is viewed as a costly distraction, wasting time and resources—best quelled.

Enlightenment thinkers championed the virtues of knowledge. Nowhere is that perspective better expressed than in the Mesmer Commission Report. The Commission, led by Benjamin Franklin, had been charged with investigating the legitimacy of Mesmerism, which led ladies to swoon and others to attest to healing effects of electric shocks. In a series of imaginative investigations, the Commission documented (from our perspective, at least) the psychological power of suggestion. But they did not dwell on that explanation. Rather, they characterized the subjects' beliefs as just wrong. Godwin's introduction continued:

> Truth is uniform and narrow; it constantly exists, and does not seem to require so much an active energy, as a passive aptitude of soul in order to encounter it. But error is endlessly diversified; it has no reality, but is the

> pure and simple creation of the mind that invents it. In this field the soul has room enough to expand herself, to display all her boundless faculties, and all her beautiful and interesting extravagancies and absurdities.

That is, one need not explain the emergence of knowledge. That happened naturally. Or automatically, without effort. Error was an aberration of nature.[12]

That view of error and the "true" path of science was echoed in John Herschel's virulent 1830 criticism of the history of phlogiston, what the 1771 *Encyclopedia Britannica* described as the principle of fire:

> The phlogistic doctrine impeded the progress of science, as far as science of experiment can be impeded by a false theory, by perplexing its cultivators with the appearance of contradictions, ... and by involving the subject in a mist of visionary and hypothetical causes in place of true and acting principles.[13]

Herschel's malignant assessment has cascaded through history, and phlogiston remains one of the canonical examples in the philosophy of science of substances-that-never-were, mistaken natural kinds, and susceptibility to error and self-delusion. For Herschel and others, errors are *transgressions*. In these perspectives, errors—like phlogiston—reflect errancy, or science gone astray and, equally, credulous scientists who succumb to some kind of cognitive weakness in ever accepting an erroneous claim, and hence are blameworthy. For them, errors can never be reasonable. They implicitly indicate lapses of science, where science is construed only as "good" science. Accordingly, errors are wholly outside science proper. A true scientist never errs. If they do err, then they were never a true scientist. If one can render error as "external" to science proper in this way, science—and scientists—conveniently retains a pristine image.

Of course, these commentators do not engage cases of error among the most respected scientists. Yes, famous scientists and even Nobel Prize winners can err. The effort to portray science as error-free (and thus, in their view, its authority as unimpeachable) is misguided. As the many examples in this book will show, this somewhat romanticized view is artificially blinkered and incomplete. Error is far too pervasive in science to frame it as exceptional or extraordinary. One must come to terms with error as an integral part of ordinary scientific practice.[14]

How do scientists themselves interpret the occurrence of error? Sociologists Nigel Gilbert and Michael Mulkay interviewed a group of scientists in the midst of a controversy in the mid-1980s, asking them why their opponents were wrong. The responses were a good glimpse into the informal gossip among scientists. Other scientists' errors reportedly stemmed from:

succumbing to charisma
"strong personality"
a rhetorical "aura of fact"
"intellectual inertia"
confrontation with "unorthodox" views
being "tenacious"
being "dogmatic"
"pig-headedness"
"fear of losing grants"
dislike
"prejudice"
"subjective bias"
personal rivalry
"emotional involvement"
"threats to status"
"naiveté"
"thinking in a woolly fashion"
"sheer stupidity"
an "ostrich approach" (of willfully disregarding the facts)
a whole generation simply "unequal to the task."

"Thinking in a woolly fashion": that's my favorite. Errors seem uniformly attributed to "unscientific" factors that interfere with sound reasoning. At least among those who disagree with you. Ironically, perhaps, few described their own errors in these ways—or even admitted errors. So, there is a tendency to equate error with irrationality, and irrationality with the abnormal interference of psychological and sociological factors.[15]

Others certainly echo these sentiments. Walter Gratzer in his book *The Undergrowth of Science* ascribes error to self-delusion, or credulity. Or *feckless* credulity. Or *unfathomable* credulity. For him, error is fueled by a fragile ego, ambition, desire for fame and advancement, envy, jealousy, and political ideology. Similarly, Robert Youngson, in his survey of *Scientific Blunders*, cites the roots of error in carelessness, plain stubborn wrong-headedness, arrogance, willful and culpable ignorance, moral frailty, preoccupation with rewards (fame, status, and wealth), and political, religious, or other ideologies (at least). In characterizing *The Scientific Attitude*, Lee McIntyre castigates those with self-delusion or willful ignorance (not just ignorance, no!—*willful* ignorance), along with the tiresome, gullible, and self-righteous charlatans. One might well imagine that honest error was impossible: an inherent oxymoron.[16]

Unfortunately, these all-too-common assessments focus on accountability, not on understanding error. Their primary purpose is to assign blame. Responsibility. Fault. *Culpability*. In many ways, they are no more than declarations that an error has occurred, coupled with an adverse judgment against an individual associated with the claim. Political in spirit, not epistemic. And not fruitful for addressing the problem.

The vacuity of these epithets about error may be highlighted by imagining that one could magically remedy them all. Suppose we ensured universal security about getting grants, softened personalities, quelled jealousies, rivalries, ambitions, and personal antipathies, instilled universal open-mindedness—and graciously pulled people's heads out of the sand and de-woolified their minds. Would we thereby solve the various errors at stake? Likely not. One does not find these factors in the arguments of professional discourse because they do not, ultimately, address the evidence or the arguments that have been advanced. They are not relevant to assessing or reconfiguring the justifications toward a more reliable claim. One must focus instead on the epistemic dimensions. While the psychological and social factors may provide important *motivational* context and shape the tenor of the discourse, they are not the sources of error themselves. That is, epistemic analysis shifts the focus about error in science from the *behavior* of the scientist to the structure of the *justification*. If our ultimate aim is producing reliable knowledge, we need to attend to the factors that, if addressed, would remedy the erroneous claims and support its trustworthiness (some of which may, indeed, have a cognitive or social dimension—as discussed in Chapters 3 and 4).

Thinking clearly about error in science challenges one to distinguish carefully between idealized science and authentic science. One may wish that science was error-free. But that does not make it so. We need to guard against confusing *normative* and *descriptive* accounts. Philosophers tend to envision science as it *ought* to be. Historians and sociologists, on the other hand, generally depict science as it *is*. Both may be important. But we should not mistake the idealized (normative) version for how science really happens.

Philosophical preconceptions can sometimes shape how history is written, as in Herschel's case. Ordinarily, it seems, we seem chiefly interested in how a discovery happened, not how it nearly did not happen. Even if these two processes were intertwined at the time. We thus tend to disregard the error. It seems "wasted effort," not part of the relevant story. We streamline the history. We erase the dead ends, blind alleys, errors, and other "hiccups" along the way. We reconstruct the past as it "should" have happened, rather than how it *did* happen. Or, blinded by hero worship, we create "myth-conceptions."[17] We may not even be aware that we are thereby distorting an understanding of the nature of science. Unfortunately, such biased histories can foster the view that science proceeds without error.[18]

Ultimately, fully rendered cases of error in authentic scientific practice are important philosophical benchmarks. In particular, they can help one assess proposed norms of science. What is realistically possible in context? Perhaps our expectations of science are not only idealized, but also utopian? Historical cases of scientific error can help us frame "realistic ideals." In the same way, the historical remedy of errors can help us discern and articulate concrete methods or strategies for finding and resolving errors. It may be worth remembering that our current scientific knowledge is the result of "real" science, not idealized science. We should not discount it, errors included. We should thus study errors in history to inform our understanding of how scientists effectively manage error.

Indeed, the relation between philosophy and history may benefit from reassessment. Later, we will see how many methodological principles did not emerge from some a priori or abstract philosophical reflection. Rather, the norms in question are by-products of dealing with errors historically. Philosophy emerges from, and is informed by, history. As David Hull reminded us, philosophy of science is often meta-science in disguise.[19]

•

The second step toward naturalizing error is circumscribing the epistemic efficacy of methodological norms. Conventionally, philosophers have championed science by characterizing its distinctive methods and disciplined protocols for establishing reliable knowledge. Heeding these norms is paraded as the foundation for trust in science. The assumption is (crudely):

right method → right answer.

Namely, adhering to all the prescribed methodologies is supposed to ensure reliable conclusions. Errors therefore presumably signal a lapse in doing science properly. In this view, there is no allowance for following the "right" methods and reaching a "wrong" conclusion. Indeed, that admission seems antithetical to the very reason for articulating methodological norms.

But a judgment that any error, by default, betrays an ineffective method in any particular instance is not fully warranted. Consider the case of incomplete evidence. It is commonplace—even among science educators at the elementary level—to say that "scientists revise their theories in the light of new evidence." That was a fundamental insight that emerged from the historical turn in philosophy of science in the 1960s and 1970s. It fueled the fruitful study of conceptual change by Kuhn, Hanson, Lakatos, Laudan, Toulmin, Shapere, Giere, Hull, Darden, and others. In such cases, one cannot realistically fault earlier scientists for absence or lack of evidence, which only appeared later. The historical scientists may have followed "the rules," but arrived at erroneous conclusions all the same. The emergence of new evidence led to a different pattern in the overall justification, and to the discarding of old theories in lieu of new ones. It was not a matter of gradual accretion of knowledge, or of one theory reducing to a special case of its successor. Erroneous conclusions could develop even without methodological flaws.

Errors may also arise simply from "bad luck." As history shows repeatedly, the development of knowledge inevitably involves an element of happenstance, or contingency. Chance meetings. Accidental events. Unanticipated new technologies. Blind trial and error. In those cases (many detailed throughout this book), one cannot fault failed methodology.

More generally, we must regard that many methodological norms are heuristic in nature, not algorithmic. They do not *guarantee* a solution each and every time. Rather, they are *strategies* that prove effective *often enough to warrant adopting them*. Yet they exhibit inherent limits to their effectiveness. We need to embrace a slightly deflationary view of methods. They need not be foolproof to be effective tools for guiding investigations.[20]

Incomplete evidence, happenstance, and heuristics all help decouple methodology from justification. That is, *sometimes*:

right method → wrong answer
wrong method → right answer.

Namely, historically, "right" method and "right" answer do not always align. Hence, the apparently paradoxical title of one collected volume exploring many such historical cases: "Wrong for the Right Reasons."[21] More exceptions of this type will be explored in Chapter 7, where one finds that perspectives that fostered great discoveries on one occasion also led, on another occasion, to notable errors.

The limits of scientific reasoning have been documented for many cognitive dispositions. Unchecked, humans are not formal logical reasoners. They tend to exhibit a set of known biases. The many forms of reasoning, linking stimulus to appropriate response, are evolved mental strategies, presumably adaptive in some contexts. Functional in one context, they are not always fully functional in another (say, in science). They are the source of many logical fallacies, so familiar to philosophers. These reasoning errors frequently reflect systematic failures in our mental engineering. We cannot fully "unlearn" these inherent cognitive tendencies. At best, we may learn to recognize them and regulate them in our thinking. Accordingly, we need to set *realistic* norms for conducting science.[22]

In summary, just because someone has followed the standard methodological norms does not mean that they have escaped the possibility of error. Even if you institute controls, there may still be other confounders (e.g., Eijkman's and Vorderman's studies on beriberi—see Chapter 2). Just because you have collected a large sample and used a low p-value as a statistical cut-off does not mean that you have dodged a statistical fluke (e.g., the "oops-Leon" particle or the clinical testing of Avastin—Chapter 2). Just because you have studied a wide scope and broad range of variable values, does not mean that you have exhausted the relevant scope of a prospective universal law (e.g., Boyle's law or Newtonian mechanics—Chapters 3 and 6). Just because you have exercised great precision and limited the measured uncertainty, does not mean that you have escaped systematic error from hidden variables (e.g., the Hubble constant—see later this chapter and Chapter 2). Just because you have replicated an experiment and attained the same results, does not mean that you have eliminated all artifacts (e.g., mesosomes, polywater, or cold fusion—see Chapter 6). Just because you have checked someone's credentials and reviewed the data, does not mean that you have safeguarded yourself from fraud (e.g., Henrik Schön or Victor Ninov—see Chapter 4). Just because you have fixed one error, does not mean that you have fixed all errors on the same topic (e.g., the high-energy intermediates of oxidative phosphorylation—see Chapters 2 and 5). These cases help remind us that methodological norms, while valuable in most cases, have contexts and limits.

•

To summarize, naturalizing error means simply acknowledging that error is a "natural" part of doing science. We cannot hope to eliminate error entirely. But we can learn to *manage* error (Chapter 8). Methodologies help safeguard researchers from many misleading conclusions and unjustified conclusions, yes. But they are not foolproof. Error happens. Ironically, to err is science. That is why correcting error is also part of science.

What counts as error

Once you begin to discuss error in science, you soon find that error means a great many things to different people. So it seems appropriate at the outset to clarify the term and the concept, at least for the discussion in this volume.

If one adopts an etymological approach (as many do), one will trace the term "error" to the Latin *errare*: to wander, or to go astray. Namely, someone who errs has followed a "wrong" path. The linguistic lineage (through Old French) also carries a second meaning: "mistake, flaw, defect, heresy." The judgmental tone resonates strongly with the Enlightenment view of scientific error as a transgression, or *moral* failing. However, from an epistemic perspective, an ethical lens seems out of place. The behavior of the scientist—whether virtuous or blameworthy, even their concordance (or not) with sound methodology—is itself peripheral. What matters foremost—for public trust in scientific claims, say, or for guiding the further work of other scientists—is the reliability of the conclusion itself. More particularly, is the claim in question *justified*? So, for the moment at least, I invite the reader to forbear concerns about fraud or misconduct or "questionable research practices," often raised in contemporary commentary on error. Let us limit our orientation to "honest" error, perhaps? Our chief focus here will be the claim and its *justification*, perhaps the quintessential problem for philosophy of science.

This does not wholly resolve the problem of conceptualizing error, however. The meaning of error proves to be quite slippery—confusing, and sometimes even paradoxical. For example, imagine characterizing an error as "a lapse in knowledge." If so, and you *know* something is an error, then it cannot be an error any more, because you know it is false. Perhaps you can only say that once upon a time you were mistaken. At that earlier point, you could not possibly have called it an error, because you were unable to identify its flaws, and did not know any better. So when did (or does) the error exist? The problem seems to be in the *transition* of justification, and in mistaking false claims for true ones (or vice versa).

Consider the case of the planet Pluto. Or *former* planet Pluto? Or the *dwarf planet* Pluto, once classified as a planet? Gravitational calculations in the early 1900s indicated that there should be another planet orbiting beyond Neptune. An active search by Clyde Tombaugh revealed in 1930 the presence of such a massive body, which was named Pluto. In 1978, it was even discovered to have its own moon. Not all planets remained planets. In the early 1800s the first asteroid to be found, Ceres, was considered a planet. With the discovery of other asteroids, its status changed. In a similar way, in the 1990s, with more powerful telescopes, more observations and the aid of computers, astronomers found more objects not quite planet size in the outer Solar System. Then more. Meanwhile, the original calculations used to hypothesize Pluto were discredited, as based on incorrect values for Neptune's mass and position. Pluto's status was challenged, then formally demoted by an assembly of astronomers in 2006. Was Pluto never a planet, even though it was called one for decades? When did the error occur? In 1930, when it was "wrongly" designated a planet? Or only in 2006, when the International Astronomical Union voted to strip it of its planetary status? Was the misstep in the original classification of Pluto (which seemed to follow all the rules), the "failure" to be cautionary about it, the faulty calculations, or perhaps the very concept of a planet, being too inclusive or vague in the early 1900s? Even such a relatively simple case illustrates the puzzle of defining errors.[23] The puzzling—indeed troublesome—aspect of errors is that they were not perceived as errors at the time (or from the claimant's perspective). Even if the error now seems in retrospect (or from another vantage point) glaringly obvious to us. To resolve this dilemma, we need to conceptualize an erroneous claim—along with its justification—in *diachronic* terms. We must couple a fully contextualized historical view, embracing one justification, with another justification (developed only later) that we regard as discounting the former, and providing a more fully justified alternative claim.

Accordingly, one may establish a (provisional) benchmark definition:

> *An error is a scientific claim whose justification is later found to be unwarranted.*

That is, an error is a claim once deemed "reasonable" and later found to be unsupported, or unreasonable. An error is necessarily a retrospective (or external) attribution. One major question here will be: How could the interpretation of good reasons change? And change so dramatically? Namely, in what ways might our *justifications* be incomplete or ultimately unwarranted, but completely hidden to our initial awareness?[24]

Framed in this way, the topic of errors in science opens some profound, and possibly disturbing, questions. How can there be holes or blind spots in our reasoning? How can we trust scientific claims if the justifications for them may later be found to be unreliable? (How can scientific claims hold any authority in public policy or decision making if they are subject to unanticipated change, especially in wholly unpredictable ways?)

Clearly, we need to understand the nature of the changes in justification. Why do they occur? *How* do they occur? How were earlier scientists "handicapped" in their abilities to notice the errors (if that is an appropriate designation)? Also, how might scientists deal with the prospect of errors? Are there any methodological tools to limit the frequency or lessen the scope of errors? Can history guide us? Can we make scientific reasoning more complete or secure? How? Ultimately, how do we develop trustworthy, or reliable, knowledge? These compelling questions are why we need a well-developed Philosophy of Error.

•

Justifications in science encompass a vast spectrum of practices. So, if we acknowledge that each is susceptible to error, the scope of a philosophy of error becomes, accordingly, quite broad.

First, philosophers of science will be most familiar with the conventional focus on the relation of theory and observations (T/O), or theory and evidence (T/E)—the stuff of introductory textbooks: for example, confirmation of predictions, consilience of inductions, probability of belief. Here, the analyst of error has occasion to probe the broad range of fallacies in deductive logic (also stock fodder in teaching critical thinking skills): affirming the consequent, mistaking correlation for causation, underdetermination of theory by data, overlooked *ceteris paribus* clause, and so forth. Equally, let us not be blind to the many other forms of reasoning, each with their own distinctive patterns of logic: induction, abduction, analogy, and probabilistic and error-statistical reasoning (at least). We may surely also attend to the use of more sophisticated combinatorial strategies: such as the method of multiple working hypotheses, natural selection-type search, hypothetico-deductive inference, or Mill's method of difference. Each step in reasoning marks another occasion where justification may go awry. All the errors in scientific argumentation—familiar territory for philosophers—fit comfortably in a philosophy of error.

But error is hardly limited to the conceptual realm. Work in philosophy of science over the past several decades has expanded well beyond the simple T-O framework. Most notably, the New Experimentalists have underscored the epistemics of laboratory and field work and the microreasoning about the very process of observation (broadly construed). Perceptions matter. Experimental design matters. Apparatus matters. We now talk comfortably about instruments—microscopes, thermometers, spectrometers, and such—as having *epistemologies*. Using them—and using them *properly*—is integral to justifying the meaning and validity of their outputs. Lab equipment is susceptible to malfunctions and artifacts: opening more opportunities for errors, contributing to a now-burgeoning image of the justification structure. Meanwhile, measurement theorists have underscored the epistemological complexities of measuring systems and the simple act of recording a single value for voltage or distance, say. At the same time, more attention is being afforded to the material culture of science, and how that plays into conclusions and their justification. One may thus also consider the significance of expertise and competence—does the technical execution of experiments indeed correspond to their intended design? The whole realm of observation (again, broadly construed) involves a rich tapestry of forms of justification that also command the attention of someone interested in error.[25]

At the other end of the epistemic spectrum, perhaps, philosophers of science have also delved into the aspects of justification at the social level: namely, the process of vetting, revising, and filtering claims through the critical discourse in a scientific community. For many years, such dynamics, while well known and hardly discounted, were generally taken for granted. Peer-reviewed journals, evidence-based debate, resolution of disagreement, consensus, and such were all largely assumed to function unproblematically, as a matter of course. However, these processes, too, are vulnerable to allowing erroneous justifications to slip through and be widely endorsed. The crucial significance of these interactions, along with their institutional frameworks, are now more widely appreciated. In particular, Cultural Studies of science and many sociological, feminist, and Marxist case studies helped lay bare the potential for systematic gender, racial, class, and/or political bias. Homogeneous or exclusive communities can easily fail to achieve Merton's norm of "organized skepticism." Complementary standpoints or contrasting theoretical (or political) perspectives seem essential for an effective system of checks and balances. So, we may add several social practices governing critical interchange as essential to the justification process. For the philosopher of error, the system of discourse and how it manages the development of a scientific consensus—namely, the domain of social epistemology—is also fundamental to the overall structure of justification, and thus to the study of error.[26]

Surveying the broad reach of justification beyond conventional logic might allow us to reconsider the conceptual level and probe its details further. For example, philosophical focus on theories has been supplemented with study of models and model-based reasoning. Focus on formal (hierarchical) logic and law-like reasoning has been complemented with exploration of "lateral" reasoning based on cases, narratives, exemplars (or paradigms), simulations, and clinical case studies. Likewise, the study of heuristics, proxy variables, and such—all admittedly imperfect, but demonstrably fruitful—has blossomed. One can add reasoning about the suitability and limits of model organisms or generalizing from in vitro studies, or the reliability of statistical analyses and other forms of reducing data. Already aware of the psychological tendencies of confirmation bias, one can easily expand the repertoire of individual cognitive biases to include cultural biases, such as gender, racial, religious, political, or class biases. Having ventured into the information age, we might now be concerned with the quality of curation of large databases (e.g. Human Genome Project and/or protein sequence repositories), as relevant to justifying any conclusion based on sampling from those data. Again, all these steps of reasoning from one set of data or claims to another, from one concept to another, require justification. And all are occasions where error may occur, and all fit within the purview of a philosophy of error.[27]

All these dimensions of justification are familiar to philosophers of science. But perhaps they have not been fully addressed in terms of the errors that may be associated with them. One element of a new perspective may be seeing them all as fully networked or integrated. A flaw in any one step can ripple through the whole larger structure of justification with adverse consequences, in the same way that even a point mutation of one nucleotide in an organism's DNA can prove lethal. Indeed, the complexity may be a bit overwhelming—far more is at stake than some simple comparison of theory and evidence. One might stand in awe and wonder of any claim that survives the epistemic scrutiny of many successive filters, when one concatenates all these dimensions together. They may sketch a view of justification in science that is extraordinarily complex: both fragile and resilient. And humbling, perhaps. And that would be central to respecting a role for a philosophy of error.

I propose a holistic view for interpreting the structure of justification (based on mapping) in the next chapter, and the three major layers of justification just mentioned form the basis of more detailed discussion of error types in Chapters 2–4.

Epistemic posture

Another key factor is a scientist's level of commitment to a claim's justification. (Note, again, that our focus is the justification, rather than the "truth" of or "probable belief" in the claim itself.) In a particular historical context, was the claim accepted as fully endorsed, or only with qualifications? Was an epistemic commitment made?

So, if a researcher entertains an explicitly tentative hypothesis, while never embracing it fully, can that ever be an error? If it is published, but never defended as fully demonstrated, is it eligible as a genuine error? If the scientist's reasoning is sound, but just one assumption later turns out to be misplaced, does that count as a mistaken justification? If one scientist advocates for or defends an idea, but other relevant experts in the field dismiss it, does that count? For epistemic analysis, the relevant benchmark seems to be when the claim is deemed appropriately justified. (In a sense, nothing is either true or false but thinking makes it so?) *Speculating* creatively, or *entertaining* concepts that are considered only *plausible* or *partly (incompletely) justified* should not be confused with error. Accordingly, an essential element is one's *epistemic posture*, or epistemic stance.

For example, between 1645 and 1761—namely, for over a century—there were more than 30 claims to have observed a moon of Venus. That included such notable figures as Cassini, Lagrange, and Montaigne. Yet none of the claims were defended with any vigor. None were endorsed by a second observer or accepted by other astronomers. Nor was the prospective moon ever given a name (as scores of erstwhile chemical elements were). That is, no report was ever considered truly justified to begin with. So, none can truly be considered an "error" (in the sense of a justification upended). Eventually, all the reported observations were abandoned, regarded either as artifacts of the telescope or of human vision or as other astronomical objects near Venus visually. But even these dismissals were based on suppositions, never robustly confirmed. In 1761 and 1769, multiple observations of the transit of Venus across the Sun, done with reliable optical instruments, failed to reveal any planetary satellite, and that seemed to settle the issue. The degree of effort devoted to a purported moon of Venus certainly warrants historical interest (as superbly documented by Helge Kragh). But despite the scale of the work, the epistemic posture never involved a justification that would have warranted a retraction: There were no formal errors here.[28]

It is commonplace, of course, to say that *all* claims are "tentative"—or provisional or fallible or subject to change—even when presented as justified. But such a disclaimer is not very helpful, really. This popular expression is really no more than a rhetorical "get out of jail free" card, tucked up one's sleeve, to be played only when needed. The epistemic commitment is missing. Just as if it was mindfully withheld. Under such cautious contentions, *no* justification can ever be considered complete or acceptable.

Still, an epistemic posture may certainly be conditional. Thus, a scientist who is aware of particular qualifications, or "known unknowns," may explicitly state them. Stipulating context or provisions indicates the scope of such residual uncertainty. It acknowledges the known limits of one's research and of the corresponding justification.

The notion of varying "levels" of commitment to a claim is hardly new or unfamiliar. For example, Latour and Woolgar referred to various "modalities," and Bayesians tout a notion of "degree of belief." In studying error, however, we are concerned with just those cases when some definitive commitment was made and *later* withdrawn. Some assumption becomes suspect, or some new evidence offers fresh context. Based on the change in justification, the commitment *for the very same claim* is now *reversed.* Our goal should be to understand *why* the conclusion itself changed. Surveying and organizing the spectrum of possibilities is the primary aim of the next three chapters.[29]

How, then, should we characterize epistemic posture or epistemic commitment? Perhaps the most explicit would be to circumscribe the scope of the justification with respect to the available evidence. It may also be "scaled" to how one plans to apply the justification. A scientist's "level" of commitment may thus be signaled indirectly by their actions. I will note three significant shifts in level of commitment.

First, does the scientist or lab group in question *pursue* the claim? Are resources invested in investigating it empirically? Are its implications or assumptions explored theoretically? These actions reflect a judgment that the idea is at least *plausible*. A "tentative" posture, here, is nonetheless a positive stance. That is, it reflects a posture about burden of proof: that the claim can be regarded (albeit modestly) as warranted until further evidence indicates otherwise. Thus, even for something nominally labeled a "hypothesis," an observer can measure concrete commitment in terms of time, lab resources, and other effort. For example, before research-granting agencies fund a research project, they typically require some justification that the endeavor itself has merit, or that the target concept is already partly justified. That is, even an assessment of theory promise involves justification, as nicely elaborated by Laurie Anne Whitt. Accordingly, it is common for scientists who are simply pursuing a hypothesis to say that it "failed" an experimental test, or that it proved "wrong"—indirectly signaling a former expectation and that an epistemic commitment has changed, although we need not regard these as genuine errors.[30]

So, consider Charles Darwin's proposal that modern chickens evolved from red-footed jungle fowl. Recent genetic analysis has indicated that the ancestors seem to be *gray*-footed jungle fowl, a different species. (Well, critics of evolution had a field day parading that "error" as another fatal flaw in evolutionary theory!) But Darwin had never presented his claim as definitive or fully justified. His primary intention was to suggest how a domesticated species could be traced to wild progenitors, illustrating descent with modification. So he offered a plausible example. Speculation, especially when acknowledged as such, is not error.[31]

A second level of epistemic commitment occurs when the claim, along with its justification, is *published*. Using publication as a benchmark expresses an assumption that introducing an idea into a public forum, where the justification is open to scrutiny and criticism, reflects a significant level of confidence about, or commitment to, a claim. The claim moves from the security of the private realm to the vulnerability of the public realm. Publication also takes effort. That reflects a practical level of commitment. And in cases of peer review, publication generally (although not always) reflects that the proposed justification has met at least professional standards of epistemic responsibility—and has not exhibited some glaring reason for being discounted entirely.

So, this may help us assess Linus Pauling's published claim in 1953 that DNA was a triple helix. That claim was dramatically eclipsed only a few months later by Watson and Crick, with their double-helix model. Should we view this as a prime example of error in science, as suggested by at least two historians? The commentators drew attention, in particular, to Pauling's professional stature and credibility. He was a Nobel Prize-winning chemist. Earlier he had successfully elucidated the alpha-helical structure of proteins. So even Pauling's opinion would have carried some persuasive weight. Yet Pauling and his co-author never presented the triple-helix idea as more than a "promising structure." The evidence was all suggestive. As a claim, it was nowhere near justified, even for the authors. We can easily imagine that Pauling was eager to stake a position that would establish priority of discovery—*if* it turned out to be correct, while hedging against embarrassment if it later turned out to be wrong. There was no substantive epistemic commitment. It was not an *error*. Again, speculative claims or hypotheses that are merely entertained for consideration, even if published, do not reflect errors, precisely because the justification is regarded as incomplete.[32]

Scientists adopt a range of postures toward publication, with some being much bolder (and more willing than others) to present ideas that might eventually be rejected. They may be more willing to advance overstated claims. In addition, standards of quality may vary among different publications. Still, publication is a helpful heuristic (other indications notwithstanding) for supposing that there is some form of epistemic commitment behind presenting the claim formally to others. For my analysis, it is of considerable interest when the justification for a published claim—even a cautious, plausible one—is later rejected, or negated by further evidence. We should want to know why.

A third important benchmark is *consensus*. This shifts the level of the claim, and the epistemic commitment embodied in "acceptance," from the individual to the community. Consensus can be notoriously slippery when one tries to establish it formally. Still, if one takes the system of checks and balances within the scientific community as critical to the epistemic process, then consensus among different perspectives reflects a depth of epistemic commitment not available to an individual. What appears in textbooks (as an expression of consensus) matters. These are the kinds of claims that one expects to serve as a stable benchmark in public policy or personal decision making. Again, if we are interested in science as a source of reliable claims, we should be interested when scientific consensus changes—and more importantly—why.[33]

For example, Fred Hoyle famously rejected the concept of the Big Bang when everyone else seemed to accept it. By contradicting the consensus of the community—and disagreeing with the theory which ultimately prevailed—Hoyle is considered by some commentators to have erred. But for Hoyle, and even for most of the community, the justifications for the hypotheses of a stable versus an expanding universe remained indeterminate at the time. The apparent consensus was just a shared best guess. No one knew for sure, and they acknowledged this. Of course, this did not stop them from staking their positions and parading their favored theories. Hoyle's dissent is surely interesting, but due to the acknowledged uncertainty, one cannot consider it an "error" in justification.[34]

Error and uncertainty

A focus on epistemic posture opens questions about the status of uncertainty. Typically, we tend to say that uncertain claims are susceptible to error. Namely, we label them uncertain precisely because we acknowledge that they may be "wrong." But if the commitment to the justification is never made, then we cannot be caught wholly off guard if the evidence or the reasoning changes and with it the conclusion. An epistemic posture of uncertainty is thus *not* error (as conceptualized here).

In cases of uncertainty, we withhold commitment from a conclusion that we know would place us in epistemic jeopardy. Indeed, uncertainty is an explicit recognition of the absence of a secure justification. We openly admit the indeterminacy. In cases of error, by contrast, a faulty justification has gone unnoticed. Indeed, understanding the difference between visible and invisible dimensions of a justification is perhaps one of the core challenges of interpreting error.

Historically, the concepts of error and uncertainty have been closely related and sometimes not always effectively distinguished. That confusion originated in the early history of statistics and the origin of the modern concept of measurement "error." Geographers and astronomers mapping the Earth and sky in the late 18th century noticed that successive measurements for the position of a star or fixed landmark (such as a mountaintop or church steeple) did not always match. However, such features could not shift on the landscape or in the heavens. Some measurements, they reasoned, had to be *wrong*. To correct the apparent "errors," the scientists endeavored to improve their instruments. They refined telescope optics. They machined their surveying instruments and scales more carefully. Yet despite their efforts, the inconsistencies persisted. How could they identify which measurements were "correct," and which were "incorrect"?[35]

By considering the measurements as an ensemble, however, they found that a pattern emerged. When they were depicted graphically, the scattered measurements clustered around a central value. Those who collected the data decided that the mean value must represent the "true" value. In a sense, statistical analysis was born. Calculating simple average values solved the problem at one level. But the distribution was also generally consistent from case to case, forming a bell-shaped curve (today's familiar normal distribution). Measurement "errors" followed a uniform pattern, which could be expressed mathematically. Here, arising out of apparent chaos, was one of those striking regularities scientists strive to discover. It became a new "law" of nature. They called it the *Law of Error*.

It was not long before they realized that the "Law of Error" was related to probability distributions. The field of mathematical statistics developed over the 19th century from that notable insight. Today we know the historical "Law of Error" as a Gaussian distribution. It describes a probabilistic, or random, variation: a form of *uncertainty*. It is *not error*. The astronomical and geodesic "mismeasurements" were not wrong per se. They were *imprecise*. Inherently "fuzzy." That is, they were *not inaccurate*. They *were* indeed the "correct" value, but also inescapably coupled to an additional factor of characteristic variation. The "Law of Error" was not really about *error*, then. Rather, it was about *uncertainty*. Historically, the momentous discovery of the statisticians was, ultimately, to realize this distinction: that imprecision is not the same as inaccuracy, and *that measurement uncertainty is not error*.[36]

Sadly perhaps, the term "error" is still used to describe the *uncertainty* of measured values. For example, investigators use the term *standard error* to denote the numerical range that encompasses 68.2% of a quantitative uncertainty. A *margin of error* (such as reported in public poll results) does the same, typically at a 95% level. Researchers graph their experimental results with mean values and an *error bar*. That error bar is a shorthand version of the bell curve, or the former Law of Error.

Yet the modern term of "measurement error" is, I claim, a misnomer. There is no error involved. The "error" does not refer to a false claim, nor to any purported failure in justification. As the history of the Law of Error indicates, measurement "error" refers to *uncertainty.*

Ideally, then, experimentalists should modify their terminology. They should refer not to measurement "error," but instead to *measurement uncertainty.* "Error bars" should be labeled, more plainly and clearly, as *uncertainty bars.* Standard error is, more properly, *standard uncertainty.* Margin of error is really *margin of uncertainty.* Indeed, the whole field known as Error Statistics seems, in my view, mislabeled. It should rightfully be called *Uncertainty Statistics.* This simple change in terms will, I contend, help promote a clearer focus on error itself, as separate from questions of uncertainty. This is simply another way to acknowledge that precision and accuracy are distinct concepts.

A similar confusion of terminology occurs with the general concept of experimental error. Experimentalists conventionally talk about two types of "error": systematic error and random error. Systematic errors are, in a sense, the true errors. They encompass all the ways the recorded values or observations may be *uniformly* biased. Many factors presumably affect or distort each measurement equally. Some such factors researchers can identify and quantify. Hence, they can make corresponding adjustments or corrections to the measurements. For those that remain unknown, they use the term experimental error. The sources of error might be incorrect reference values in calculation formulas, properties of the measuring instrument, or underlying assumptions about how the observational setup relates to the target quantity being measured. That is, all these factors may affect the *accuracy* of the values determined. Once these errors are discovered, they can be corrected (just as the early astronomers and geographers tried to do). Until then, they remain possible, but ultimately unknown errors.

Not so for the second type of experimental errors, conventionally called random errors. As their name might indicate, they exhibit a probabilistic distribution. Again, the sources cannot be fully specified. But these "errors" are visible, in a sense, in the variation of values from a set of "identical" measurements. Again, there is observed *uncertainty.* Some values seem too high, some too low. The effect of the variation is thus not to alter the *accuracy* of experimental values, but to limit their *precision.* To borrow Ian Hacking's apt phrase, these uncertainties can all be "tamed" by mathematical analysis.[37]

Again, systematic (true) error relates to accuracy; random "error" to uncertainty and precision. And again, terminology in experimental work can help distinguish between *sources of error* and *sources of uncertainty*.[38]

The interplay of (systematic) error and (random) uncertainty may be illustrated with the history of the Hubble constant, the rate of the universe's expansion. In 1928, Edwin Hubble had evidence for expansion of the universe from red shift in the light from distant galaxies. But determining the *rate* of expansion was more difficult. One needed stellar distance and age measurements, which were intertwined with values for the relative brightness of a star and its absolute brightness. Hubble solved the dating problem by using a special type of star, known as Cepheid variables. They pulsate, and their period of pulsation is related to their absolute brightness. The relative (observed) brightness of the Cepheids could thus help establish the distance in galaxies where they were located. The series of calculation thus fell into place. Hubble calculated the rate of the universe's expansion and, with it, its age.

One may easily appreciate the many *uncertainties* in Hubble's value. There is the sensitivity of the telescopes to brightness (especially for the low relative brightness of very distant galaxies). There is the number of distant stars you can find: too few and you risk an unrepresentative sample. There is the distance calibration using the Cepheids' relative brightness, itself an estimate. Indeed, around the same time other astronomers used similar techniques, but arrived at different values for Hubble's constant, which ranged from 450 to 600 km/s/Mpc. It is not that one value was correct and the others were in error. Given the acknowledged sources of variation, the range of values were just different approximations. In the 1930s, then, the different values for the rate of expansion reflected scientific *uncertainty*.

However, in 1952, using more powerful telescopes, Walter Baade discovered that the Cepheid variables, once considered a uniform group of stars, actually combined two different sets of stars. Stars that Hubble once thought had a certain brightness became *clusters* of stars, each with a lower luminosity. That change cascaded into the determinations of distance. Distance into stellar ages. Stellar ages into the rate of expansion, and the rate of expansion into the age of the universe. Suddenly, the universe seemed twice as old. The evidence changed; the justification changed: an *error* had been exposed (and, simultaneously, remedied). Notably, the new value for Hubble's constant was well outside the range of previous calculations—beyond the former appraisals of uncertainty. Baade had exposed a *systematic* error: an undetected assumption about the uniformity of Cepheids.

Then in the late 1950s, again using more powerful telescopes still, Allan Sandage observed that Hubble had misidentified another key group of stars used in the calculations. They were nebulae (HII regions), not stars. The rate of expansion was cut again, to about 75 km/s/Mpc, and the age of the universe nearly doubled again. Another systematic error.

In the past half-century, new calculations have continued to appear. But they now fall within a range of 50–100 km/s/Mpc. Different methods for estimating distance have emerged, allowing astronomers to cross-check them. The eponymous Hubble Telescope has also provided more highly resolved images and additional data that help provide higher precision, or a decrease in the distribution of measurements. Thus, there is still observational (measurement) uncertainty. But as the range of values narrows, there is increasing confidence in the values obtained. The history of the Hubble constant thus exhibits both error ("systematic error") and uncertainty ("random error"), which may help clarify the distinction between them.[39]

The historical pattern of systematic errors versus precision can be found in the history of the values assigned to many physical constants, such as the value of the electric charge (*e*) or the atomic mass of chlorine.[40]

In summary, by articulating the role of epistemic posture in error, one can articulate more clearly the distinction between error and uncertainty. Errors result from changes in justification, and "positive" knowledge is converted to "negative" knowledge (but still remains knowledge). New evidence resolves alternatives in support of a new claim. Uncertainty, by contrast, is a form of unknown, an indeterminacy, where justification is *unresolved* (even if its scope is circumscribed). This distinction helps inform the analysis of the "natural history" of errors and error resolution (Chapter 5) and further contributes to a novel conceptualization of knowledge itself (Chapter 8).

Taking stock

Having set forth the central theme of error, having sharpened the focus by conceptualizing error in terms of epistemic posture and upended justifications, and having established the scope of the problem (both inclusively and exclusively), it is now time to take stock of earlier works on error and what insights, resources, and background they provide.

Comments on error can be found scattered throughout the modern era, certainly since Francis Bacon (see "Error types" in Chapter 2)—we have already encountered the Enlightenment sentiments of Herschel and Godwin, quoted earlier.[41] Many observers simply compile lists of errors and conclude that error is inevitable, stemming from human frailty and such, and they often chastise scientists for their self-delusion or credulity (or even "feckless credulity").[42] In 1936, psychologist Joseph Jastrow suggested they could all be subsumed in a study of "Errorology," probing the mental origins of the "impediments or inexpertness of the thinking process"—that is, any failure to exhibit proper logic.[43] Attention to securing rigor in scientific reasoning is essential, of course. But, as many of the examples already cited indicate, deviation from idealized logical reasoning does not exhaust or even typify the cases of error in science.

A more fruitful vision was introduced by Deborah Mayo in her 1996 *Error and the Growth of Experimental Knowledge.* She considered how scientists could *learn* from error and how they could *argue* from error (using Karl Popper's notion of severe tests). She also sketched how they could cope with the errors that inevitably occur. For example, investigators may minimize errors or limit their adverse effects. That includes, first, working to avoid repeating past error types—say, through experimental controls ("before-trial planning"). Second, one may monitor for the occurrence of error ("after-trial checking"). That might include measuring the chance effects of sample size: a vital role for statistics. Third, for errors that are unavoidable, one can try to accommodate them by "subtracting" their influence from the results.[44]

I hope to continue and further develop Mayo's pragmatic approach here, especially in expanding it beyond the realm of experimental reasoning, to also include the conceptual and social-level forms of justification noted earlier.

Many commentators express the problem of errors in terms of "failure."[45] Yet this framing depends, foremost, on reference to prior expectations and values (often assumed or unstated). Namely, an experimental finding or observation is a "failure" only if you expect clear results that unambiguously convey the answer you sought. (Why should we *expect* every replication to "succeed" or every model prediction to come true?) The normative behavioral dimension, I have argued, distracts us from the epistemic core. To resolve errors, we need to regard them more as puzzles about justification than as evaluative judgments of inadequacy or "failure."

Yet another group has described errors as forms of "bias"—generalizing, perhaps, from the problem of biased samples in clinical studies (a statistically unrepresentative set of evidence). They have expanded such biases to include ways evidence is (mis)interpreted. Viewing all errors in terms of bias seems to imply that conclusions are never wrong; they are just warped or distorted—a view that seems strangely at odds with episodes from history where scientists outright rejected many theories or concepts as false, not merely misleading. In addition, one may easily confuse the source of error (the biasing factor) with the error itself (the conclusion with a mistaken justification). As I will detail in Chapter 7, sometimes bias can be fruitful to discovery—in some cases, ironically, based on the very same biases that lead, on other occasions, to error. So, the concepts of bias and error, while related, should be kept distinct. Biases can be *sources of* error, but are not errors themselves.[46]

Others have tried to sidestep the negative connotations of error entirely, it seems, in favor of a diffuse category of "going amiss" (echoing, perhaps, the notion of "failure"). That view has prompted a long ad hoc list of experimental mis-steps—malfunctions, dead ends, puzzlement, confusion, noise, artifacts, monstrosities—but without any systematic organization.[47] This approach, in particular, seems to miss the significant distinction (described earlier) between error (blindness to a flawed justification) and uncertainty (anomalous findings or incongruent conclusions). Research that is stymied by puzzling or ambiguous results is not the same as error, although an unknown error may be lurking nearby.

Yet other commentators note the multitudinous ways of being wrong and cite this as a reason to doubt that any unified concept of error will ever be found.[48] For them, no general theory of errors is possible, apart from their individual instances. In my view, that remains to be seen.

All these approaches—going astray, failure, bias, going amiss, and error nominalism—can be informative. Yet they do not focus on the misplaced justifications. Indeed, none of these various characterizations and frameworks seems to have struck a chord or provided a seed crystal around which a philosophy of error has coalesced.

Mayo, for her part at least, explicitly pinpoints error to faulty experimental inference. She identifies three basic categories in the structure of their justification: (1) validating the conditions and the relevance of the data themselves (essentially, theories of data or evidence), (2) establishing the theoretical assumptions and ruling out alternatives (experimental design), (3) and statistical reasoning.[49] But these are limited (again) to experimental knowledge. In the next chapter, I will articulate a broader framework that spans the three layers of justification noted above.[50]

Another major task in conceptualizing error is classification. Can we catalogue the many types of errors, organize them coherently, and construct a functional taxonomy? This will be the primary focus of the next chapter, so only a few initial remarks are needed here. Historical approaches to taxonomy often stem from the fundamental conceptualization of error. Hence, one set of taxonomies is based on logical fallacies and flaws in argumentation.[51] Another set views the errors as cognitive pitfalls, incorporating a broader spectrum of psychological processes.[52] Again, there is a subtle tension between concepts alone and the more complete structure of a full justification. Another frequent approach to taxonomy is largely procedural: namely, where in the process did the scientists go "astray"? Hence, some taxonomies dissect the experimental environment and classify error according to assumptions, design, materials and instruments, execution, data analysis, and so on.[53] Others reflect more explicitly the structure of the typical scientific paper, essentially using its sections as categories.[54] Yet another approach highlights the varieties of "questionable research practices" (QRPs), such as truncating data or *p*-hacking. These procedure-based perspectives are helpful, too, although perhaps not complete when one considers the social dimension of discourse, criticism, and reconciling conflicting claims. In almost all these taxonomies, fraud and falsified results are problematic outliers: Where does honesty (or documentation of integrity) fit within a conventional scientific argument? It is, rather, an element of context, involving the social structure of the discourse—and this is another reason why many proposed taxonomies seem limited in scope. A more comprehensive (or synoptic) and systematically organized perspective on error in science is mapped out in the next chapter.

Finally, we may look to history for ideas about how to *manage* errors. That is, let us study errors not to malign science for its past errors, but to orient it toward more effective practices in the future. This habit of error-analysis and updating one's principles is already customary among engineers, at least.[55] Mayo's approach of learning from error definitely underscores the notion that understanding errors is intimately related to guiding, and sometimes improving, scientific practice. If indeed errors are inevitable, then the primary challenge becomes not eliminating errors altogether, but managing them when and where they occur.

Here, the recent flurry of concern over failed replications of famous experiments and the so called "Reproducibility Crisis" is relevant.[56] While there has been substantial concern that, according to one researcher, "most published research findings are false,"[57] and that science is "broken," the posture has nonetheless been one of "fixing" science. If fraud or QRPs are on the rise, then the appropriate response is seen as developing new methods, social practices, or publication norms that provide disincentives and checks against "scientific malpractice."[58]

A healthy response to error is thus to find remedies and to improve the process of science. For example, in her focus on learning from error, Mayo has introduced a handful of useful concepts (some mentioned only briefly) that I adopt here and elaborate on (see Chapter 8). First, *error repertoires* are lists of errors based on "the history of mistakes made in a type of inquiry."[59] They are important diagnostic tools. Second, experimentalists develop important knowledge of the specific "effects of mistakes," namely, "the kind and extent of the effect attributable to different errors or factors," what I call *error signatures*. Finally, *error probes* are diagnostic tests, in the spirit of Popper's severe tests, that seek to expose a certain error type, if it is present.[60] These are tools in error diagnosis and in testing the resilience of findings at the experimental level. In a similar way, Lindley Darden[61] has developed methods at the conceptual level: chiefly, for anomaly resolution and for diagnosing and fixing faults in theories. These perspectives for developing fruitful strategies form a background throughout the book, and the results will be summarized in Chapter 8.

The philosophical study of error is also greatly enriched by a wealth of historical work. The continuing fascination with the topic of error has yielded a series of edited volumes with numerous relevant case studies.[62] While these collections are mostly ad hoc—and lack a comprehensive philosophical framework for interpreting error—they nonetheless provide plenty of raw material for study. In addition, many historians and sociologists have invested in extended book-length studies of particular episodes of error, providing valuable detail at a fine-grained level—for example, on caloric and chemical affinities; the fifth force; polywater and cold fusion; the planets Vulcan and Pluto, and the moons of Venus; over 100 chemical elements; AIDS research; and Einstein's claims.[63] Many individual papers, while shorter, are also valuable. I have supplemented these with my own historical research. It is part of my task in this book to integrate and synthesize the insights from these many well-informed studies.

In introducing a recent volume on error and uncertainty in scientific practice, Marcel Boumans and Giora Hon contended that "the kind of error to be corrected differs from case to case." That is, they "are idiosyncratic, case-specific and local misapprehensions." Their dismal conclusion (stated twice, lest the Reader miss it) was that "there is no general theory of error, and there never will be one." "Indeed," they continued," the task of theorizing error is insurmountable."[64] This book challenges that dire verdict and offers one vision of a comprehensive philosophy of error in science.

A guide to the book

What lies ahead in subsequent chapters? What needs to be addressed?

First, what are the various sources of error? How do scientists make (well meaning) mistakes? One may suppose at first that there is very little reason, philosophically, for scientists to err (echoing the Enlightenment view). Yet errors do occur. That leads us to history as an indispensable resource. What do we learn, not from abstract theorizing or idealizations but from the concrete experiences of the past, that can be appropriately generalized? Our first task is to compile an inventory of the various error types. And then a way to organize them. By documenting errors from the past, one builds an important reference for recognizing or diagnosing possible errors in the future—and for understanding more fully the very epistemic processes of science (Chapters 2–4).

Second, how are the errors found? How are they then isolated to particular error types, and then fixed? Again, we profit from historical analysis. What processes helped resolve error? In what contexts? Inevitably, we appeal to more evidence. But how does that evidence arise and what makes additional evidence significant? (Chapters 5–6)

We seem to prize error-free knowledge. So we might strive to purge error from the process of science itself. Having identified the many sources of error, what is the prospect of eliminating errors from science entirely? What is the relationship between error and discovery? How might insights and blind spots ironically be entangled? (Chapter 7)

Finally, where does all this lead? Can the negative knowledge of error have a positive role? How are the methods of science shaped by past errors? If science errs, what can we say about scientific progress? Can characterizing previous errors at an abstract level assist us? How? Can we manage error in scientific practice more effectively with checklists and/or checkpoints? Where might a more sustained study of the philosophy of error lead? (Chapter 8).

2
Observational Errors

Sources of error • error types • mapping and the layers of justification • observational errors • material errors • framing errors — sampling bias — small sample size — incomplete sampling — proxy variables and heuristic gaps — confounders • instrument errors • human observers • observer bias • artifacts and observer effects • summary of observational errors

Having characterized the problem of error itself, and established basic definitions, scope, and background (Chapter 1), one may be prepared to plumb the causes of error in actual scientific practice. Just how do errors in science happen? If we intend to find them and fix them, we should have some idea of their origin. Not generally or vaguely, but specifically. Scientific researchers often encounter pitfalls in their equipment or experimental design. And they work to minimize them, and they remain on guard to spot mishaps. They refer to them as *sources of error*. In this chapter, I expand this notion from the laboratory to the whole process of science.

So, how do scientists err? Or—more properly (keeping our provisional definition of error in mind)—how do the justifications of scientific claims go awry? If they are initially accepted as unproblematic, how do they later fall apart? I have already noted the broad reach of faulty factors in justification, from dirty glassware or improperly calibrated instruments to premature generalizations or blindness to the gendered nature of one's very questions (Chapter 1, section "Error naturalized"). But how might we characterize specific error types and assemble them in an informative *error inventory*?

First, I review earlier efforts to identify and classify error types. Next I review the popular map metaphor of scientific representations as a way to organize the types across multiple levels of justification. Then I begin to enumerate and provide brief case examples of error types: in this chapter, at the observational level; and in the succeeding two chapters, at the conceptual and social (discoursive) levels.

Toward a Philosophy of Error in Science. Douglas Allchin, Oxford University Press. © Douglas Allchin (2026).
DOI: 10.1093/9780197827703.003.0002

Error types

Here, I continue a project that I began in 2001: to document the various sources of error that scientists encounter in their practice.[1] The cases come from history. And from more recent, contemporary experience. As we collect more and more cases of error, we are able to ask: What similarities or patterns are there? Can we identify general *error types*? How should we organize them coherently? Might we also be able to identify the most frequent error types, and/or perhaps the most severe ones?

First, we may first review similar historical efforts to identify and classify various error types. Awareness of errors certainly seems as old as science itself, as expressed, for example, by Francis Bacon in the early 17th century. Bacon is well known for his grand vision of an inductive science, which helped inspire and guide how others built modern science. Yet alongside the ideal, he cautioned his readers that potential errors might lead one astray:

> The illusions and false notions which have already preoccupied the human understanding, and are deeply rooted in it, not only so beset men's minds that they become difficult of access, but even when access is obtained will again meet and trouble us in the instauration of the sciences, unless mankind when forewarned guard themselves with all possible care against them.

Forewarned is forearmed. Bacon described how our thinking can become muddled in several ways. He was essentially identifying psychological pitfalls centuries before there was any formal study of psychology.[2]

Bacon named four general sources of error in how humans think. He referred to them with fanciful imagery as "phantoms" or "illusions," or occasions where imagination might be taken as reality. They impair our ability to think clearly and effectively. Today, we might characterize them as cognitive biases.[3]

First, there are illusions based on how humans tend to think (as a species, or "tribe"). We see patterns where none exists. We adhere to first impressions. We let wishes guide our perceptions, or attend unduly to the unusual or unexpected. Thus, some conclusions will be incidental by-products of our thinking process, not genuine depictions of reality. The modern reader will easily see the relation of "illusions of the tribe" to cognitive biases and heuristics,[4] or the role of motivated reasoning.[5]

Second, some errors stem from our personal idiosyncrasies, unique backgrounds and tastes: illusions of the individual (or, in Bacon's cryptic metaphor, of a person's "cave" or "den"). As a result, some people focus on similarities, while others emphasize differences; some are oriented to details, while others are more holistic; some are preoccupied with the past, while some with the future. These dispositions diminish the objectivity of our conclusions. Here, the modern reader may think of research on personality types and thinking styles.

The third type of error is illusions of discourse (or of the "marketplace," where humans gather and exchange thoughts). Communication itself, along with the ambiguities of language (as well as, perhaps, the use of persuasive rhetoric or "marketing"), can confuse us. They can "lead men away into numberless empty controversies and idle fancies." Here, a modern reader will likely see a crude description of "spin," and the work on persuasive tactics by advertisers and public relations consultants.[6]

Finally, there were errors of dogma, or of comprehensive "systems" or overarching ideologies (in Bacon's exotic terminology, illusions "of the theater"). Bacon targeted a few cult-like theories familiar in his own time. But we still encounter grandiose theories and fads that tend to receive fervent support. Bacon certainly did not view those grand systems as conducive to good science. I would like to think that Bacon would approve of the modern psychological and sociological analyses of conspiratorial thinking.[7]

So, for Bacon, thinking could be adversely influenced by four error types: (1) inherent cognitive deficits, (2) individual bias, (3) language, and (4) unchecked adherence to programmatic ideas. One can certainly find resonances of Bacon's observations in modern cognitive science, as detailed in Chapter 3.[8] However, Bacon did not provide many details about his four "illusions." Nor did he articulate strategies for remedying them. Indeed, his categories seem almost as puzzling now as the arcane metaphors he used to label them. Nor does his obscure, somewhat haphazard organization seem very systematic, or helpful for informing contemporary scientific practice. Rather, they are mostly antiquated curiosities. Still, his effort was an important historical landmark in thinking about types of errors.

By contrast, if one walks into a lab today, one will likely hear researchers talk about systematic versus random errors. Random errors arise from the "noise" in the apparatus, the limits of measurement, and other variations in conditions. These are perhaps best characterized as *uncertainty*, rather than as true errors (section "Error and uncertainty" in Chapter 1). Identifying them in this way allows one to manage them mathematically through statistical analysis or through more precise measurements. Systematic errors, on the other hand, are trickier. They arise from flawed theoretical assumptions, poor experimental design, or just plain ignorance of relevant variables. They tend to alter all the results in the same way—not randomly. But by their very nature, these sources of (true) error are unknown, hence very hard to detect. Not surprisingly, they concern experimentalists the most. So, it is instructive to differentiate between uncertainty and error, but unfortunately, there is no commonly accepted or comprehensive list of the systematic errors, thus the value of this categorization does not extend further. [9]

There have been various other efforts to conceptualize error types. Each offers a glimpse of the challenge and the issues involved.

For example, Giora Hon approaches the problem from the perspective of an experimental physicist. He describes the errors as one might encounter them sequentially in the course of an investigation, sorting the errors to the various stages in which they occur: (1) background theory (laying down the theoretical framework of an experiment), (2) assumptions concerning an actual experimental setup and its working (constructing the apparatus and making it work), (3) observational reports (taking observations or readings), and (4) theoretical conclusions (processing the recorded data and interpreting them). These largely coincide with Mayo's framework for experimental inference, although Mayo refers to errors in the value of a parameter (taking observations) in a different place. Yet, while this categorization is surely convenient for the practitioner, it provides little in the way of philosophical insight on the nature of the justification problem. For example, these error categories do not seem specific enough to clarify a particular source of error, such that one may consider how to remedy it. A process-oriented approach is also vulnerable to the fact that errors may emerge even with proper methodology (see section "Error naturalized" in Chapter 1).[10]

Mayo expands on the pitfalls of experimental inference by suggesting, more specifically, four types of recurring errors, which she calls *canonical errors*: They are as follows:

(1) Mistaking chance effects or spurious correlations for genuine correlations or regularities
(2) Mistakes about the quantity or value of a parameter
(3) Mistakes about a causal factor
(4) Mistakes about experimental assumptions.

As Mayo herself notes, this short list is hardly exhaustive. (In this chapter and the next two, I will add substantially to this brief and somewhat ad hoc list.) But the concept of a "canonical error" is informative in the sense that it acknowledges that scientists repeatedly encounter instances of *particular* sources of error, and that one might profit from trying to characterize *general* error types based on them.[11]

The problem of interpreting experimental sources of error has gained additional prominence recently in the context of the so-called Reproducibility Crisis. What happens when a replication "fails"? What went wrong? Is the original experiment in doubt based on some source of error that escaped notice initially? Or is the replication itself perhaps unreliable (failing, perhaps, to adhere to "the original recipe")? Or is it all just a matter of statistical uncertainty from one data set to the next? Stephan Guttinger and Alan Love[12]—melding views from biochemical benchwork and philosophy of science—have provided one "taxonomy" for interpreting the possibilities. First, they distinguish between theoretical failures (identified merely as overgeneralization) and methodological failures—notwithstanding fraud as a third possibility. They then partition the methodological failures into finer groups (which echo Hon and Mayo). This is a very broad scheme, and largely incomplete, but again reflects the effort to put some concrete structure to the problem of error. Earlier, in a short essay in *Nature*, Glenn Begley responded to the crisis from his perspective as a senior scientist at a biotech pharmaceutical firm. He identified "six red flags of suspect work." The list appealed to some familiar standards: blinding; basic repetition; full data disclosure, positive and negative controls; validated reagents; and appropriate statistical tests.[13] It seems informative that someone in the field, scouting for promising results, finds that the six safeguards against error are violated often enough to profile them as problematic in a major journal. More recently, Vazire and Holcombe have outlined a similar kind of error list: their "observable self-correction indicators" (OSCI).[14] While these ventures are oriented to the current controversies about replication, they exhibit specifics beyond the earlier crude categories. But, as I will show below, they are still not exhaustive enough. Nor systematic enough.

Another major effort has focused on promoting evidence-based medicine and reducing the errors that can sabotage clinical research. Originally sketched by David Sackett in 1979, it has since grown into the Catalog of Bias Collaborative (catalogofbias.org). The focus has been on "biases," defined as "any process at any stage of inference which tends to produce results or conclusions that differ systematically from the truth."[15] The biases (as sources of error, in our terms) are sorted (again) into the procedural stages of a research study: (1) conceptualization (study design), (2) selection (obtaining a fair sample), (3) conduct (patient and doctor behavior that conforms to the study design), and (4) reporting (various forms of miscommunication). In accord with the focus on clinical medicine, the biggest category identifies various forms of ascertainment bias (also known as sampling bias)—namely, "when data for a study or an analysis are collected (or surveyed, screened, or recorded) such that some members of the target population are less likely to be included in the final results than others." That is, the epistemic goal is to compare equivalent populations with and without a treatment in question. But there may be ways for one group to have some confounding factor (not being tested), but which influences (or "biases")—completely unnoticed—the results of the targeted treatment. It may be in how patients are enlisted or recruited, perhaps in the timing or location. It may be in the presence of coincident drugs or treatments more in one group than the other. Or unknown correlations based on patient types and their doctors. These biases were all discovered and documented through the history of clinical research and now function in a cautionary role. This catalog is a fine example of a collection of error types, in this case oriented primarily to the domain of clinical medical research (see also discussion of "Error repertoires" in Chapter 8),

The various efforts just noted (by Hon, Mayo, Guttinger and Love, Begley, Vazire and Holcombe, and Sackett, et al.) focus primarily on the limited realm of experiments in a lab setting or in clinical research. But errors certainly extend more broadly to theoretical debates and the critical discourse among scientists. For example, the concepts of phlogiston or caloric (or others entities in Figure 1.1) were not merely based on a few experiments, but served a major theoretical role in synthesizing, modeling, or explaining a broad range of phenomena. The relevant epistemic factors in those cases also belong in our error catalog or error inventory.

Here, we may be reminded, for example, of the strategies for theory change articulated by Lindley Darden (developed in the case of the history of Mendelian genetics, and applied also to challenges to the central dogma of molecular biology and other cases).[16] Philosophers of science do not customarily or uniformly view theory change in terms of error. But when one theory succeeds another, the former is typically regarded as misleading or wrong. Or the scope may ultimately be circumscribed, in cases where earlier incautious overgeneralizations were found to be wrong. Certainly, the network of justification can change with new evidence, spurring scientists to abandon one theory in lieu of another. Theory replacement (or theory displacement) casts a shadow of error behind it. It is strange, perhaps, that according to education theory many forms of learning involve replacing one conception with another. Kuhn's view of scientific revolutions was certainly shaped in part by Piagetian models of learning. The educator's discrepant event is paralleled by Kuhn's anomaly, a student's exploration is akin to the proliferation of scientific theories in response to a science community's crisis, and consolidation is analogous to the resumption of normal science under a new paradigm. So, characterizing the piecemeal elements of theory change—as forms of conceptual errors—is no less important than enumerating the experimental ones.

These conceptual reconfigurations are where human cognitive patterns—or cognitive biases (Bacon's "illusions of the tribe")—can foster misleading perceptions or distort reasoning from one claim to a supposed equivalent. For example, one may mistake one familiar case as being more representative than it is: a bias of salience. One may mistake the scope of a concept as larger than it ultimately turns out to be: overgeneralization. One may amplify extremes and neglect the mundane: sharpening and leveling. There has been plentiful research in psychology identifying these various heuristics, their biases and occasions of failure. Indeed, these pitfalls have become so familiar and exhibit such common currency that they are enumerated on Wikipedia. One impressive catalog of these errors was assembled based on clustering and organizing 188 of these thinking heuristics/biases registered there: the Cognitive Biases Codex.[17] But we need to trace their occurrence in science more thoroughly.[18]

As previewed in Chapter 1 (section "Error naturalized"), errors may also occur at the social level, in the discourse among scientists. The field of social epistemology is still relatively young, and the articulation of errors at this level seems to be under development still. I hope to address this deficit in Chapter 4. As we will see, however, Robert Merton's norms for "the growth of certified knowledge," first proposed in 1942, seem to anticipate various problems at this level, and prove informative.[19]

All these error types—from contaminated samples or instrumental artifacts to logical fallacies or cultural biases to misjudgments about a peer's credibility or the stifling of fruitful critical discourse at the social level—were part of my own beginning effort to organize error types over two decades ago. We need to craft a synthesis of experimental, conceptual, and social perspectives and make sense of errors at all levels, and show how they fit together in a unified structure—to be considered next.[20]

Mapping and the layers of justification

To convey the relation between scientific inquiry, justification, and history, as a framework for interpreting error, I want to draw on a conventional (some may say old-fashioned) conception of science and scientific theories: that scientists regularly generate representations of the physical world. Here, we should interpret representation in the most liberal way, as *re*-presentations: to include models, causal mechanisms, mental schema, theories, diagrams, graphs, computational simulations, and so on. For the sake of visualization, one may consider them all *maps*. The map metaphor for science has a venerable heritage—adopted by John Ziman,[21] Bas van Fraassen,[22] Horace Freeland Judson,[23] Henry Bauer,[24] and Ron Giere,[25] at least—and has been explored most recently by Rasmus Winther,[26] although I find the earlier account by David Turnbull[27] more cogent. Namely, scientific theories (or models, or *maps*) are selective, conventionalized, and contextual indexical representations. They are never intended as exhaustively realistic replicas.[28] Perhaps, for our purposes here, we can settle for an intuitive image of maps as depicting (with qualifications) some corresponding empirical landscape or territory (its domain)?

The next question is: How are such representations, or maps, produced and justified such that we may trust them (as able to inform effective interaction with some aspect of the world)? To articulate the epistemic core, here, we may briefly revisit Ian Hacking's analysis of microscopes and the apparently benign question, "Do we see through a microscope?"[29] Namely, is the image that strikes our retina a reliable representation of the specimen under view? Hacking poignantly debunked the naive impression of a microscope as a lens system that simply enlarges the image. Most microscopic images are produced by diffraction, as light passes through the specimen and is deflected by some of its internal structures. What we observe are shadows, in a sense. Nevertheless, through experience, a microscopist learns to read the image in a way that is faithful to the original specimen: a re-presentation, or map. The image is surely transformed along the way. For example, refraction bends different wavelengths of light differently, leading to chromatic aberration: a potential error. However, knowing how a microscope works, we have been able to accommodate the aberration by devising achromatic lenses or by using monochromatic light sources. The key is understanding the nature of the transformation, and ensuring that the process preserves the relevant information about the essential structures and the relationships between them. The upshot for Hacking (and for us) is that we do not see *through* a microscope, we see *with* a microscope. How do we know (or justify) that what we are seeing is genuine? Hacking discusses several strategies, but the key element is understanding the *mapping*. "When an image is a map of interactions between the specimen and the image of radiation, and the map is a good one, then we are seeing with a microscope." Ultimately, this is the foundation for interpreting maps or representations of any kind. Is the mapping a good one?[30]

A microscopic image, a telescopic image, a sonogram, or an ocean-floor sonar profile are all simple *maps*. So too are oscilloscope screens, spectrometer readings, a thermometer's mercury level, bioassays, and seismographs. In the past, some may have regarded these as "simple" observations, involving no element of justification. But Hacking's analysis of the microscope, along with the work of the New Experimentalists and measurement theorists, has underscored that even the most basic observation, measurement, or data collection is fraught with epistemic peril and the outcome—the observational map, say—must be justified if we are to escape misleading error.

But this is only the beginning in constructing a fact, concept, or theory. The process of mapping is compounded at each stage of a scientist's work. Data points are assembled into graphs (also maps). Graphs are compared and statistical differences analyzed (more mapping on another level, entailing another layer of justification). Results from one sample population are contrasted with another, with correlation coefficients, say. Patterns emerge. Models are constructed to reflect the interaction of multiple components. Is the selection of relevant components and the nature of their interactions justified? And so science develops. Maps of maps of maps. The justification structure in science certainly seems more complex than the conventional focus on reasoning between just theory and observation or theory and evidence. Scientific papers are complex achievements of argumentation.[31] A synoptic view of science involves layers and layers of justification, each one susceptible to error (see Figure 2.1).[32]

A complete image of the structure of justification in science thus involves a trajectory, of sorts, multi-layered and highly integrative, evoking Bruno

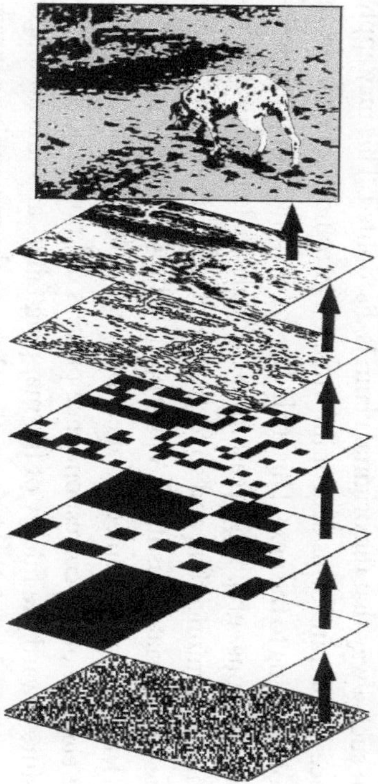

Level	Stage	Justifications
Social-Level (Discoursive)	scientific claim informing public policy or personal decision making	faithful communication of the consensus of relevant experts; credibility of media
	consensus	robustness of alternative reference frames; consensus of diverse contextual perspectives; reviews and meta-studies
	public claim	publication; fair & critical peer review; professional networks; credibility filters
Conceptual / Cognitive	theories	explanations, meta-models; consilience of inductions; families of models
	modeling	models, idealizations, simulations, predictions, causal concepts
	patterns of data	generalizations, law-like formulas; interpretive statistics: t-test, chi-squared, ANOVA; correlations, ratios, indices; curve-fitting; comparisons of graphs; taxonomies & classifications
	organized data	data reduction; descriptive statistics: averages, medians, ranges, sample size; graphs, trends, percentages; uncertainty bars; data files
Observational	observations / raw data	measurements, images, instrument readings & output, "inscriptions," data points; lab notebooks
	"real world" phenomena	material events & causes

Figure 2.1 The layered structure of scientific justifications, via successive stages of reasoning, from local observations to more global claims and generalizations.

Latour's image of long networks, the serial enlisting of allies, and centers of calculation. The path originates in the documentation of some natural phenomenon—measurements and observations, which yield "raw" data—and extends through many transformations to the ultimate forging of agreement across a community of scientific experts. As a claim coalesces, it traces a trail of longer and longer chains of justification. As the evidence is synthesized in successive stages from disparate clusters of observations, the justification builds a wider and wider branching network. At the same time, it knits together more and more local phenomena and observations into successively larger and more encompassing global perspectives. Quite an extensive process, when dissected fully. The resultant series of maps may thus be arrayed along three dimensions simultaneously: from *local* to *derived*, from local to *networked*, and from local to *global*. This is the overarching structure of justification exhibited in the "natural history" of a scientific claim, from shards of data to community consensus.[33]

And any of the successive steps may go awry. The process of justification may falter anywhere along the way. Namely, any single step in this vast multilayered structure may be a source of error. That helps dramatize the task of assembling an inventory of error types.[34]

In Chapter 1, I provided a provisional definition of error as "*a scientific claim whose justification is later found to be unwarranted*." Here, using the map metaphor, we may further characterize the unwarranted justification in terms of

> *an error is a (faulty) mapping that does not preserve the original structure of the world, at first unnoticed.*

That is, an error is fully discovered/constructed when the justification map changes in such a way that the original claim is discounted. This may apply to any process that transforms data or information into another form—from one layer of mapping to the next. This includes, for example, sampling, well-framed observations, graphing, reasoning through analogy, causal inference, faithful communication, and so forth. Is each such transformation, or remapping, appropriately justified? At one point in time (or from a particular viewpoint), the mapping seemed justified. Later more information indicates that some element of the correspondence process did not work. The epistemic posture toward the fidelity of the mapping falters and it is no longer seen as preserving the original information through the remapping process.

Observational

- material errors
- framing errors
    - sampling bias
    - small sample size
    - incomplete sampling
    - skewed proxy variables and heuristic gaps
    - confounders
- instrument errors
- fallible human observers
- observer bias
- artifacts and observer effects
- flawed experimental assumptions
- flawed experimental design (other cases of misframing)
- lack of competent execution by technicians or other laboratory personnel
- systematic failures of heuristics ("short-cut" methods)

Conceptual

- overgeneralizations
- faulty assumptions
- theory-laden judgment and confirmation bias
- cultural bias
    - religious bias
    - gender bias
    - racial bias
    - class and political bias
- heuristic gaps
- cognitive lapses
    - biases in causal reasoning (e.g., correlation/causation; post hoc fallacy)
    - biases in probabilistic and statistical reasoning (e.g., base rate fallacy)
    - others (e.g., fluency bias; salience bias)
- unaddressed alternatives
- cryptic alternatives
- conventional logical fallacies
    - (e.g., affirming the consequent; denying the antecedent)
- inappropriate analogies
- inappropriate choice of models or model systems

Social-Level (Discoursive)

- communal confirmation bias
- communal cultural bias
- lapses in communication
- lapses in peer review (pre-publication)
- publication bias
- credibility bias and fraud
- conflict of interest

Figure 2.2 A partial inventory of potential sources of error, or general error types.

Echoing Hacking: The mapping is *not* a good one (any more). These particular factors are the *sources of error* that we want to catalogue and characterize as more general *error types*.

Again, one may sort the structure of a scientific justification into three rough layers of mapping: observation, conceptual interpretation, and social-level (discoursive) assessment. The first is intimately connected to the material processes of the lab or field site, documenting concrete phenomena: observation. The second is largely the cognitive work transforming that empirical information into ideas, typically organized in a publication or "report" for presentation to peers: conceptualization. The last layer involves the interactions of scientists in a community of fellow experts who assess, discuss, revise, filter, validate, and share those ideas: social-level discourse. These layers form very general categories for conceptualizing error types, or general sources of error (Figure 2.2), as described in more detail in this chapter and the next two.

Observational errors

The potential for error begins at the scientist's first interface with the empirical world, what one may generically call *observation*. What are the epistemic dimensions of mapping at this most basic level, between territory and map?[35]

Measurements and other forms of documenting phenomena are foundational, whether in the lab, the field, the clinic, or the observatory. Or in mining data from any vast reservoir of digitized information. These basic "raw data" are vulnerable to misinterpretation in many ways, based on accuracy, precision, clear specification of conditions, relevant controls, appropriate methods, consistency, theory of the instrument, calibration, standardization, replication, adequate sample size, experimental skill, the craft of stabilizing phenomena, detecting technical artifacts, placebos, blind and double blind studies, concurrence among alternative methods, choice of model organism, curation of large data bases, and more. A philosopher is well positioned to appreciate the shortsightedness of popular appeals to "the simple facts of the matter." Accordingly, the challenges of merely "observing" now command the attention of many philosophers: the New Experimentalists and mathematicians in measurement theory, for example. As philosopher Ian Hacking once mused, "experimentation has a life of its own" (1983, p. 150). All these processes are susceptible to error.[36]

"Observation" may be complex. For example, one may be appropriately humbled that the "Observation of the Top Quark," published in 1993, was the product of 403 authors (the members of the D0 Collaboration) and several years' work. All that, for one observation. Similar complexities, undertaken by large collaborations, have been involved in detecting solar neutrinos, the Higgs boson, and gravity waves. Nearly all observations are indirect, using telltale material or transduced energy traces as vicarious selectors (namely, proxy indicators, adopted heuristically) of some remote event. Even human vision is highly mediated by neural processing (in various layers of the retina, the lateral geniculate nucleus in the thalamus, and the visual cortex). The notion (recently defended) that "naked-eye observation" is somehow the purest and most basic form of perception and the bedrock of science just seems misplaced. Every step in the transmission of a "signal," every transformation is open to error.[37]

One may also take note, perhaps, of the remarkable challenge behind what may seem ordinary and mundane: How does one bridge the physical and symbolic worlds? How does one convert an event, a position, a time, or a physical property into language, a sign, a number, a graphical trace, or a mental entity? Each observation is an empirical benchmark, of sorts, from which a conceptual map can be assembled. Once the "inscriptions" or "data" are created, we may easily forget their contingent and tenuous link to the territory. A close examination of observational practice may help us remain alert to the epistemic processes involved and their potential fragility. For convenience, we may group concerns here into at least four types: materials, framing (including experimental design), instruments (including human observers), and interference with the phenomenon being investigated (including spurious artifacts).[38]

Material errors

Perhaps the very first challenge is to ensure that the thing that is being observed is indeed the thing that is being observed. Are the concrete physical conditions correct, or interpreted appropriately? That is, the material dimension of the territory itself, or the empirical landscape, may deviate from the planned observation. It is no secret that published reports of research typically include a separate section titled "Materials and Methods," allowing others to evaluate the conditions of observation. Philosophers have, perhaps, been too concerned with the intellectual dimension of science—its theory, its "theory-laden" observations, it controversies—thereby eclipsing attention to the material dimension of practice and the concrete dimensions of the laboratory. A focus on error may help restore the balance?

Robert Crease has imagined that lapses such as impure materials, defective cables, poorly machined parts, unanticipated factors, and the like are exceptions. "The process of experimentation itself requires that cases in which one has been misled completely by the experiment be exceptions, the stuff of lunch table anecdotes."[39] Nevertheless, material lapses do occur. They are common enough—or the consequences great enough—to warrant sustained study. This is true, even for errors that may seem trivial at first. They can have profound consequences, as illustrated in the case of a viral contaminant in a PCR kit that led to a paper in *Science* (later retracted) that claimed that chronic fatigue syndrome was caused by the XMRV mouse virus.[40]

Consider the dilemma of Antoine Lavoisier, tasked with monitoring the purity of the Parisian water supply in the late 1700s. Following custom, he evaporated samples of river or canal water and measured the earthy residue—presumably its impurities. However, suspecting the glass vessel, Lavoisier measured the mass of the vessel before and after, as well as the earthy residue after boiling a sample of rainwater. Verdict? The glassware itself was disintegrating and introducing impurities into the purity test.[41]

Water purity still matters today. For example, consider the case of the tainted Perrier water in 1990. Imagine a relatively routine workday at the Environmental Health Department in Mecklenburg County, North Carolina. The lab there was responsible for testing samples for water quality. Dilute the sample. Run it through the mass spectrometer. Look for any traces of possible organic pollutants. Go on to the next sample. But on January 19, 1990, a novice employee encountered a blip on the screen. It appeared even when the sample was just water. That was odd. An error somewhere, probably. But where?

That spurred a search for the source of error (see also section "Isolating errors" in Chapter 5). The lab checked the machine. They checked the utensils and then all the sources of potential contamination they could imagine. After a few days, they began to wonder about the water used to dilute the samples. Earlier, they purified their own water. "We used to take deionized water," the laboratory director explained, "boil it, purge it with nitrogen, seal it in a container, then cool it down. But this was fairly time-consuming for the amount we were using." So, for five years they had been going to the grocery store and buying bottled water, "two or three bottles at a time." Not just any bottled water, though. Perrier. Premium, "all natural" water. "There were no organics in it, and," the director admitted plainly, "it worked." For a while, at least. Suddenly they were testing the "pure" Perrier itself, like one of their normal unknown samples. Surprise! It had trace quantities of benzene. Well, so much for the hallmark purity of Perrier! Having isolated the error, the county testing lab could resume normal work. However, the contamination was reported to relevant federal agencies. Perrier recalled bottles worldwide. They lost sales for several months. The error was tracked to a failure to change their water filters on schedule. That was easily fixed, but not their lost market share.[42]

So, a contaminant can muddy the water. Other times it is contaminated (or mislabeled) cell lines. Or commercial antibodies. Or modern human DNA in a Neanderthal DNA sample (see section "Error repertoires" in Chapter 8). Or a sample of hafnium that is not as pure as the manufacturer advertised on the label (see section "The case of the mislabeled hafnium" in Chapter 5). Or arsenic that has not been sufficiently washed from a bacterial sample, thereby appearing to be incorporated into its DNA.[43]

The notion of a contaminant can be quite broad. In 1883, Heinrich Hertz studied whether cathode rays were charged particles or electromagnetic radiation. Passing the rays through an electrical field and seeing no deflection, he concluded they were waves. About 14 years later, J. J. Thomson performed the same experiment, but was more careful to evacuate air from his tube and secure a better vacuum. Now, the rays bent. Namely, ion-filled air had "contaminated" Hertz's cathode ray tubes.[44]

Material errors can be extraordinarily frustrating. They can sometimes be the most unassuming factors. One group studying isolated cell lines in breast tissue was conducting work in two locations. Their results differed. When they finally worked in tandem, they realized that they used different mechanical methods for separating cells, both standard protocols. One gently rocked a suspension of cells to and fro; another used an automated stirring rod. One is reminded of James Bond's classic martini specification: "shaken, not stirred." Yes, it can make a difference, and possibly lead to an interpretive error.[45]

Perhaps, as Crease contends, material errors are exceptions. But when they do occur, they can nonetheless be significant. One dismisses their relevance at one's peril. It is not without cause that investigators and their lab technicians fret about contamination and exactitude in the physical conditions of their observations—even fussing about clean glassware. If material errors are few, it may be because lab techs know their importance and work so diligently to prevent them and to cross-check for unintended mistakes. Such work on checking for and preventing errors will be revisited again in Chapter 8.

Errors in framing

As noted above, one wants to be sure that the thing that one is observing is indeed the thing that one thinks one is observing. Is the "territory" appropriately delineated by the methods of observation? Borrowing from the map metaphor, we may call this an issue of *framing*. Namely, is the domain of observations properly aligned with what one intends to "map"? It is about articulating boundaries, or a field of view defined by placing a "frame" around it. Here, it describes what phenomena or events are regarded as relevant—and thus included in the observations—and which are irrelevant, and thus excluded from consideration. In terms of error, I am interested in cases where an observation (or set of observations) yields an error not because the observation itself is "wrong" or deceptive, but because one misconstrues what it is an observation of.*

Several error types involving misframing (or misalignment) are possible. I discuss five in this section: sampling bias, (small) sample size, incomplete sampling, proxy variables, and confounders. These error types in (mis)framing are, in general, already familiar to experimentalists, although perhaps with different names or labels. By grouping them together here, I hope to emphasize their thematic relationship. A set of "plain" observations is not always so simple as it may seem. What is included and, equally, what is excluded matters.

Establishing the appropriate frame of observation is closely associated with experimental design and controls. The goal is to include and document

* The term *framing* (or *misframing*) is susceptible to many meanings. Here, the key issue is relevance, not contextualization. An observational frame, in the sense I adopt here, does not refer to what physicists call a "*frame of reference*," used to describe a position that establishes the set of coordinate axes. That is about context. Nor is "framing," here, about a conceptual *framework* of background assumptions and theory, which also designates a whole context through which interpretation proceeds. Nor is an observational frame about "framing" a message, used to analyze how the contextualizing of information in communication or media has measurable psychological effects. Rather, an observational frame is similar to the philosophical concepts of *scope* (of a theory or model) or its *domain* (of application). The "frame" circumscribes what phenomena or events are deemed relevant, and which irrelevant. It separates what is included and what is excluded as an "observation."

all the relevant factors or variables, while excluding possible distractors, confounders, or coincident variables (or, if unavoidable, monitoring and subtracting their effect). This design goal is found whether one is finding the appropriate conditions in nature or creating those conditions somewhat artificially in a laboratory.

A nice analogy is provided by Nelson Goodman in his musings on the classic problem of induction. When are we justified in generalizing or extrapolating from a subset to the whole? Goodman asked, "What makes a good (or fair) sample?" Namely, when does a part justifiably *exemplify* the whole? Goodman's everyday analogy was a tailor's swatch of fabric. When does the swatch convey the pattern fully and "honestly"? In a similar way, one might ask: When does a music clip allow you to interpret the whole musical work? When is an excerpt of a movie or book sufficient to characterize the whole? The same challenge applies to observational frames in science. When does a limited set of data or measurements fairly represent the aspect of the natural world we are trying to characterize or understand?[46]

Sampling bias

Unrepresentative samples, even if "authentic," can be misleading. The observational frame can be metaphorically askew, relative to the intended target. One may misinterpret the selected cases as representative of all cases, even in large, well-organized experiments.

This error type is widely familiar, so few examples are needed, here. The classic historical case is the Lanarkshire Milk Experiment. In the spring of 1930, the Department of Health in Lanarkshire, Scotland, studied the effect of milk supplements on the growth of students. About 20,000 children participated. For four months, half received ¾ pint of milk per day, the other half none. Of course, children from wealthy families were already well fed, while others were poorer and weighed less. To measure the effect of the milk alone, the two groups needed to be comparable. Various strategies were designed to ensure that the selection of the individuals would be appropriately balanced. But the safeguards were not adequately administered. The local school teachers (many admitted afterwards) "adjusted" the selected groups. In the end, more students who appeared undernourished were assigned to the group that received milk. The average weight among the group that received no milk was actually higher at the beginning. Nor had they controlled for the weight of the children's clothing. Children grew, of course. But by the end, it was impossible to know if the milk itself truly had any effect. Namely, the researchers did not observe what they had planned to observe. The samples had been biased. Not long after, mathematician Ronald Fisher articulated statistical rules for experimental design, to circumvent or at least minimize such problems in selecting samples and arranging treatment groups (a new methodological norm, arising from articulating a novel error type; see section "Methodological norms" in Chapter 8).[47]

However, securing the appropriate samples is not always easy. For example, how does one track whether there is an intergenerational "cycle of violence" contributing to child abuse?

For many years, the accepted wisdom was that physical child abuse was perpetuated through families, whether genetic or learned. Numerous studies had shown that parents who abused their children had themselves frequently experienced such abuse earlier at the hands of their parents. But this approach traced causality "backwards." It was blind to the behavior of others who also experienced such abuse but did *not* "inherit" or later exhibit abusive parenting. That is the link between cause and effect that was misframed: the samples were not fully comparable.

To study such long-term effects, one needs prospectively oriented longitudinal studies—difficult to fund and to pursue institutionally over many years. However, a study published in 2015 was able to identify cases of abuse in the late 1960s and follow the individuals across the succeeding decades. While there were some cases of generational repetition, overall these was no correlation or increased risk of parental violence in the next generation. The earlier studies falsely indicating a "cycle of violence" had been misleading because they looked at an unrepresentative sampling of all the relevant cases. In the big picture, the observations were misframed. That is why sampling bias is another important error type.[48]

Small sample size

Another significant way that samples may be unrepresentative is by being too small (or too narrowly framed). On occasions, simply by chance (or mischance), the limited observations that one makes may not reflect the overall phenomenon. The resulting mapping, from sample to generalization, is not faithful. Sample size is (again) a familiar dimension in contemporary study design. But just because we know about the problem does not mean that is it always noticed or avoided.

Such was the case in 1976 while searching for the bottom quark, a subatomic particle. Evidence for such particles requires creating billions of collisions in a particle accelerator, monitoring the scatter for certain kinds of by-products, and determining their telltale mass. It is like trying to discern a clear voice amidst lots of static. One is trying to distinguish a "signal" amidst substantial amounts of "noise." In this case, 27 events were detected. In all 11 events formed a marked cluster in a narrow mass range, easily noticed when graphed. That was the apparent "signal" of what the team at FermiLab was looking for. There was the possibility, of course, that such a cluster could occur randomly—not due to any "real" particle at that energy level. The group calculated the statistical probability: less than 2%. So the team promptly published their findings in *Physical Review Letters*, announcing the preliminary evidence for what they called the Upsilon (ϒ) particle. Investigations continued, seeking further evidence and detailed information. More bombardments. More events. But now, the threshold for measurements was higher, and thus more sensitive. The clustering of events dissolved. It had been coincidence, after all. No momentous discovery (yet). Accordingly, the particle was renamed, tongue-in-cheek, honoring the project leader (and later Nobel laureate) Leon Lederman. From Upsilon particle to "Oops-Leon" particle. A lasting tribute to the risks of small sample sizes.[49]

Small samples may also be viewed *qualitatively*. That is, a limited number of observations may not tell the whole story fully or accurately. For example, the fossil relatives of the chambered nautilus, ammonites, puzzled paleontologists for a long while. Some pairs of species seemed to be found together in the same rock with startling regularity, even across different geological formations. In many case, the two species seemed to appear and disappear in the same geological periods. Eventually, with enough cases, paleontologists had to admit that the species pairs—sometimes assigned to separate genera—had been misclassified. They were actually male and female forms of the *same* species. Now, the potential error is a guide to taxonomy, with rules for deciding about sexual dimorphism in extinct species, where mating cannot (obviously) be observed directly. (Again, the regularity of an error has led to methods for managing the error type; see section "Methodological norms" in Chapter 8).[50]

So, adequate sample size, whether viewed quantitatively or qualitatively, is important to establishing a sufficiently representative subset, or fair "swatch," in Goodman's terms. The observation frame needs to exemplify the intended target. Small sample size is another error type.

Incomplete sampling

Sometimes, scientific reasoning may be misled because the full spectrum of relevant cases has not yet been observed. The sampling of relevant variants is incomplete. They may thus fail to exhibit the ultimate overall pattern of the targeted observational frame—not because there are too few samples but because the full diversity of features has not yet been exemplified. Here, the unrepresentative set of samples arises through the contingency of history. That is, the sample's limitations are not based on any inappropriate methodological filter (systematic sampling bias), nor based merely on the random probabilities in collecting a small sample (small sample size). One may generalize prematurely, in a sense, before the sampling should be considered complete (see also section "Overgeneralization" in Chapter 3).

For example, consider the history of the term "oxygen." The term—meaning "acid-former"—was coined by Antoine Lavoisier, and arose in the context of his theory on acidity. He observed that the acids whose chemical composition was known contained oxygen. Burning sulfur with oxygen yields SO_3, which combines with water (H_2O) to yield sulfuric acid (H_2SO_4). Burning phosphorus likewise yields phosphoric acid (H_2PO_3). Burning nitrogen yields nitric acid (HNO_3). Carbon dioxide in water yields carbonic acid (H_2CO_3). So, oxygen seemed the common element. Hence, he renamed the gas, earlier referred to as "eminently respirable air," as the new principle of acidity, or *oxygène*. At the time, the elemental compositions of "marine acid," or "muriatic acid" (hydrochloric acid) and "acid of fluorospar," or "fluoric acid" (hydrofluoric acid), had not yet been determined. When those were finally determined, it was clear that they do not contain oxygen. Lavoisier's sample of acid types—his "swatch"—was, unfortunately, unrepresentative. Our term "oxygen" today is a powerful reminder of the risks of incomplete sampling. A reliable scientific map needs a complete-enough sampling of benchmarks in the relevant domain.[51]

So too for Casimir Funk's characterization of certain micronutrients as "vital amines," or vitamins. The ones he examined had nitrogen, yes. But there were others that did not. The nitrogen was not relevant to their essential role. We still call them vitamins, though—a poignant reminder of the incomplete (unrepresentative) sampling that guided Funk's terminology.[52]

The incompleteness of the sample may not be obvious at first. For example, the gradual emergence of fossils of human ancestors has provided false impressions of the human lineage. As the 20th century unfolded, each new human fossil was dated and placed on the timeline. Paleoanthropologists dutifully connected the dots, reading the human lineage from the fossil record as a progression from "primitive" to modern forms. Then some fossils from very different human types were found coexisting in the same time layer. That challenged the idea (suddenly exposed as an implicit, undeclared assumption) that humans constituted a single lineage. That view had to be jettisoned as an error. Over time, more human fossils revealed even further branching. We now see human evolution as a tree, not a gradual progression. The incomplete record of human fossils had for many decades fostered a misframing of our evolutionary history and, in a sense, of our human identity.[53]

Some philosophers highlight the importance of accuracy as a way to avoid error. Others stress precision. These examples help underscore the critical (but perhaps underappreciated) role of *completeness* of evidence in the relevant observational frame(s).

Proxy variables and heuristic gaps

In some cases, the relevant phenomenon may be difficult to access, or the critical variable difficult to measure. Undeterred, researchers find an indirect route: using a *surrogate* or *proxy variable.* But the proxy may prove to be an inadequate equivalent for the intended observational frame.

Sometimes, a substitute measurement is just a matter of convenience, cost, or time. It streamlines the research process and makes limited funding stretch further. For example, width of tree rings is used as a proxy for weather during a particular year: thicker rings indirectly indicate more rainfall and higher temperatures or longer growing seasons. Petroleum geologists use the color of microfossil conodonts to reconstruct historic oceanic temperature conditions and their suitability for oil and gas formation—and thus have been valuable tools in fossil fuel exploration. The widely known method of radiocarbon dating is based on measuring the ratio of carbon isotopes in a sample. They assume certain levels of carbon dioxide in the atmosphere and relative rates of radioactive decay. That is, the measurement is based on the residue of a historical process and certain assumptions, rather than on an actual observation or determination of time. Proxy variables are all indirect indicators, or "vicarious selectors" in Donald Campbell's term. (By this, he meant that learning, as a form of natural-selection-type process, can be based on secondary information rather than on the relevant factor itself.) They may seem shaky, but scientists rely on them extensively, with varying degrees of confidence.[54]

Proxy variables are a form of heuristic, a "short-cut" that is more efficient, especially for complex operations. Heuristics work most of the time. But not always. What is being tested and what it indicates will be generally correlated, but not perfectly. So heuristics tend to have systematic points of failure. The limits of heuristics may be illustrated through such familiar things as home pregnancy or COVID-19 tests. The potential for either false positives or false negatives is well known. Those results occur in the event that the test's proxy variable does not match perfectly the phenomenon one is ultimately trying to measure (Figure 2.3). For example, the home pregnancy test measures the amount of a hormone, human chorionic gonadotrophin, that is generally only produced when a fertilized egg cell has been implanted in the uterine lining. The test is not a direct observation of pregnancy, but of a side effect. So, the test can deliver false negatives (predictably) if the test is performed too early, the urine sample has been diluted, or the results are read too soon. False positives may arise when the subject has been undergoing fertility treatment or taking certain drugs, or when there is a "chemical" or ectopic pregnancy. Accordingly, the tests come with appropriate disclaimers that they are only 99% effective. There are known exceptions. Those exceptions underscore the general principle that proxy variables shift the target observational frame slightly and can thus sometimes yield errors.

One notorious case of proxy variables involved an apparent link between childhood leukemia and electromagnetic fields (EMFs), widely reported in the media in the 1980s and 1990s. That is, those who lived near high-voltage power lines were presumably at greater risk for cancer. It was alarming, especially as amplified by veteran journalist Paul Brodeur and by the public's vague fear of "radiation" of any type. Yet the original study announcing the correlation did not actually measure the EMFs(!). As a later review from the U.S. National Research Council explained: "Measuring residential fields [EMFs] for a large number of homes over historical periods of interest

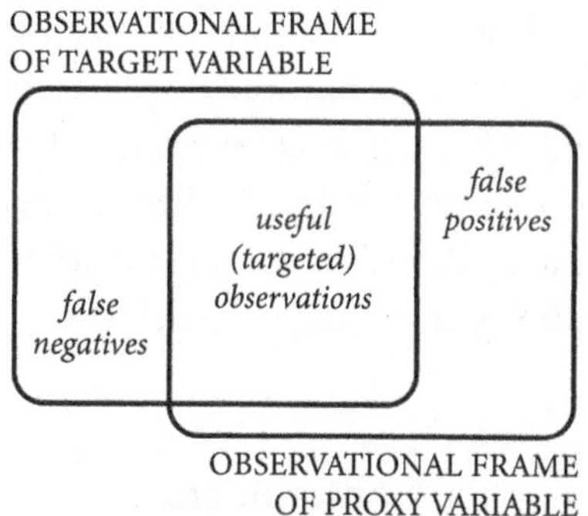

Figure 2.3 Errors may arise from misframing, stemming from the use of proxy (or surrogate) variables.

is logistically difficult, time consuming, and expensive, so epidemiologists have classified homes according to the wire code (unrelated to building codes) to estimate past exposures." The wire code is a composite index based on the size of the electric wires and their distance from the residential buildings, which can be determined easily from public maps. That is, the investigators did what sensible scientists do. They found a proxy variable to represent the exposure to EMFs inside the homes. It was the wire codes that were correlated to a 1.5% increase in childhood leukemia.

Given the significance of the study for public health, others followed, checking for methodological errors. Yet most adopted or adapted the original methods and many reached similar results. After persistent incongruences, however, the National Research Council swallowed the expense, visited the relevant homes, and measured the actual EMFs inside. When those measurements were finally done, they did not match the values estimated based on the wire codes. The correlation with cancer disappeared. The "wire-code paradox" persists. Even now, certain wire codes can be correlated to a modest increase in childhood leukemia, but what other factor they may indicate (besides EMF values) remains unknown.[55]

Another common form of proxies in biology—especially increasingly in toxicology—is the use of in vitro testing as a surrogate for in vivo testing. Namely, tests are done with cell cultures "in glass" as opposed to in living organisms. There are obvious advantages in reducing the cost and ethical burden of working with living animals. Still, the cell system lacks context, such as interactions with other cells and organs. Results, especially dosage effects, do not always transfer well.[56]

Heuristics are valuable tools in scientific investigation. But they may also fail and lead to error. Researchers ideally need to be alert to the limits of proxy variables and other surrogate measures, to know when caution regarding results may be warranted and when further research is needed prior to stable conclusions. We will encounter them again when considering conceptual and discoursive errors (sections below).

Confounders

In other cases, an observation frame focused on one particular variable coincides (unnoticed) with a second variable. The observational frames overlap. Unfortunately, that correlated variable—a *confounder*—may sometimes account for the observations one attributes (unwittingly) to the target variable. That is, a coincident variable can dictate what happens during an experiment, yet go unnoticed. While the notion of confounders is widely familiar, here I want to underscore its relation to framing. Observations may need to be subtly reframed, to exclude or accommodate the confounder.

Most notably, some influential factors may typically coincide with the phenomenon or with the relevant variable being observed. The observer, unaware of their influence, may unwittingly mistake one factor as responsible, when another, hidden variable is ultimately key. Observations need to be mindfully framed, so that the effects of any confounder is "screened off" (see also section "Isolating errors" in Chapter 5).

For example, in the 1890s Christian Eijkman (future Nobelist) tried to document the cause of beriberi in Java. In lab studies with chickens, he had observed the causal role of rice diet. Polished white rice led to the disease, while unpolished rice (still with its cuticle coating) would cure it. Still, the sample size was limited and used a model organism as a heuristic. Both were potential sources of error. Eijkman thus enlisted public health official Hans Vorderman to survey the incidence of beriberi in the local prisons. In all 100 prisons. Nearly a quarter million prisoners. No problem with small sample size there. The results impressively confirmed the role of rice diet.

But Eijkman's claims were more specific. He blamed a bacterium in the starchy rice and credited a curative anti-toxin in the coating of the "unpolished" rice. But perhaps there was a confounder? Could the bacterium have originated from another source? Perhaps it was airborne, and infection resulted from poor ventilation in the prisons? Perhaps it was waterborne, with permeable floors as the culprit? Perhaps it was released from the aging prison walls? Or perhaps it was transmitted person-to-person, with population density as the key causal variable? A skeptic might have alleged that any of these factors was responsible. Thus, Vorderman's study dutifully collected statistics on them, to show that these coincident variables did not exhibit a correlation with beriberi, and could be ruled out as confounders. Here, he endeavored to be specific and clear about the observational frame to ensure the reliability of the conclusions.

Split frame: Beriberi was not bacterial at all. Eijkman's successor at the military hospital in Java, Gerrit Grijns, imagined a different causal scenario. He attributed the disease to a *lack* of some vital nutrient in the rice cuticle, absent in the polished white rice. Both interpretations fit the observations from the chickens and from the prisons. Grijns's challenge was to reframe the observations to focus specifically on what he saw as the relevant variable: the missing nutrient, rather than any bacterium or toxin in the rice (for more, see section "Isolating error as epistemic work" in Chapter 5). Using the alternative observational frame, Grijns eventually showed that the pattern of the data no longer fit Eijkman's casual "map" based on bacteria. To justify beriberi as a nutrient deficiency, not a microbial disease, one needed to properly frame observations (see Figure 2.4).[57]

Confounders lurk everywhere, it seems. In the 1910s, there was a similar controversy over the cause of pellagra. Was it a microbial infection or a nutrient deficiency? Charles Davenport, a noted biologist from the Cold Spring Harbor Laboratory, was called in to adjudicate. His conclusion? Pellagra was

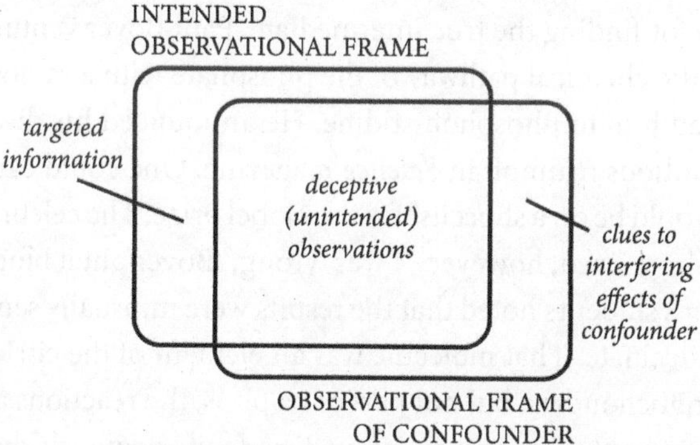

Figure 2.4 The misframing arising from the presence of confounders (spurious coincident variables).

neither! It was genetic! Davenport supported his conclusion with numerous pedigrees, each showing the incidence of pellagra across multiple generations of the same family. But Davenport's pedigrees were highly misleading. Pellagra was, in fact, nutritional—a deficiency of nicotinic acid (vitamin B_3). The poor diet stemmed from poverty, not inherited metabolism. And where economic opportunities were lacking—as in the southern United States at that time—people born into poverty remained poor. No wonder that Davenport's pedigrees could exhibit a familial pattern of pellagra. But what they really documented was families trapped in economic hardship. Pedigrees were not an effective observational frame for isolating the cause of pellagra. They hid the coincident—but more relevant—variables of poverty and malnutrition. Davenport's 1916 report was not the first nor the last time that pedigrees would be used inappropriately to try to justify genetic explanations for what were ultimately social patterns.[58]

The appearance of a confounder does not necessarily discount the observation itself. Rather, it concerns how the observation is linked to the empirical "landscape," or what one interprets as the "domain." The observation is *misframed.* Ironically, a misframed observation may still be valid, but meaningful only in another (and perhaps less significant) context.

Consider the anti-climactic "discovery" in 1963 about energy processing in the cell. The central energy reactions take place within the membrane-bound mitochondria and biochemists were searching fervently for a key intermediate that would link the final reactions in the membrane—oxidative phosphorylation, or ox phos—to the end product, the unit energy molecule ATP (adenosine triphosphate). Already eight intermediates had been proposed, and each found to be mistaken. That had sharpened the sense of tense competition for finding the true intermediate. Paul Boyer ventured in, hoping to track the chemical pathway of the phosphate with a radioactive label. That soon led him to phosphohistidine. He announced his discovery with a sense of cautious triumph in *Science* magazine. One could easily imagine that Boyer would be on a short list for the Nobel prize. The celebratory atmosphere was short-lived, however. "I was wrong," Boyer put it bluntly in 1981. One of Boyer's students noted that the results were unusually sensitive to the amount of succinate. That molecule was an element of the citric acid cycle, also in the mitochondria, but not part of ox phos, the reactions at issue. The labeling experiments, Boyer soon learned, had not adequately discriminated between the two sets of energy reactions, one embedded in the membrane and other in the internal fluid. Separating them was not that difficult at that point. As a result, Boyer's lab traced the histidine to an enzyme (succinyl CoA synthetase) that harvests energy from the citric acid cycle. Phosphohistidine was, indeed, part of energy transfer in the cell. But not involved in the high profile ox phos reactions that formed ATP. Boyer's lab had indeed discovered something, but not the prestigious high-energy intermediate of ox phos. While the claim about phosphohistidine was an error in one context, it was a novel claim in another. The domain of Boyer's observation of phosphohistidine had originally been misframed.[59]

Instrument errors

Scientific observation frequently involves instruments. Technological devices often substitute for our senses, adding mechanical rigor—and precision—to the process. They can also extend the reach of human perception—to infrared or ultraviolet radiation, to microwaves or X-rays, and to a full spectrum of sound frequencies. They can detect electrical, magnetic, or gravitational energy, or physical properties such as temperature, humidity, pH, or dissolved oxygen. But does the instrument function reliably? Sometimes they fail. When they do, they may introduce observational errors. For all their opportunities, instruments introduce a new layer of epistemic concern. They can jeopardize the reliability of the observational mapping that they are designed to help. Can we trust their measurements, recordings, or "inscriptions"?

Ideally, an investigator knows exactly *how* their instrument works, to help them check that it is functioning properly. That is integral (as noted above) to Hacking's conclusion about seeing *with* a microscope. That may mean establishing a valid "theory of the instrument" itself before one can even proceed. As Daniel Rothbart and Suzanne Slayden note, even a spectrometer has an epistemology. Namely, do we understand how the light of different wavelengths is converted into electrical signals, which are converted to a dial meter or a digital counter? Have we tested the instrument itself, before we use it in testing other things?[60]

Even so simple a device as a rain gauge can hold surprises. Yes, just a simple cylinder to catch and measure the amount of rain. In the mid-1700s, early meteorologists noticed that rain gauges placed at ground level did not register the same amount of precipitation as the very same gauges placed on roofs or several feet above ground. The puzzling discordance earned its own special name: the height-catch problem. After more than a century of attention to the problem, they determined that elevated gauges and the structures that held them aloft created wind eddies that perturbed the falling rain. Even gauges placed next to each other might collect different amounts of rain, depending on the turbulence. That led to standardizing the methods of rain gauges, so that at least there would be consistency and measurements would be comparable. (Again, understanding the error led to a methodological solution—here, a standardized observational protocol; see section "Methodological norms" in Chapter 8).[61]

Knowing about one's instrument and its behavior can help one detect errors in its measurements or images—and to selectively discard them as unreliable. A classic example is the 1919 telescopic observations used to validate Einstein's theory of relativity. The aim was to measure the bending of light rays by the Sun's gravity, observable during a solar eclipse. To measure the deflection, one must compare telescopic photographs taken at different times—one during the eclipse, of course, but also another before (and possibly after) as the reference standard for where one should expect to see the stars (without the gravitational effect). When the international team assembled the equipment on an open field in Sobral, Brazil, they tested it. They discovered that a mirror on the primary telescope exhibited a surface irregularity—an astigmatism—leading to distorted images. Even small changes in the reflected angle would be amplified into significant differences, where very precise measurements were essential. They proceeded anyway, hoping that they could accommodate any error. But when they developed the photographic negatives from the eclipse day, the stars were blurred. Charles Davidson recorded in his diary the sad news: "It was found that there had been a serious change of focus, so that, while the stars were shown, the definition was spoilt. This change of focus can only be attributed to the unequal expansion of the mirror through the sun's heat.... It seems doubtful whether much can be got from these plates." Basically, with inexact "points," one could not align the two images precisely, necessary for (re)scaling and orienting the images for comparison. Despite all the expedition effort and expense, then, the team leader Frank Dyson ultimately decided to scrap the data. In a sense, it did not matter whether the calculated measurements supported or challenged Einstein's theory. The results were relatively useless because the instrumental setup had not functioned appropriately. Thus, when the team announced the findings of the two expeditions—in surprising accord with Einstein's somewhat outlandish prediction—they relied only on images from a second, smaller telescope at Sobral and one in the Atlantic.

Much later, in 1978, with the advent of computers, the staff of the Royal Observatory in Greenwich was able to reanalyze the Sobral images, now taking into account the apparent instrument errors. Those revised values were, fortunately, commensurate with the others already published. In a sense, then, one could put to rest earlier doubts about whether the data had been discarded inappropriately, simply because they did not yield the "expected" results.[62]

So, as illustrated in the cases, the height-catch problem of rain gauges and the Sobral eclipse telescope mirror, instruments may be a source of observational error. But mindful attention to these possible errors can avert potential disaster, and even open a path to significant new discoveries. That posture of actively checking for—and then eliminating or ruling out—possible sources of errors proves to be an important strategy in managing and limiting error and in fostering confidence in the reliability of scientific knowledge (for more, see section "Error probes" in Chapter 8).

Human observers

Human observers are instruments, too. They see, hear, smell, taste, and touch. They record their perceptual data in lab and field notebooks—more "inscriptions" of a different kind. Aware of the misreadings of mechanical instruments and measuring devices, we might well reflect on the epistemic status of human observers as experimental "instruments," too. They are certainly subject to various perceptual sources of error. But rather than refer simply to "human error," one may perhaps specify how their perceptions fail, with an eye to prospective remedies.

When Hacking described how a microscope maps a specimen to an image, he did not extend his analysis to the human peering into the microscope. But perhaps one should. One may certainly regard the biology of human perception as analogous to the operation of physical instruments, and equally subject to flaws or biases. Eyes, for example, are not simple pixilated cameras. The retinal receptors have various context-dependent thresholds. Ambient light levels matter. After responding to one light stimulus, the molecular transducer in the cells (rhodopsin) needs time to recharge (refractory period). So sensitivity to the same light intensity is not strictly continuous. Next, nerve cells converge in another layer of the retina, integrating impulses into various patterns, such as "on centers" and "off centers." These are further integrated along the neural pathways to the brain, defining edges, lines, corners, and so on. For example, contrasts of light and dark are more readily perceived. The "image" in the visual cortex of the brain is not a one-to-one mapping of sensory cells.

Neural processing of observations continues. For example, new perceptions are influenced by earlier ones. Based on memory, certain sensory signals become salient, others less relevant. That is, some things are more likely to be noticed than others. Also, the brain may redirect the eye's gaze to examine one thing in more detail, while directing it away from something that is ultimately just as important. Perceptions that match earlier patterns may be processed unproblematically, while anomalous ones may trigger caution or doubt. Expectations matter. They can subtly filter or even "correct" perceptions to fit anticipated results. The "take-home" caveat for science is that human observers are complex instruments susceptible to error.

Philosophers of science, at least back to Norwood Hanson and Thomas Kuhn, have drawn attention to the vagaries of human perception. Many provide examples of optical illusions as analogies. Many opine blithely, but vaguely about "theory-laden observation." However,a meaningful understanding of error types in science will be better nurtured by concrete examples of such perceptual errors, culled from the history of science and from contemporary practice.[63]

For example, we may turn to the notorious case of N-rays, a purported new form of radiation reported in 1903. Just a few years later, they were discredited as illusory. Were they ever "observed"? What *was* observed? What was the role of human perceptions? Many commentators interpret this episode moralistically, in terms of "nationalistic fervor," "feckless credulity," and "submissive assistants ... perpetuating delusions." Here, I treat it more modestly, in terms of "normal" observational errors. Initially, distinguished French physicist René Blondot was studying the polarization of X-rays when he noticed an altered brightness in some electric sparks. He pursued the anomaly further and found evidence of another form of radiation, which he call N-rays. He continued his investigations, using spark brightness, diffusion screens, calcium sulfide residues, and phosphorescent screens to manifest the rays. Over the next several years, over 40 individuals confirmed and extended his observations, leading to over 300 papers. Yet others failed. Critics noted how the observations seemed to invariably depend on the observer. They highlighted the subjective nature of the detection. In rejecting one paper submitted to the prestigious *Revue Scientifique*, the editor advised the author: "Your results can be explained simply by supposing that your eyes do not have sufficient sensibility for appreciating the phenomena." Efforts to photograph the N-rays were not uniform and did not resolve the observational problem. Were N-rays real? How might one assess the observers as scientific instruments?[64]

We have noted that experimentalists regularly confront the challenge of instrument reliability. They habitually need to check their instruments to ensure that they are in working order. That includes calibrating them. For example, they can use a blank, or null sample, to check that there is indeed a "zero" reading: no background "noise." They might follow that with a standard sample (of known value), to check that the instrument's reading is correct. Small adjustments can sometimes be made to bring the instrument back into alignment. That strategy was essentially applied in the fall of 1904 to gauge if Blondot and his assistant, as human instruments, were functioning reliably. Robert Wood, from Johns Hopkins University in the United States, went to Blondot's lab in France to experience the N-ray phenomenon first hand, and to benefit from the observational guidance of the premier source. Wood saw nothing. Apparently he did not have the requisite observational skills. But then Wood tested his hosts, rather than the N-ray phenomenon. He periodically blocked the rays with his hand, while Blondot observed the rays. With Wood's hand obstructing the ray's supposed path, Blondot reported how the luminosity of the electric spark varied, as if the hand had been moving in and out. When Wood's hand was kept away, Blondot reported the same variation. What Blondot saw was apparently unrelated to the whether the N-rays were transmitted or not. Later, Blondot demonstrated how an aluminum prism refracted the rays into a wide spectrum, with peaks detected at specific wavelengths as the screen swept along the spectrum. He then repeated the exercise, but in the darkened room he was unaware that Wood had surreptitiously removed the prism. There was no spectrum, yet Blondot reported seeing the transmissions at exactly the same spots. Wood quietly replaced the prism, but now the assistant was suspicious of foul play. On the next trial, the assistant observed. In the dark, Wood made noises as though he was again tampering with the prism, but touched nothing. The assistant could see nothing now, and accused Wood of a disturbance. But of course, Wood had left everything as the assistant had arranged it. In this way, Wood was able to "calibrate" the two human observers. Whether N-rays existed or not, the tests of the testers indicated that these two individuals were not reliable detectors. Their published observations could not be trusted.[65]

The whole episode seemed to be echoed a few decades later, when Fred Allison described a powerful new method for chemical analysis, based on comparing polarized light transmitted through ionic solutions in a magnetic field. Like the N-rays setup, Allison's apparatus also involved a spark generator. And like the N-ray case, some observers seemed able to repeat the observations, others not. Other labs introduced new methods for assessing the brightness of the light, and used photographs or photometric technology to measure the difference in light intensity. They documented variation in the intensity of the spark that could easily have been mistaken as caused by the samples. Research on the Allison magneto-optic effect was gradually abandoned, attributed to undependable and limited human observation.[66]

Observer bias

The vulnerabilities of human perception can be a significant source of possible error—notably, when the phenomenon is at the physiological edge of sensation or a conspicuous subjective judgment is involved. An additional concern is *observer bias*, where experience or theoretical expectations can actively shape apparently unmediated sensory perceptions. Again, it is helpful to remember that the human mind is itself an instrument: a complex instrument that selects and edits neural impulses at every level of processing. The effects can be very subtle. As illustrated in the two episodes just described, they are generally unintended, unconscious, and largely hidden. But the problem recurs historically.

Nowadays, medical researchers are duly concerned about observer bias in clinical trials of new drugs and treatments. That is, a doctor's knowledge of a treatment and beliefs about its efficacy influence their "objective" diagnosis or assessment of the patient's condition—the very data that will be used to test that very question. General awareness of this psychological dilemma in medicine emerged early in the 20th century, and efforts to forestall it soon followed (another nice example of how knowledge of error types can grow and be productive—see section "Progress through error" in Chapter 8). One early landmark study was designed by Adolf Bingel in

1912–1913 in Germany. For decades, diphtheria had been a major scourge across Europe. Serum therapy (a method recognized in the very first Nobel Prize in Medicine in 1901) had certainly ameliorated the situation. Bingel acknowledged its effectiveness, but was skeptical about the serum working because of a specific anti-toxin. He wondered if the serum itself—any serum—might be equally effective. (Namely, was the special "anti-toxin" preparation a needless confounder?) By this time, the notion of controls for experimental comparison was widely appreciated. So Bingel established two groups to compare the contrasting observational frames. Some patients received the conventional "anti-toxin" serum, while others the ordinary horse serum. In an effort to avoid a biased sample, he assigned every other admitted patient to alternate groups. Moreover, Bingel was aware of the importance of sample size, so by the end of his study he had 477 individuals in one group, 461 in the other.

Still, Bingel was aware of an *additional* potential source of error: observer bias by the physicians treating the patients. Given the controversial nature of his idea, their preconceptions posed a special danger. Bingel reminded his readers that it is "extraordinarily difficult ... to evaluate the influences of therapy on disease unless they are obvious, as for example, the success of a surgical operation or cure of syphilis with mercury or Salvarsan. The therapeutic optimist very easily sees improvement, and the skeptic sees nothing." He thus wanted "to achieve an objective overall assessment," rather than the doctor's informal, possibly biased, "impressions." So, he explained later: "To make the trial as objective as possible, I have not relied on my own judgment alone, but have sought the views of the [at least six] assistant physicians of the diphtheria ward, without informing them about the nature of the serum under test (namely the ordinary horse serum). Their judgement was thus completely without prejudice. I am keen to see my observations checked independently, and most warmly recommend this 'blind' method for the purpose." Here, Bingel used the term still familiar today: "blinding." An observer who remains "blind" to the test conditions remains "without prejudice," as Bingel noted, and has no expectations to exercise. That method gave stronger credence to Bingel's contentious conclusion: *any* serum was effective. The theoretical "how" assumed by the Nobel Prize winner was in error. Of course, the method of blinded analysis that Bingel helped pioneer is now a standard of responsible clinical research.[67]

Observer bias equally haunts a variety of fields in modern science. It has now been documented in forensic science;[68] in animal models of human health conditions;[69] in assessing aggressive ant interactions between nestmates and non-nestmates, trying to test the theory of kin selection;[70] and even in tropical herbivory,[71] to name a few.

Observer bias is insidious, surely. One would not necessarily predict it, philosophically. Yet it can be documented in practice—subtly increasing the frequency of certain effects and amplifying the effect sizes. It occurs unconsciously and can easily remain hidden, often only realized in retrospect. While such bias is unintentional, it can nonetheless be managed intentionally—for example, through Bingel's strategy of blinding. The topic of observer bias (among many biases in science) is currently a matter of active discussion among researchers. How extensive is it? How damaging is it to the conclusions of science, perhaps to studies that have already been published and which may form a historical bedrock of the discipline? Regardless, what can be done to minimize it? The custom of blinding, familiar in medical and psychological research for over a century, is now informing more fields of science. We will return to the historical development of methods for regulating errors in Chapter 8.

Artifacts

When instruments function well, they surely enhance scientific investigation. Sometimes, however, rather than facilitate observing some facet of nature, they *create* the very phenomenon being observed. The resultant "observations"—images, measurements, and signals—are artificial and entirely spurious. That is, instruments are susceptible to introducing *artifacts*, meaningless data that may be mistaken as "real." An artifacts is, in a sense, a confounder present in the experimental apparatus or observational method itself.[72]

One case that has received substantial attention by philosophers of science is the production of bacterial mesosomes. These spiral membranous structures first appeared in 1953 in images using the new methods of electron microscopy. By the mid-1970s, however, they were dismissed as artifacts: caused by the treatment of the cells with osmium tetraoxide or other fixatives that damaged the cell membranes. Ironically perhaps, they are perfectly reproducible.[73]

Another would-be "discovery" was the report of "anomalous water" in the 1960s. A team of distinguished chemists in Russia condensed water on the inside of a glass or quartz capillary tube. It mimicked the kind of narrow space found in porous rocks, tiny blood vessels, and plant phloem cells. Under these conditions, the water seemed to exhibit extraordinary properties: higher viscosity, lower melting point, and lower vapor pressure. Theorists responded by proposing that the water molecules formed long polymer chains. They dubbed it *polywater*. The possibility of impurities in anomalous water, explaining its remarkable properties, was considered from the start. Precautions were taken. The preparation was refined and spectroscopic analyses were made. Given the promising military and commercial applications, investigations flourished. For a few years. Then claims about contamination became more frequent and more concrete. Several labs published new spectroscopic analyses, comparing polywater with proposed contaminants and with "polywater" prepared with methods that specifically excluded the suspect materials. The verdict? Anomalous water was tainted with silica (from the glass or quartz tubes), as well as (in various labs) sodium and other minerals and organic compounds that could easily have come from sweat. Anomalous water was not so anomalous, after all. It was not a chemically unprecedented polywater. It was, basically, dirty water. In 1973, the original advocate, Boris Deryagin, acknowledged the error with characteristic scientific understatement: "We have established that there are no condensates free of impurity atoms and simultaneously exhibiting anomalous properties. ... Consequently, the anomalous properties of condensates may be explained, not by the formation of a new modification of water, as was previously supposed, but by the peculiar features of a reaction taking place between the vapour and solid surfaces in the process of condensation." (Yes, even now one can still produce "anomalous water" and observe its special properties. It is entirely reproducible. However, it is an irrelevant artifact of the preparation procedures.)[74]

Artifacts can easily deceive experimenters. A 2017 paper in *Science* presented results indicating that an electrical current could induce a magnetic effect in a certain type of insulator. However, as research continued, the team learned that the current had other, unintended effects. It generated heat in the sample. That, in turn, heated the sample holder. And it was *the sample holder*, not the sample itself, that exhibited the measured magnetic effect. As the authors stated in a retraction three years later, "it became clear that a large part of the reported diamagnetic signal arose from a mechanism we did not anticipate. This signal is attributable to localized heating of the sample holder, caused by the unavoidable Joule heating in the sample." That is, they had unwittingly *created* the misleading data with their instrument—the very definition of an artifact.[75]

Glitches can hide in the most unassuming—and difficult to find—places: even in computer software. For many years in the 1990s, the U.S. Environmental Protection Agency sought to determine the number of deaths associated with certain levels of soot in the air. After five years of generating results, one investigator found an error in a default setting in the statistical program that crunched the raw numbers. That resulted in slightly "overcounting" the relevant mortality rates. In another case, in the late 1960s Joseph Weber reported detecting gravity waves using two large aluminum drums. But, after years of controversy, the apparent signals were traced to effects of the computer program that analyzed the data from multiple locations. Another elusive artifact.[76]

As noted earlier, scientific instruments can fail. But sometimes they fail specifically by producing artificial "effects" that are mistaken for real ones, or for the normal course of nature. Artifacts are another potential source of observational error.

Observer effects

Knowing that instruments may yield artifacts, one may inquire into how human observers or experimenters themselves may disturb what is observed. Sometimes, the act of observing can change the very thing that the scientist is hoping to observe: an error type appropriately called *observer effects* (not to be confused with observer bias, above).[77]

For example, in the mid-1840s, James Joule's work on the mechanical equivalent of heat was based on quite sensitive measurements, involving very small amounts of heat. Joule set up his apparatus in the cellar of the brewery where he worked, where he could expect a stable temperature. Furthermore, he insulated the vessel. Even so, he needed to ensure that the water started each experimental run at the same temperature as the room, lest it absorb or lose heat and distort the results. Even with all these precautionary measures, Joule found that *his own body heat* could affect the temperature readings. Namely, the observer was disturbing the phenomenon being observed. As a result, Joule had to shield himself behind a wooden screen, well away from the apparatus. To read the thermometer, he had to install a small telescope, viewing the water tank through a small window in the screen. Sometimes, the act of observing can unwittingly change what is observed.[78]

Sampling of atmospheric carbon dioxide (CO_2) is also subject to observer effects. Even the breath of an unwary person taking a sample contains enough carbon dioxide to "pollute" the sample. Accordingly, Charles Keeling developed strict protocols in 1962 to prevent this observer effect:

> Five liter spherical glass flasks, previously evacuated to a pressure below 1 micrometer of mercury, were exposed by opening a greased stopcock so that air expanded into the flask. ... Although this procedure is simple to execute, special precautions must be consistently observed to avoid contaminating a high proportion of the samples. The sample taker, to minimize contamination from his own breath, was instructed to sample only when the wind was at least 5 knots. After first breathing normally near the site for some moments, he exhales, then inhales slightly, and finally without exhaling again, walks 10 steps into the wind, where he takes the sample. ... Only one member of the South Pole field party was designated each year to take samples. Prior to arrival in Antarctica, he received two days of instruction from Scripps personnel. The results of his practice sampling were determined by gas analysis while he was still undergoing training.

Namely, hold your breath, walk into the wind, wait, and take the sample. Only valid on a windy day. And only if you have been properly trained and have passed your practical exam.[79]

The effects of the observer may be subtle, indeed. In 1873, Charles Darwin was investigating insectivorous plants. He wanted to know if the leaf glands of his pitcher plant secreted the acidic fluid that collected at the bottom of the tube. In a letter to a colleague, Darwin sketched a possible experiment. First, wipe the leaf surfaces clean. Then try to elicit a secretory response by applying a small sample of fibrin—an animal extract that Darwin had used successfully to elicit a digestive response in another carnivorous plant, the sundew. But perhaps the mere physical irritation of the leaf's gland might produce the fluid: an unwanted observer effect. Darwin was keenly aware that he might create an irrelevant and misleading phenomenon. So (he advised his correspondent) he planned to apply a bit of equally damp cotton or moss on another point. That would enable him to test and rule out any unwanted interference. Darwin's method allowed him to check against an prospective observer effect: another fine, if modest example of managing potential error (see Chapter 8).[80]

Summary of observational errors

To review, then, observation can be a complex process, subject to many sources of errors—material errors, such as wrong substances or contaminants; framing errors, such as sampling bias, small sample size, incomplete sampling, unchecked proxies, and confounding variables; misleading data from malfunctioning instruments or from instruments whose operation is not fully understood; perceptual errors from human observers; data biased by human observers with expectations; artifacts introduced by the experimental apparatus; and human interference significantly altering what is being observed. Any single one might lead to the unraveling of trustworthy observations.

But this list is hardly meant to be exhaustive. One might easily add to the list other observational error types, such as the following:

- Flawed experimental assumptions
- Flawed experimental design (other cases of misframing)
- Lack of competent execution by technicians or other laboratory personnel
- Systematic failures of heuristics ("shortcut" methods).

The purpose of the survey in this chapter (and the following chapters on conceptual and social errors) is to illustrate the concept of error types and their diversity, and perhaps to provide impressions of at least the major (most frequently encountered) error types. We may look forward to developing a more complete inventory or error types in the future.

One may also acknowledge the conceptual context of "observation" or the collection of raw evidence. In some cases, the data that is collected is the direct expression of the question that is asked. And that, in turn, may reflect assumptions or biases that have not been acknowledged or sufficiently probed. For example, if one does not inquire about women's health as distinct from "human" health (using a historical database of male-only patients or animals), then one's data may be misframed, and one's conclusions skewed. Assumptions can be slippery and their role is addressed again in the next chapter.

3
Conceptual Errors

Conceptual errors • overgeneralizations • faulty assumptions • theory-laden judgment and confirmation bias • cultural bias—religious bias—gender bias—racial bias—class and political bias • heuristic gaps • cognitive lapses • unaddressed alternatives • cryptic alternatives • epistemic hubris

A second group of error types occurs at the conceptual level. That is (using our model for mapping justification, Figure 2.1), when observations or data are deemed reliable, scientists can establish meaningful patterns and relationships among them. This creative work is largely achieved through individual mental processes. The cognitive tools are many and varied. For example, repeated associations may lead to tentative generalizations. Correlations provide clues to possible causation. Differences allow one to isolate causal factors. Similar cases invite exploration through correspondence. Of course, researchers apply the many well-known basic forms of reasoning (again: deductive, inductive, abductive, analogical, probabilistic, and error-statistical, at least), as well as more complex forms of reasoning (again: the method of multiple working hypotheses, natural selection-type search, hypothetico-deductive inference, or Mill's method of difference). Through these, observations are mapped and remapped into simple "laws," taxonomies, schemata, formal exemplars (or paradigms), models, families of models, causal mechanisms, explanatory structures, and more comprehensive theories—all concepts of varying scope and levels of generality or abstraction. Over time, philosophers, psychologists, historians, and others have documented an extensive set of scientific reasoning methods (to use a general, if somewhat vague label) for developing conceptual interpretations.[1]

Any of the reasoning patterns may fail, even if the evidence itself is reliable. Assumptions may be misplaced. Relevant contexts may be misconstrued. Boundary conditions may be ignored or completely unknown. A limit of generalizability may be overlooked. The reasoning process itself may be open to irregularities. For example, analogies that are fruitful in one context

Toward a Philosophy of Error in Science. Douglas Allchin, Oxford University Press. © Douglas Allchin (2026).
DOI: 10.1093/9780197827703.003.0003

may be misleading in others. (They are heuristics only, like proxy variables.) Probabilistic reasoning may be susceptible to the rare event. Statistical models can be misapplied. And given real human minds, cognitive processes may not consistently exhibit the epistemic ideals. Fallacies abound. Biases intrude. Miscalculations or other mistakes enter. Human cognition regularly fails to meet the exacting normative standards of "rationality" articulated by philosophers. So the epistemic concerns are many.[2]

Conceptual errors might be organized in many ways. There are, of course, numerous fallacies, both formal and informal, logical and argumentative, copiously documented by philosophers and teachers of critical thinking (the *Internet Encyclopedia of Philosophy* lists 229). They form the foundation for any inventory of error types, of course.

Here, however, I focus on a few conceptual errors that history indicates are especially relevant to science: overgeneralizations; mistaken assumptions; confirmation bias (or "theory-ladenness"); coherence with one's cultural beliefs (including religious, gendered, racial, class, and political perspectives); analogies, models, and other heuristics; other cognitive dispositions related to memory, attention, and causal thinking; failure to fully address alternative interpretations; and blind spots based on the limits of imagination, technology, or relevant collateral knowledge (see again Figure 2.2). The emergence of conceptual errors should not be surprising. To err is human; to err is science. And that includes being oblivious to one's own shortcomings. Still, one may characterize how and where these sources of error occur—as general error types—in order to inform practices of checking for or mitigating errors (Chapter 8).[3]

Overgeneralizations

One aim of scientists is to notice, describe, and then explain regularities in the data: general patterns that apply beyond the particular set of source observations. Usually, one assumes that the relationship holds uniformly across similar cases. However, sometimes the pattern may break down. It may appear only in limited circumstances, which may not be obvious at first. "Similarity" may be misconstrued. Researchers may err by *overgeneralizing* their findings. It is a widespread (and, I contend, underappreciated) error type in science.*

* Overgeneralization relates to the new riddle of induction, as described by Goodman (1965). That is, the problem is not that generalization itself is risky, but that we should be concerned about

Note that overgeneralizations differ from "hasty generalizations" (or "jumping to conclusions"), another familiar error type (see section on "Unaddressed alternatives"). In the latter case, one accepts a conclusion before due consideration of the evidence or before an adequate sample size has been examined (last chapter). The pattern that one thinks one sees in the data ultimately does not exist at all. The premature conclusion reflects the cognitive disposition to attribute patterns even to random data: *patternicity*. In the case of overgeneralizations, by contrast, the data do indeed justify a general conclusion. However, its proper scope is bounded. One cannot legitimately apply the concept as widely or as broadly as initially imagined.

The challenge is analogous, perhaps, to the problem of properly framing observations or of securing a sufficiently complete and representative sample, discussed in the last chapter. In both cases, the aim is to align the map with the relevant domain. In the case of overgeneralization, one applies the pattern of the map too broadly. The domain, or scope—the frame of the concept, now—is more circumscribed, or more local, than one had assumed.

Overgeneralization errors can occur even with the most modest scientific concepts. Perhaps the most basic work of science is finding a systematic relationship between a pair of variables. Two sets of observations may be correlated, perhaps. Or inversely correlated. That relationship can often be expressed in simple mathematical equations or formulas. From the late 17th century onward, they avidly sought to establish these basic "laws" of nature as a foundation. Many laws have since become widely familiar through introductory science courses. Laws such as these are conventionally presented as universal and invariant. Namely, they apply everywhere and without exceptions. But their histories tell a different story. Many of these simple mathematical expressions hide complexities that were discovered only after they were well established and had gained reputations as "laws." For example, Newton's laws were found to be inaccurate at relativistic velocities and small masses. Mendel's law of independent assortment was limited with the discovery of chromosomal linkage, and what was once considered a "law of dominance" has been abandoned entirely. Snell's law of refraction does not apply to calcite (or Icelandic spar), which exhibits double refraction, nor does it apply at angles of total internal reflection, or materials with negative-refraction indices. Galileo's law applies only to pendulums of very small angles (and so corrections were needed in the design of pendulum

induction classes, or what we deem a warranted basis for generalization. Here, scientists are too generous in characterizing the relevant inductive class.

clocks). Boyle's law met successive restrictions (see Figure 6.1 and section on "Expand scope, widen domain" in Chapter 6). Most "universal" and "invariant" laws are, ironically, contingent and deeply contextualized. Elements of context may include what may be variously characterized as assumptions, provisos, qualifications, ceteris paribus clauses, or boundary conditions. Originally, these laws were overgeneralized.[4]

Not all overgeneralizations are expressed in lawlike terms. Consider, for example, the fate of the "central dogma" of molecular biology, as expressed by James Watson: DNA replicates as DNA, and is also transcribed into messenger RNA, which is translated into proteins. The flow of genetic information was considered one-way only. Then Howard Temin and David Baltimore discovered reverse transcriptase, violating the assumption that DNA was only synthesized from a DNA template. Finding that error was significant enough that it earned a Nobel Prize. Similarly, for the discovery of prions by Stanley Prusiner: proteins that (at some level) can "reproduce" themselves (or their alternate conformation). That was the occasion for a Nobel Prize, too. In awarding the prize for the discovery of interference RNA, the Nobel organization described it, too, as an exception to the central dogma. Discovery of errors—here, major overgeneralizations—can be significant (see "Negative knowledge" in Chapter 8).

Problems of overgeneralization also plagued assumptions about the genetic code, the mechanism for generating proteins from DNA/RNA sequences—specifically, mediating between the shape of a trio of nucleotide bases and the corresponding amino acid. Once molecular biologists deciphered the encryption pattern (the code), it seemed consistent across all species, from bacteria to humans to tobacco plants. All life seemed fundamentally and elegantly unified. Most biologists figured that once the translation system had evolved, it was so central and so complex that any further change would damage the protein products so severely that the change would be lethal. "The genetic code, once established, would therefore remain invariant." That is, the "primitive machinery" and the "primitive code" seemed "frozen"—and hence universal.[5]

However, studies of mitochondria indicated that they contained their own DNA, distinct from the nuclear DNA, and that these genes were translated using a different code: namely, where different amino acids were paired with different RNA codons. Then a different set of variants in the code was found in a group of one-celled ciliates. And more were identified as studies continued. Currently, there are 33 known variants. The genetic code is not "universal." One must circumscribe its scope. Moreover, biologists have now found counterexamples to the notion that the genetic code was "frozen" with the 20 well-known "essential" amino acids. There are two such nonstandard amino acids—pyrrolysine and selenocysteine—that seem "hard-coded" in variants of the genetic code. Even more remarkably, in one species of archaebacterium, there is a novel enzyme—meaning a completely novel gene at another layer—that links the RNA to the amino acid (pyrrolysine aminoacyl-tRNA synthetase). The "universal" amino acids have turned out to be not strictly universal. The idea of a frozen, universal genetic code was overgeneralized, too.[6]

Another realm of overgeneralization is simply overstating what the evidence warrants. Hence, an investigator may claim that detecting a correlation in the data justifies a conclusion of a causal connection that would explain that correlation. Or they may regard population-level data (an average, say) as sufficient for making specific predictions about individuals in that population (thus disregarding any inherent variation). Or they may inappropriately interpret data indicating a causal link as sufficient evidence for a certain causal model or mechanism. For one pair of statisticians, at least, this failure to heed the inferential boundaries of one's evidence "is the most common error in data analysis."[7]

As an error type, overgeneralizations indicate that conceptual errors are not measured exclusively by being "right" or "wrong." One must simultaneously consider the intended domain, or scope of application. That is, the assessment of concepts is not just about the relationship of theory and observation, or theory and evidence. There are numerous other factors involved in how we frame the concepts that map our world—most notably, context.

Faulty assumptions

As noted in the cases above, context matters significantly. Assumptions strongly shape reasoning and are subject to error, even if the reasoning proper is sound. In a sense, they are akin to contaminants in observation. Sometimes, the assumptions are explicit. Often, they are not. A researcher may be completely unaware that any critical assumption has been made. Thus, a major challenge for science is to expose these implicit contexts: counterintuitively perhaps, another form of "discovery." Frequently, the only clue to the relevance of a hidden idea is when one encounters some anomalous finding, which is eventually traced back to the missing assumption (see section on "Incongruences" in Chapter 5).

One may easily recall, for example, the series of erroneous assumptions in determining the Hubble constant, and hence the age of the universe (see section on "Error and uncertainty" in Chapter 1). First, Baade found that the Cepheid variable stars, used as benchmarks for calculating astronomical distance from Earth, were not a homogeneous group, as earlier presumed. Later, Sandage discovered that another group of reference objects were nebulae, not stars. The calibration of distance was revised again. In 1956, Elizabeth Scott noted how the ability to discern clusters of stars varied with distance (based on the cluster size and luminosity). That led to a distance-related bias in interpreting benchmark stars (now called the Scott effect). And the Hubble constant changed by a factor of two (again). These details are typical of the challenges faced by scientists. Observations are frequently indirect. Assumptions merely about what one is observing, or how, are commonplace, simply to get the job done. But they are subject to error.[8]

As another example, significant assumptions also lay buried, unnoticed, in William Thomson's influential calculation of the age of the Earth in the late 1800s. Thomson (later, Lord Kelvin) applied his basic principles of thermodynamics to the history of the planet, assuming that a once-molten mass had cooled at a regular rate, based on the formula for heat conduction. Of course, some basic data were needed and Thomson duly measured the heat conductivity of rock directly, and the current temperature gradient was observed using deep boreholes and roughly matched data from mines. The resulting age (between 20 and 400 million years) did not leave much time for evolution. Thomson was just fine with that. In his original 1862 paper, however, Thomson had provided a brief caveat, almost as an aside. He defended the method as valid "unless sources [of heat and light] now unknown to us are prepared in the great storehouse of creation." In retrospect, that proved blindly prescient. Decades later, radioactivity was discovered. Ernest Rutherford made the connection to the age of the Earth in 1905. Perhaps with an additional source of heat, the Earth was not cooling quite as rapidly as Thomson had claimed? The *New York Times* ran the reassuring headline: "Doomsday Postponed." Without explicitly acknowledging defeat, Thomson quietly ceased promoting his argument. In the long run, radiogenic heat changed the calculated result only by a factor of about two. However, revealing the role of the assumption, overlooked for decades, had a historically devastating effect on the rhetorical authority of the argument, once treated as irrefutable.

Yet another implicit assumption proved far more critical. Thomson's former student, John Perry, realized that heat flow in the Earth's interior would likely differ markedly from conditions at the surface. Indeed, the interior was surely hotter and thus probably fluid, even generating convection currents, as another geologist had suggested a few years earlier. Thomson's formula for heat diffusion had been far too simple. Using a few more provisional assumptions, Perry found that the age of the Earth could well be tens or hundreds of times greater than Thomson originally said. Over half a century later, Perry's models emerged again—by which time, geophysicists had found other, surer ways to measure the age of the Earth. The thermal structure of the Earth is now considered much more complex than Thomson assumed, including the effects of a cooling crust on heat retention.[9]

Anyone familiar with science can surely describe numerous other cases where assumptions got a researcher in trouble. One might imagine that we should strive to eliminate such assumptions from science. If we could. Yet not all assumptions lead one astray. Some can open the way to discovery. They can even lead to insights and blind spots at the same time (for more on this conundrum, see Chapter 7). Reasoning is inherently a risky endeavor.

But where do assumptions come from? Particular kinds of assumptions pose particular problems, as elaborated more fully in the next two sections.

Theory-laden judgment and confirmation bias

Assumptions are context. And the most prominent context in science is theory. Theoretical assumptions are a major guide to reasoning. For example, theory helps knit disparate observations together as meaningful. It helps fit any new interpretation of evidence into a network of established claims. Accordingly, theories can fruitfully guide research. They may point to new phenomena lurking in the shadows. They may profile gaps in prospective patterns that need to be explored. They may identify relevant information worth collecting or what features are especially worth observing.

At the same time, however, theories can also mislead. They may misdirect primary attention. They may frame ultimately important details as apparently irrelevant. They may support and perpetuate illusory patterns. They may lead investigations down unfruitful pathways. Paradoxically, then, theoretical assumptions can play contrasting roles, sometimes fostering insights, sometimes creating blind spots (see Chapter 7 for a fuller discussion). Here, I note a few cases where theoretical assumptions contributed to errors.

A simple but striking example is found in the anatomical drawings of Leonardo da Vinci. Da Vinci dissected corpses to study anatomy. He was a consummate draftsman, of course, and rendered his subject with great detail and accuracy. So, how do we interpret his drawings of the abdomen, where we find a duct connecting the liver with the spleen? According to the ancient Greek theory of humors, there was such a duct. It carried "black bile," the humor responsible for melancholy, from the liver to the spleen. Yet no such vessel exists. Still, da Vinci drew it, and there it remains in his drawing still: an ironic tribute to the power of erroneous theoretical assumptions.[10]

Theory can powerfully guide reasoning. Recall Christian Eijkman's interpretation of beriberi (see section "Confounders" in Chapter 2). Eijkman went to Java to find the cause of the disease following a triumphant period for the germ theory. In only a decade, microbiologists had discovered the bacteria that cause tuberculosis, cholera, pneumonia, typhoid, tetanus, diphtheria, anthrax, and a few others. Indeed, Eijkman was studying with Robert Koch, the world's leader in germ theory, when the occasion for the commission to Java arose. In addition, outbreaks in institutions had provided initial support for assuming a microbial infection. During the first year in Java, Eijkman and his colleagues used the powerful new culturing techniques to try to isolate the bacterium and to establish animal-to-animal transmission. But with no success. Eijkman persisted in the following years, until accidental events indicated a role for diet. Even so, Eijkman configured the findings in terms of the initial assumption of germ theory. While he carefully established the outer coating of the rice as a curative factor, he attributed the disease itself to the presence of a still unidentified and unisolated bacterium in the starchy kernel (and a neurotoxin it produced). Eventually, Gerrit Grijns, who was less committed to the germ theory assumption, more warmly entertained the notion that beriberi was caused by a nutrient deficiency. True, it was an unprecedented kind of nutrient, needed only in very small amounts. That had made discovering its role more challenging. Ultimately, the assumption that beriberi was microbial had been an assumption only, and apparently blinded Eijkman to the ultimate cause.[11]

When philosophers first intimated in the 1960s that observations or reasoning about evidence in science might be "theory-laden," it was something of a scandal. Shocking and titillating both. "How dare anyone impugn the objectivity and rationality of science!" Yet historians have long been comfortable with how biographical and cultural contexts have influenced the ideas of particular scientists. Scientists are human, after all. Still, as historians might note, science seems to get by and is able to achieve even some monumental discoveries. Many philosophers now endorse a more deflationary posture, where science is "naturalized" in the context of human cognition.[12]

The influence of theoretical perspectives should not seem all that surprising. In terms of how our brains work, early ideas form a context for later ones. They shape what one notices. They highlight certain observations as relevant, others as less so. They filter perceptions and the collection of evidence. Confirmatory evidence will tend to be easily accepted, rather than scrutinized closely. At the same time, prior concepts may lead to discounting counterexamples. They will activate skepticism. Countervailing evidence will tend to be subjected to higher standards of demonstration. Or such cases will be dismissed as irrelevant exceptions. In common parlance, the evidence is "cherry-picked"—albeit unconsciously. That is, as cognitive scientists have now amply documented, we all exhibit *confirmation bias*. What appears as a well-formed argument may not be well informed. The evidence may be (unwittingly) selected to accord with the original conclusion. A genuine effort at justification may become no more than an exercise in rationalization. The argument may seem logically complete, but it is not based on a fair view of the relevant evidence. The resulting sample, like one limited by size or sampling bias or incompleteness (see Chapter 2 on "Observational errors"), is unrepresentative of the whole. It is a *conceptual* misframing (Figure 3.1). Theory-laden perceptions or judgments addressed by philosophers of science are merely manifestations of a general cognitive syndrome. So, perhaps Bacon was right about illusions of the "tribe," or the cognitive dispositions of our species. Understanding how human brains work is relevant to understanding how science works.[13]

Confirmation bias will typically emerge in assumptions. Some assumptions are adopted in the course of an investigation, while others reflect prior knowledge or beliefs. In either case, there is an interesting asymmetry. Early concepts contextualize the reasoning about later concepts. New concepts are less likely to be the basis for re-examining earlier ideas. Unless scientists are mindful about it. That is one topic for a fuller consideration (in Chapter 8) of how scientists ideally manage errors.

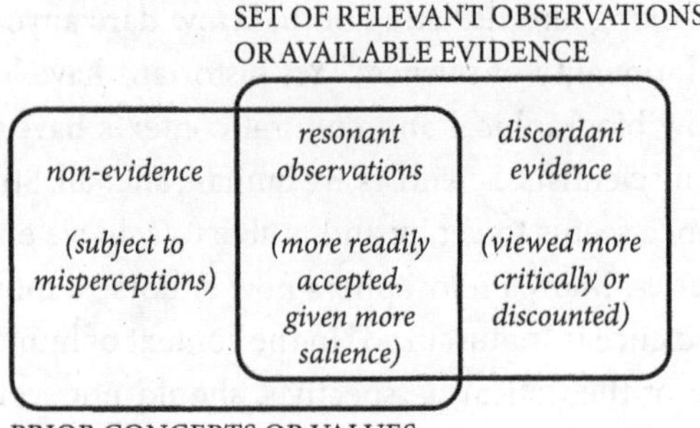

Figure 3.1 How confirmation bias distorts framing of evidence during the mapping of a justification.

As noted above, assumptions and theoretical context—or "bias"—can both promote insights and foster blind spots, depending on context. Thus, the epistemic design of science is not necessarily best served by trying to eliminate such "bias" (again, see Chapter 7). Equally important, confirmation bias can also be found at the level of the scientific community—addressed more in the chapter.

Cultural bias

Confirmation bias in science is not limited to scientific theories and assumptions. Scientists' minds are filled with ideas, values, and norms from their personal lives and cultures, as well. These, too, form an inescapable—although all too often unacknowledged—background to reasoning. That is, scientific concepts may be shaped by perspectives that are conventionally regarded as "external" to science. The deep psychological processes of confirmation bias do not distinguish between different sources of assumptions or intrinsically filter out the "non-scientific" ones. So, scientists may be subtly but indelibly influenced by the cultural norms or ideologies that permeate their everyday lives. Stepping into a lab, alas, does not radically alter how our brains function. Accordingly (and it should really surprise no one), scientific reasoning can exhibit *cultural bias.* It is just another, more indirect form of theory bias.

In the 1970s and '80s (and beyond), sociologists of science documented the influence of culture on science. In case after case, they exposed how various historical claims embodied unmistakable gender bias, racial bias, class bias, or other forms of politics or ideology. That scholarship challenged the cultural authority of science. Indeed, that often seemed the intent of the sociologists, who embodied their own political contexts. One widespread response was to denounce all science, embrace relativism, and even promote abandoning science or the search for reliable knowledge. However, another response—followed throughout this book—is to adopt a suitably humbled, deflationary view of science and seek remedies. We can honestly acknowledge the capacity of science to err without abandoning its aims. Let us appreciate the sociological studies, then, as valuably identifying sources of error. These errors might well be regulated through appropriately informed (and possibly new) epistemic practices (see also discussion of "Merton's norms" in Chapter 4 and "Methodological norms" in Chapter 8).

Indeed, many of the culturally based errors were ultimately corrected historically. (Albeit far too belatedly: one should not interpret these corrections as a triumphant demonstration of "self-correcting" science.[14]) The first step in transforming science, perhaps, is to articulate and understand these errors and to characterize them as a discrete error type. In this section, I survey a diverse but all too limited sampling of cases of cultural bias, where confirmation bias in science was oriented variously by religion, gender, race, class, or political ideology. These types of errors are rarely listed alongside the experimental errors in the previous section or the lapses of reasoning elsewhere in this section. Indeed, they tend not to be conceptualized as sources of error at all. Yet doing so might allow us to think more mindfully about how they might be addressed more methodically and perhaps minimized. Articulating error types is the first step in exercising better regulation of error in ordinary scientific practice. There are many variants, so this section is especially long for the sake of completeness.

Religious bias

If we venture back to the early 1800s, we will find William Buckland being named the very first university professor of geology. That was remarkable for an institution like Oxford University that primarily prepared clergy for their profession. Buckland himself was an ordained minister. Indeed, Buckland's inaugural address celebrated the theological role of his science: *Vindiciae Geologicae; or the Connexion of Geology with Religion Explained.* Later, he would author one of the renowned Bridgewater Treatises, arguing from geological observations to the existence of God and his beneficent nature. So, one may not be surprised how Buckland interpreted his discovery of fossils in a cave in northern England in 1821. The cave floor was littered with bones from hyenas, rhinoceroses, and elephants, half buried in mud. Buckland carefully reconstructed the series of events from the geological clues. First, the cave had formed. Then, the larger animals—too large to fit through the narrow cave opening—must have been torn apart by a predator and dragged into its den. Finally, all these animals (including the predator hyenas)—none of which still inhabited England—were killed by floodwaters, which carried silt into the cave and half-buried the remains. In the spirit of his earlier religious advocacy, Buckland presented his findings as evidence not of any flood but, in a Biblical context, of Noah's flood. Buckland's religious theme appeared prominently in the title of his study: *Reliquae Diluvianae*, or "Relics of the Flood." The work was good science: enough to earn the prestigious Copley Medal from the Royal Society. But it was steeped in religious assumptions. To his credit, perhaps, Buckland later qualified the Biblical dimension of his conclusions. His work inspired study of fossils in flooded caves throughout Europe, which would presumably confirm the worldwide nature of the flood. As it soon became clear, however, many different floods had buried bones at different times. Buckland acknowledged that the evidence did not support the idea of a single worldwide flood, as depicted in the Bible, as he had once assumed.[15]

Buckland was a towering figure in 19th-century geology. One cannot discount his bias. So, too, for Michael Faraday, another titan of science. Yet Faraday was deeply influenced by his religious sect, the Sandemanians. As described more fully in Chapter 7 ("Michael Faraday, Sandemanian"), their view of the unity of nature spurred both the investigation of a relation between electricity and magnetism (successful) and the search for a link between gravity and electricity (a *gravielectric* effect—ultimately, unsuccessful). To these we may add two more: William Crookes and Alfred Russel Wallace. Both applied a scientific perspective to spiritualism (the belief that some aspect of the human psyche persists after death and can manifest itself in the material world). And both believed that the phenomenon was real. Crookes applied his scientific acumen in pursuing evidence for spiritualism. As a result, he exposed nearly a hundred frauds. That is, he was not so beguiled by his beliefs that the possibility of chicanery escaped him. Still, one or two mediums seemed to "pass" his critical analysis. Using his own camera, he made one photograph that he interpreted as scientific demonstration of an authentic spiritual being. Wallace, too, was persuaded in his observations of various spiritual events and defended them scientifically as legitimate phenomena. And while Wallace was perhaps less critical than Crookes, he contextualized his beliefs more strongly in evolutionary concepts. His close work with native peoples convinced him that humans exhibited a distinctive morality that could only be explained by transcendental forces and immaterial spirit. We may surely find holes in Crookes' and Wallace's beliefs today, but their effort to apply scientific rigor to the problem indicates, in part, how cultural religious beliefs can shape an individual's scientific work.[16]

What is remarkable about these great 19th-century scientists—Buckland, Crookes, Wallace, and Faraday—is that despite their unimpeachable achievements, they did not fully escape or transcend their cultural context. They erred in ways that can be traced quite clearly to particular assumptions associated with their religious beliefs. Even the most outstanding scientists may express cultural bias by bringing apparently benign or hidden assumptions into their scientific reasoning.

Gender bias

Cultural biases and assumptions that lead to error may be gendered, as well. Examples are now plentiful in the literature, so only one thematic topic may be needed here as illustration. (Further cases are addressed in Chapter 6, where we see remedies for error in action.) However, the case exemplifies well the blind spot of cultural bias that fostered a series of errors, invisible to many scientists.

The case of female anatomy is especially interesting, as it would seem to be merely a matter of recording basic observations. Substantive study of the female skeleton began in the 1500s, embodying the growing use of investigative observation and dissection. The first drawing of a female skeleton appeared in 1583. It labeled and described thirteen particular features that differed from the male skeleton. That list reappeared unchanged over the next two centuries. But several of these features are no longer accepted as valid, nor were all the functional descriptions adequately substantiated. For example, no ossifications or structures are needed to support the weight of breasts. Yet all these errors helped characterize a woman as a mother "down to her bones." That is, the perceived social role of women in that historical period was reflected in how the male anatomists characterized the female body.[17]

Anatomical errors continued to appear in a new era of female skeletal drawings in the 1700s. Again, those errors reflected the influence of cultural norms. One popular 1759 drawing depicted a female with a disproportionately small head and exceptionally narrow ribs that helped accentuate wide hips. That was the standard of beauty for women, again based on an assigned reproductive role. "The most beautiful and womanly in all its parts," wrote Joseph Wenzel in 1788, "is one in which the pelvis is the largest in relation to the rest of the body." Samuel Thomas von Soemmerring provided a drawing more faithful to reality in 1796. But another commentator disparaged it as not as genuine as the earlier one. The rib cage was too large, he complained, given that women breathed less vigorously, in accord with their (natural) sedentary lifestyle. Gendered perspectives fueled unseen confirmation bias in transforming norms into supposedly objective observations.

The gendered nature of all these claims was further indicated in the anatomists' explicit explanations of the skeletal differences, which linked the features to their imagined social roles. They appealed to an ideological role of women as mothering (wide hips to accommodate childbirth), deferring to men on intellectual matters (who had larger brains), and yielding to a more muscular (and hence more "powerful") male. The ideal of male superiority was conveyed in Alexander Monro's claim that "the Bones of Women are frequently incomplete, and always of a Make in some parts of the Body different from those of the robust Male." These "deficits" were attributed to a weak constitution, an inactive, sedentary life, and "a proper frame for being mothers," all construed as given by nature. For Chelsden, the differences all reduced to menstruation. Jaucourt made it all very clear in the 1764 edition of the *Encyclopédie*: "All of these facts prove that the destiny of women is to have children and to nourish them." Those views persisted into the 1800s. Physician J. J. Sachs contended that "the female body expresses womanly softness and feeling. . . . The roomy pelvis determines women for motherhood. . . . The weak, soft members and delicate skin are witness of woman's narrower sphere of activity, of home-bodiness, and peaceful family life." Positivist philosopher Auguste Comte shared his view: "The study of anatomy and physiology demonstrates that radical differences, at once physical and moral . . . profoundly separate the one [sex] from the other." Cultural perspectives were not tested, but rather restated as apparent scientific truths.[18]

Later in the 1800s, the anatomical focus would shift to the cranium and its role in female intelligence. The work was neatly summarized by Theodor von Bischoff: "The laws of divine and natural order reveal the female sex to be incapable of cultivating knowledge." In the 1860s, in the shadow of a growing women's rights movement, many men sought to "protect" women from new threats to their perceived biological nature. We may also wonder that the strong cultural bias was coupled rhetorically to a commitment to be rational and objective. For example, the anthropologists relied on measurements, as a gesture toward objectivity. They invested themselves in the effort with some vigor, measuring skulls in numerous ways. They combined some lengths or angles into ratios to reflect significant features of shape or form. Ultimately, over 600 instruments were used. One researcher claimed that one needed over 5,000 separate skull measurements. There was, as one historian noted, "a Baconian orgy of quantification." Of course, whenever the evidence did not align with expectations, the researchers discounted the assumptions, rather than accept the unwanted conclusion. Error after error appeared, always to be revised (for more on the successive errors in craniology, see section on "Communal cultural bias" in Chapter 4). The rhetoric of science, guided by cultural ideology, was far stronger than the rigor in the practice of science.[19]

The gendered work of anatomists is notable in illustrating again how scientists can be blind to their own errors. The many quotes above exhibit their unguarded posture. That is, the writers seemed wholly oblivious to how cultural assumptions shaped their reasoning. That is one reason why a thorough inventory of error types is valuable: to nudge the implicit and invisible processes of biased thinking into awareness, where they can be addressed and managed more explicitly (see Chapter 8).

Racial bias

Cultural views relevant to scientific reasoning can extend to race, as well. A particularly egregious (and now widely cited) example appeared in 1851, in *The New Orleans Medical and Surgical Journal*. A group of physicians had been commissioned to report on the physiology and maladies particular to the endemic slave population. The very purpose of the project reflected the first notable assumption: namely, that this population—forcibly displaced and enslaved Africans—constituted a medically distinct group. After reviewing several infectious diseases, including yaws, scrofula, and consumption (tuberculosis), the committee named two hitherto unrecognized diseases: *dysaesthesia aethiopica*, a "hebetude of mind and obtuse sensibility of body, called by overseers 'rascality,'" and drapetomania, a mental affliction that caused slaves to run away. In retrospect, one can easily spot the flaw in assuming that these behaviors had a biological cause, rather than a social one, and, further, that avoiding an excessive workload or trying to secure liberty were pathological, rather than "healthy" responses to unjust enslavement. Nevertheless, the doctors staunchly defended their claims as scientific. They assured readers that these conclusions had "not been derived from books or medical lectures, but from facts learned from their own observation in the field of experience, or picked up here and there from others." Simple, unbiased observations, apparently. In concluding the paper, however, they betrayed their assumptions. They pined about the

> false dogma that all mankind possess the same mental, physiological and anatomical organization,

and that in documenting the new "diseases" they were

> establishing the truth that there is a radical, internal, or physical difference between the two races, so great in kind, as to make what is wholesome and beneficial for the white man, as liberty, republican or free institutions, etc., not only unsuitable to the negro race, but actually poisonous to its happiness.

Of course, the committee had not considered the "negroes" in the context of their homeland culture, nor had they observed the behavior of privileged slave owners subjected to similar enslavement—possible confounding variables essential to their conclusions about intrinsic biological causes (rather than environmental contexts). This case of cultural bias, oriented to race, is extreme, of course. Even contemporaries (who did not assume that slavery was ordained) ridiculed the publication. But it highlights an important question for error and bias in science. Did the "observation" of these new diseases lead to justifying the institution of slavery (as the authors contended), or did the assumption of slavery as justified seem to dictate the "observations" themselves? The inappropriate inversion in justification is another prime example (similar to female anatomy) of naturalizing as an error type: whereby scientists inscribe cultural notions into the rendering of nature, which presumably provides a context that justifies those very cultural assumptions.[20] That is why these error types, as outlandish as they may seem now, are so important in a complete philosophy of error in science.

Other efforts to naturalize a racial hierarchy appeared in the work of *geologists* in the mid- and late-19th century. They carried their prestigious credentials from Harvard, Princeton, the U.S. Geological Survey, and elsewhere. The central tenet was that various features of physical geography, such as peninsulas, deserts, or canyonlands, or the climate of whole continents, promoted or limited the development of "civilization." The environment did not just provide a context of what resources were available, but also strictly determined the intellectual abilities and moral character of various peoples who inhabited a region. For example, people in hot climates would be inherently inactive and lazy, not disposed to "industry" or a healthy work ethic. Thus (the geologists contended), Negroes were "naturally" subservient, and America was geographically "fit" for slavery. Or, people in arid habitats would never develop agriculture or amass a material culture. As a hunting or nomadic people, they could simply not achieve the status of a "civilized" society. "They do not build houses." They do not have large cities or museums or concert halls—hence, they are "less advanced." They are "miserable as compared with the more favored races of the world." "They cannot rise above a low state of development." The difference between savages, barbarians, and civilized societies was supposedly determined by the physical landscape. Nathan Shaler expressed the common assumption in his 1897 textbook:

> Looking back over the history of life upon the earth's surface, the physiographer is forced to the conclusion that its highest estate embodied in the moral and intellectual qualities of man has been, in the main, secured by the geographic variations which have slowly developed through the geological ages.

Such views went unchallenged in the exclusive academic communities, and then were propagated to a public that shared the racial prejudices.[21]

The geologists' purpose—explicitly stated in some cases—was to legitimize the hierarchy by explaining it. It was not merely to interpret any observed differences or to investigate unknowns. Ultimately, the assumptions provided a framework to cherry-pick facts, which were then used to support the "explanation" as objective. Distinctive traits for a race or culture (already assumed) were associated arbitrarily with whatever geographical feature could be found. There was no systematic testing of general concepts, or even concerted effort at critical analysis. These are telltale indicators of superficial rationalization through reference to nature, in lieu of rigorously controlled observations: again, naturalizing as an error type. Still, everyone involved at the time seemed to believe that it was good science. Shaler assured his readers that the project was "better, indeed, than the old faith, for it will rest on the firm foundations of our own knowledge, rather than on the trust in the opinions of our elders." The geologists were blind to how their cultural perspective shaped their selective use of facts.[22]

Nowadays, even the very concept of race as biological (especially as it might relate to skin color) is widely regarded as scientifically unfounded. Still, as historians have now thoroughly documented, science exhibits a long heritage of providing ostensive evidence for racist-infused ideas. Again (as in the case of gender), cultural perspectives can support assumptions that lead to error in science.[23]

Class and political bias

Socioeconomic class is another cultural context that may shape scientific reasoning. For example, Charles Davenport betrayed his class-oriented prejudices in his work on genetics and especially his extensive reliance on pedigrees as a form of "evidence." We noted in the section on "Confounders" (Chapter 2) how Davenport mapped the occurrence of pellagra onto family trees, concluding (falsely) that the disease was hereditary, not nutritional. Because poverty—and thus poor diet—can easily be found in multiple generations all living in the same socioeconomic conditions, pedigrees alone cannot distinguish between genetic and social causes.

The same error haunted Davenport's arguments about low mental ability. He claimed that "feeble-mindedness," too, was hereditary. By displaying a handful of pedigrees, he seemed to regard his claim as sufficiently demonstrated. However, several critics noted that the definition of the condition was so vague and susceptible to subjective judgment that the whole enterprise of study was meaningless. British geneticist David Heron observed that data on ancestors' mental ability was often attributed without evidence. He also took issue with the data collection, noting that field workers were specifically instructed to search for instances that confirmed Mendelian patterns, and to not record others. (Namely, by design, they "cherry-picked" data, creating an unrepresentative sample.) Later, American geneticist Thomas Hunt Morgan put Davenport's reasoning in perspective:

> Family pedigrees in which an unusual number of individuals [mentally] below par are present undoubtedly give the impression that something is inherited, but until all the social conditions surrounding the childhood of the individual are examined and given proper weight, serious doubts will arise as to what form of inheritances is producing the results. It is quite probable that there are extraneous factors involved in such pedigrees.

Namely, without proper controls, how could one conclude that mental ability (strong or weak) was genetic and not "inherited" by virtue of a family's educational opportunities and social privilege (or lack thereof)? Davenport acknowledged all the problems briefly, but he brushed them aside without any further scientific evidence. His dismissive behavior (in contrast to engagement with the criticisms) indicates a form of confirmation bias.

For Davenport, feeble-mindedness was simply hereditary—as were nearly all nervous diseases, which seemed to consist of behaviors he deemed undesirable. Davenport's comments, terms, and casual asides, however, tend to betray the ultimate source of his scientific errors. He characterized feeble-minded individuals as "defective," "abnormal," and "inferior." Although the condition may have been vaguely defined, it served, he said, as "a convenient group in which to put all of the socially inadequate." It was a judgment on a condition that he obviously found troublesome. He described the feeble-minded as a "danger"—or a "grave danger"—without fully articulating the nature of the supposed threat. What he did mention was the social "burden": namely, the "cost to the state." Davenport's disdainful rhetoric thus exposed his values: "industry" (vs. "indolence"), "ambition" (vs. "listlessness"), "foresight," and abstinence and self-control (vs. "licentiousness" or "alcoholism")—all of which he saw as contributing to a "safe and progressive social organization." It is no wonder, perhaps, that Davenport also listed *pauperism* as a genetic trait, closely associated with mental inferiority. Even poverty, he claimed, was constitutionally hereditary, not social. The cultural unacceptability of feeble-mindedness (as he conceived it) was thus Davenport's assumption far more than it was any conclusion reasoned from his pedigrees or other evidence. And those prejudices were precisely what he carried forward into his arguments about social policy on immigration, reproductive rights, and social welfare, after they had acquired the aura of scientific authority. Namely, Davenport tried to enlist science to establish as "normal" the lifestyle and values of his socioeconomic class—and (perhaps not coincidentally) those of his wealthy patrons. His pursuit of a science of feeble-mindedness was, simultaneously, an effort in developing a biological rationale for his own social status, while politically disvaluing others. Such behaviors were not nervous disorders and they were not hereditary. The cultural bias here was socioeconomic in nature.[24]

Cultural bias can penetrate to the most unassuming subjects. The behavior of slime molds might seem wholly unrelated to politics, for example. Yet assumptions about "normal" political structure contextualized theorizing about how slime molds organize when they reproduce. For most of their lives, slime molds consist of single cells that feed on detritus in moist forests, even if they congregate into slowing creeping masses. When resources become scarce, however, they aggregate, sometimes displaying circular waves or spirals of movement. Collectively (now as a multicellular organism), they form a fruiting body that releases reproductive spores. The scientific question was: How do the individual cells "know" when and where to come together? How do they coordinate their actions to reproduce?

In the 1950s, a molecular signal (acrasin) was identified and found to be released—like a homing beacon that established the target rendezvous point. Was there a special type of cell with certain genetic properties that originated these critical signals? In 1961, one researcher proposed that a distinct cell type formed when a spore initially colonized a new area: a "founder cell." That would later be the locus of control. Roughly a decade later, mathematical models showed how the striking waves of cell migration could be generated by signal pulses from the aggregation center, which one researcher dubbed the "pacemaker." That became the standard model.

But no one could clearly characterize or isolate the special founder or pacemaker cells. For example, if the cells at the center of an aggregating mass were removed, another center would soon emerge. Some mathematicians tried a different approach, dispensing with a role for any predetermined founder cell. What if all the cells could release, as well as respond to, the chemical signal? Mathematically, chance irregularities in the environment alone, processed through successive interactions across all the cells, could lead to the same patterns of aggregation. No need for founders. Namely, the slime molds could self-organize.

Few biologists had recognized that the notion of a "master" cell was an assumption. Positing a central authority as necessary to coordinate and exercise control ultimately reflected a deep assumption about political structures. The new modelers challenged that by asking: "Do pacemakers make aggregation fields, or do aggregation fields make pacemakers?" Well, it turned out, both. But the centers arise mostly by chance. Any cell might become a pacemaker. There is no centralized authority. The natural history of slime molds is now a prime example of the science of self-organizing systems, and the concept of emergence, a focus of much new research.[25]

Similar assumptions about centralized authority have fostered errors (or significant omissions) in other fields, as well. The cell was once considered to be controlled exclusively by the genes in the nucleus. Biologists referred to DNA as the "master" molecule, the "blueprint" of the cell. But research has now highlighted that many parts of the cell act semi-autonomously: mitochondria, chloroplasts, centrioles, at least. The principle of centralized control describes only some aspects of cell function. In a similar way, beehives were characterized as early as 1609 as a "feminine monarchie," with a "queen" bee. The different forms of adult honeybees have been termed "castes," further projecting an image of social hierarchy. However, recent work has helped detail the dynamics of social decision making. As an ensemble, honeybees make many decisions: about where best to forage, where to locate a new nest, and how to distribute labor among various competing tasks (collecting nectar, pollen, water, or wax, for example, or prioritizing hive defense versus care of the brood)—all achieved with no central "command" or brain. Honeybee hives too, like slime molds, exhibit a degree of self-organization. That understanding was long obscured, however, under erroneous political metaphors and cultural assumptions.[26]

•

In summary, cultural perspectives may bias scientific reasoning just as theoretical commitments do. As illustrated here, many contexts—including religion, gender, race, class, and politics, at least—may potentially be sources of error. One may surely wonder whether these errors warrant the extensive discussion I have provided. After all, these biases will usually lead to particular errors that might be characterized in other terms, such as faulty observations, hasty generalization, cherry-picking of data, or lax standards of demonstration—those "internal" to the conventional epistemic concerns of science. Could one not just focus on those instead? Perhaps. But I hope that the various case descriptions convey vividly just how the cultural assumptions can influence reasoning, often in ways imperceptible to the reasoner. Ideally, we should be tracing sources of errors to their—well, sources. Being precise about sources of error is integral to their effective remedy.

Another reason for highlighting cultural biases in so much depth is to underscore their importance. While the role of culture in shaping assumptions in science may not be the stuff of everyday lab talk, one can see historically how they sometimes significantly shape conclusions. In many of these cases, the errors had a further role in (mis)informing social decision making. And, often enough, they justified political power or privilege beyond the boundaries of science. When researchers project their cultural norms or ideologies onto nature through unquestioned assumptions, science seems to validate an already prejudiced concept. This significant departure from the ideal of evidence as an objective "check" one may call the *naturalizing error*. It is one of the primary reasons for considering cultural bias as a source of error in scientific reasoning and giving it due consideration.[27]

Accordingly, researchers should ideally be sensitized to cultural bias as an error type. Bias may be based on other dimensions of political disparity, as well: ethnicity, sexual orientation, indigeneity, and age, among others. Scientists should accordingly conceptualize culture-based assumptions as sources of error alongside overgeneralization, lack of controls, or glitches in experimental setups. Human reasoning, no less than a piece of laboratory equipment, can malfunction. Namely, we need to calibrate our own minds as much as our laboratory instruments. We need to ensure that our samples are not biased or misframed, even on a conceptual level.

Of course, the discussion above conspicuously omits cases where theoretical or cultural "bias" led not to error but, instead, to discovery. Without historical "controls," therefore, one cannot conclude that bias *inevitably* leads to error—more on that conundrum in Chapter 7. Here, the aim has been merely to trace the sources of these errors in each case and to characterize them in terms of general error types in reasoning. That provides a foundation for managing similar such errors in the future (Chapter 8).

Heuristic gaps

Among the observational errors (Chapter 2), I noted the use of proxy variables. These surrogates are deliberate deviations from ideals for the sake of time or economy, or to help gain traction with complex phenomena.

Such heuristic strategies are also found in interpreting observations and in conceptual reasoning. For example, one makes simplifying assumptions to facilitate reasoning (mapping) from one set of findings to a meaningful synthesis, although qualifications and uncertainty inevitably remain. Or, one reasons from an analogy. The analogy may seem plausible, but the justification is a conjecture, subject to further testing and observation. Typically, the provisional nature of such assumptions is plainly acknowledged. However, if the tentative status fades from awareness, error may well surface. The reasoning fails from a *heuristic gap*.[28]

A common application of proxies in biology is the use of model organisms. Here, the heuristic shifts from a single observational variable to an integrated system, but still used in the spirit of a pragmatic surrogate. The assumption is that a related organism shares the relevant features to investigate a particular organ system or disease, say, while accommodating the pragmatic challenges of research (and also not exhibiting any significant disanalogies!). For example, Thomas Hunt Morgan chose to work with the now-infamous fruit flies, not because he imagined them to be ideally informative in and of themselves. Rather, they were easily available, could be maintained cheaply and conveniently in milk bottles on office shelves, and had a short generation time appropriate for his prospective study of mutation and evolution. Fruit flies were a proxy for reasoning about species in general. B.F. Skinner, likewise, chose to work with pigeons on his psychological conditioning experiments because they were readily available (they assembled on the ledge outside his office window) and they were relatively cheap to maintain (compared to monkeys, say). Model organisms are heuristics. They save time and expense, even if they are not the ultimate endpoint of investigation.[29]

Medical studies and tests of various drugs and other treatments for humans, of course, are ordinarily conducted with model organisms first. And much of the time, they work. Even, perhaps, most of the time. Still, observations from model organism are appropriately viewed as provisional. And, indeed, researchers typically go on to confirm the heuristic assumptions. However, errors can emerge if the model organism is somehow unrepresentative. For example, in the mid-2000s a series of animal studies on autoimmune rheumatoid arthritis all indicated the effectiveness of a drug, TGN 1412. It altered the immune response. Yet when human clinical studies began in 2006, the drug prompted severe and unanticipated immune responses. There was a systemic failure of organ function. Six patients were catapulted into life-threatening conditions. The trials were immediately halted, and pursuit of the drug, which had seemed so promising in earlier animal testing, was abandoned. Later study indicated that the lab animals had all been raised in sterile environments. The T-cells in their immune systems had not been primed to develop as they do when in more natural environments. That difference proved critical. On this occasion, the laboratory organisms were not a good proxy for humans. A similar slip in reasoning from model organism to target organism occurred in 2000 with research on a treatment for multiple sclerosis.[30]

Heuristics may also appear in models or simulations (viewed as streamlined or selective surrogates of the target real-world phenomenon). For example, ecologists have developed habitat models, which enable them to explore the geographical distribution of a species—for assessing its conservation status, or the risks of its invasiveness or possible crop damage—without an enormous amount of (costly) field work. However, such models are susceptible to errors. It is virtually impossible to obtain enough detail about climate, soil type, vegetation, and so on, to make reliable predictions. The original data may be patchy, or correlations spurious. Being able to assess all these assumptions independently can foster better models, of course. But given the limitations of resources for research, one can hardly expect to avoid all errors. That is the conundrum of heuristics and their respective heuristic gaps.[31]

Cognitive lapses

As noted earlier, interpreting meaningful patterns in observations involves various forms of reasoning. We synthesize or transform—or remap—information into other, presumably equivalent expressions: the essence of "logic." Unfortunately, perhaps, humans do not (unaided) exhibit strictly disciplined logic. Human brains evolved, apparently, to function with more casual associations and links, but still with good-enough results and adaptive efficiency. Philosophers have documented—and psychologists studied—a long list of logical fallacies, or cognitive dispositions. One author has catalogued over 200 such lapses, with names running the gamut from A to Z, from the Actor-observer bias to the Zeigarnik effect. Sometimes they are described as biases, emphasizing their deficits. Other times they are characterized as heuristics—highlighting how they provide helpful (even if fallible) shortcuts. In either case, they typically operate below conscious awareness.

They are *cognitive blind spots.* They are also potential sources of conceptual errors.[32]

These various fallacies and cognitive biases are amply documented and widely known. They have become stock material in courses on critical thinking and informal reasoning skills. Perhaps for this reason, scientists seem well aware of them. Most have learned basic strategies to recognize them and correct them before they become significant. More typically, however, they are spotted *in others.*

Most scientists are enculturated, I think, not to trust their unguarded intuition, but to think things through thoroughly and carefully. That is, many of these cognitive lapses emerge from unreflective thinking and are easily spotted and remedied with more patient attention to the structure of one's reasoning. Initial intuition can later be corrected, or kept in check, partly by more methodical thinking—a distinction Daniel Kahneman describes as "thinking fast, thinking slow." As a result, perhaps, their impact on science seems limited.[33]

Still, one may consider a few examples. Confirmation bias (already discussed in the section above) is probably primary among them.

Many other lapses concern causal reasoning. Our minds seem to function primarily on simple associations or linkages. Thus, circumstantial evidence and innuendo, although incomplete, can be psychologically quite powerful. Or we may find it difficult to sort a necessary causal condition from a sufficient condition. We may not consider fully the context of observed associations. So, knowing that A can cause B, one may encounter an instance of B and assume (improperly) that A was the cause. This error—affirming the consequent—takes an "obvious" shortcut based on an exclusive association, without entertaining other possible causes of B, besides A. (One may consider it a case of misframing from cause to effect, similar to other "Errors in framing" discussed in Chapter 2.) Another lapse is to conflate correlation with causation: to assume that an association indicates a direct causal connection. But both A and B might be coincidental effects of some other cause, C. One could readily add to the list the post hoc fallacy, looking for a single cause in a multifactorial mechanism, and so on. Interpreting purpose as a factor in causation (teleology) is also rampant and often difficult to detect. Scientists thus need to think through causal scenarios and causal evidence carefully (of course?).[34]

Another major category of cognitive "handicaps" involves probability and statistics. For example, we may ignore background probabilities (the base rate fallacy). We also have trouble combining probabilities. Interpreting negative probabilities poses problems. We do not always distinguish clearly between an absolute measure of probability and a measure of uncertainty. We can confuse failure to reject a null hypothesis with affirming the converse, or confuse Type I and Type II errors. Again, there are many others.

Yet another set of cognitive tendencies shapes thinking merely for the sake of mental economy. They do not necessarily yield more truthful representations. Such bias favors, for example, concepts or distinctions that are essentialist, universalist, either-or, or all-or-none. That is, concepts that help reasoning be faster, easier, or smoother (more "fluent," some say) seem more immediate and clearer to the mind and invite a misleading sense of greater realism. This may well include the popular principle of parsimony, or Occam's Razor. It may be merely a psychological preference for simplicity, based on minimal mental effort, rather than a trustworthy empirical conclusion based on any systematic evidence that the world is indeed fundamentally simple. So, my list of conceptual sources of error in this chapter should hardly be viewed as exhaustive.[35]

All these cognitive dispositions—about causality, probability, economy of effort, and so on—can obviously challenge scientists, although effective learning (through instruction or just mere experience) also seems an effective remedy. For the reader interested in exploring these sources of error further, I refer them to the many books cited in the endnotes.

Unaddressed alternatives

Confirmation bias (including cultural bias as a major variant) can be further accentuated or amplified by another cognitive disposition. Basically, our minds are "lazy." We jump to conclusions. We form "hasty generalizations." And then we stop. Unless circumstances demand, we seem to minimize mental effort. "Why work harder than you have to?" Namely, as highlighted by Herbert Simon, we seem disposed to be satisficers, not optimizers. By default, our minds prefer coherence and speed, rather than completeness. Thus, when we find a "suitable" concept, we tend to regard the available evidence as sufficient. Daniel Kahneman called it WYSIATI: "what you see is all there is." That is, with a solution in hand, we rarely think more deeply or systematically about alternatives—alternative causes, alternative explanations, alternative interpretations, alternative theories. Of course, that leaves conclusions—even ones based on some evidence—vulnerable to error if those alternatives should ever prove correct.[36]

The effect of unaddressed alternatives is similar to misframed observations or limited sampling (Chapter 2). Relevant information may be missing. WYSIATI tends to generate conceptual *blind spots*. Worse, one may be relying on irrelevant information in its place. And with blinders, the scientist is none the wiser. The "moral" is that confirmation alone is not sufficient. Indeed, in some cases, confidence in confirmation alone may be *misleading*.

How ironic that the conventional image of the "Scientific Method" is susceptible to this very error. Namely, in disciplinary lore (for example, as described in introductory textbooks) science relies on the hypothetico-deductive method. One proposes a hypothesis. One then designs an experiment and predicts the results. One conducts the tests, collecting measurements and observations. If the data match the prediction (applying statistical measures), one confirms the hypothesis. If not, one rejects the hypothesis. Easy and clear, apparently. But if there is an alternative explanation for the results, confirmation is not enough. One has not ruled out the other possible conclusions. As in the case of confounders in experimental observation, you may be deceived. Paradoxically, perhaps, what you *don't* take into account *does* matter. Sadly, then, the much touted hypothetico-deductive method is a recipe for rationalization. WYSIATI.

In the 1960s and '70s, philosophers of science turned to history for fresh insights. They were impressed with how theories frequently change, some being abandoned while new ones were adopted.[37] Why? For years, theory assessment had been merely a question of the quality of evidence. But now, history showed how theories that were "falsified," or disconfirmed by some data, were not always immediately rejected.[38] Rather, the availability of an alternative theory was just as important. Philosophers realized that scientists do not always exercise absolute, uniform standards of evidence. Rather, they also compare alternatives. Theory change was a matter of "*competing*" theories and of theory *choice*. Namely, considering conceptual alternatives is central to scientific practice.

A prime example is the discovery that the noble gases can form compounds. The newly discovered noble gases (helium, argon, neon, xenon) were recognized as a group and added to the periodic table around 1900. They had been characterized as unreactive and even labeled as the *inert* gases. Somehow, the difficulties faced by early chemists in investigating these elements evolved into a general impression that they could *never* form compounds. Inert, interpreted literally. Concepts of valence and filled electron shells surely helped reinforce that image with a theoretical rationale. However, Neil Bartlett demonstrated in 1962 that the noble gases (Group VIII) do indeed form compounds. More surprisingly, perhaps, the reactions often proceed rapidly at room temperature, without any need to "force" them through extreme conditions. Bartlett's pioneering work exposing that unchecked alternative was commemorated by the American Chemical Society with a historical landmark designation in 2006.[39]

One may expect that, given a scientific community of sufficient size, unaddressed alternatives will be noticed and pursued by some perceptive individual. Not so. A striking case of neglect was ascertaining the nature of the viceroy butterfly's mimicry. Here, the neglect persisted for over a century. The close resemblance of the striking orange and black patterns in the two butterflies, the viceroy and the monarch, is easily appreciated. In 1869 noted entomologist Benjamin Walsh and his protégé Charles V. Riley presented the viceroy as an example of Batesian mimicry, introduced as a concept in 1861: namely, the viceroy was a palatable mimic of the unpalatable monarch. They drew on informal field observations of predation and the relative population sizes of viceroys compared to their closest relative. And on the remarkable resemblance, of course. Later, naturalists identified the milkweed plant as the source of the monarch's toxins. No such source of toxins seemed obvious for viceroys. All the contextual evidence was consistent with the notion of Batesian mimicry, at least. But neither Walsh nor Riley (nor many of their successors) actually explored the alternative: that the viceroy might be unpalatable, too. The two butterflies might share similar warning patterns. WYSIATI? Or, equally, "what you look for is all you will ever likely find." When the taste-test was finally done with birds in the 1990s, it turned out that viceroys were indeed distasteful, after all. The similar color patterns resulted from convergent evolution: each species enhancing its survival by resembling the other (now known as Müllerian mimicry). Walsh and Riley (and others) had not systematically checked the alternative interpretation.[40]

Accepting correlation as sufficient evidence for causation is another case of unaddressed alternatives. The possibility of such lapses in causal reasoning is widely noted, although often with contrived examples. Here is a real case as illustration in context. An important 1988 study observed that high blood levels of HDL cholesterol were correlated with reduced risk of coronary heart disease. That seemed to indicate an important strategy for treating a major cause of death. Raise HDL. The finding also seemed to fit with the role of HDL: clearing the bloodstream of "bad" cholesterol that builds up on artery walls and can lead to heart failure or stroke. So, major drug companies began testing niacin (vitamin B3), known for its ability to raise blood HDL levels. In 2011, the third such study was halted early because the treatment proved ineffective. Yes (the investigators found), HDL levels were increased. But no, there was no reduced risk of coronary disease. In three successive large-scale studies, the original correlation was not found to reflect causation. Meanwhile, several genetic studies reached a similar conclusion, independently. One statement by the U.S. National Institutes of Health offered the "moral": "It's possible that HDL itself may not directly lower the risk of heart disease but that blood HDL reflects another factor that does." That is, HDL may be a byproduct of a healthy heart, while not directly improving heart health. It remains a helpful indicator, or biomarker, to assess a patient's overall risks. But it is not a guide to therapy. That was the alternative lurking in the findings of the original 1988 heart study. It took two decades to find and articulate how the observed correlation was misinterpreted as causation.[41]

Teasing apart causation from correlation is a tricky business. The details are beyond our scope, here. But they are well addressed in a voluminous philosophical and statistical literature on causation.[42]

Cryptic alternatives

Not fully addressing available alternatives is an epistemic danger. But what if the alternative is not "available"? What if it is yet unconceived? Beyond the horizon of current imagination? One can easily appreciate the problem. Can one truly think concretely about cells if there are no microscopes? Distant galaxies, if there are no telescopes? The Earth's interior, if there are no seismographs? These alternatives may be possible, but they are "hiding," outside current technological capabilities. So, too, for alternatives based on background knowledge or conceptual frameworks that have not yet been developed. We may call them *cryptic alternatives.* That is, the problem of alternatives becomes both more acute and more poignant when the alternatives have not yet been imagined.

Cryptic alternatives pose a disturbing conundrum. How can one conceptualize an error if no one could possibly have guessed or understood the eventual answer at the time (regardless of the evidence)? Perhaps the best way to interpret and understand these puzzles is through some concrete examples.

Historians of science can offer plentiful stories of alternatives that escaped notice or were beyond the reach of the current imagination. For example, understanding the massive "misplaced" boulders known as erratics. They sit in otherwise wide-open landscapes. Some—the Norber erratics in northern England—even sit precariously on small pedestals of a different kind of rock. They certainly command attention and wonder. Their name (from the Latin for "wandering") indicates the implicit puzzle they pose. Where did they come from? For example, while the Norber erratics are composed of dark "greywacke" sandstone, they rest on a bedrock of sedimentary limestone. They seem not to have originated in the local area. One geologist traced an erratic in Monthey, Switzerland, to the Mont Blanc massif, some twenty-five miles away. How did the erratics get to their current location?

In the late 1700s, Horace Bénédict de Saussure expressed the then widespread view that the huge rocks had been borne by powerful floodwaters. That explanation also fit well with other geological features often observed near the erratics: coarse scouring of the bedrock, beds of gravel far from any river, polished rocks, deposits of soil or clay mixed with rocks of different sizes (even boulders), and large U-shaped valleys. All apparently bore witness to the violent action of great volumes of water. That perspective was most vividly expressed by William Buckland in his 1820 *Relics of the Flood*, which contextualized all the evidence in the Biblical story of a giant worldwide flood (see earlier section on "Religious bias"). But even by then, geologists were doubting the global nature of any flood, contenting themselves with explanations based on regional floods, tidal waves, broken earth dams, or other forms of catastrophic water movement. In 1833, Charles Lyell attributed the erratics and related phenomena instead to the drifting of ice-floes, bringing reindeer bones and birds from the Arctic as far south as southern France, where they were deposited when the iceberg melted. Drifting offered a popular alternative to the diluvial and alluvial theories. Geologists had plentiful options to debate (which, of course, they did). However, all were errors. There was another, cryptic alternative.

The eventual answer was only cryptic, perhaps, to those who lived in England and northern Europe. The experiences of these scientists—and hence, perhaps, also their imaginations—were limited by the landscape and geology around them. Geologists living near the Alps and in Scandinavia, by contrast, were familiar with a quite different set of observations: glaciers. Ice, not water. That idea was introduced by Jens Esmark from Norway in 1824 and echoed by Ignaz Venetz in Switzerland—and by numerous uneducated Swiss peasants interviewed by Jean de Charpentier. Local perspective mattered. But would this help explain the erratics in England? Robert Blakewell noted that "in the midland counties of England, for instance, there are beds of gravel, and fragments of rock, scattered over hills, that are . . . far distant from the rocks which have supplied the fragments." But, he added, "in England, we have not the glaciers to assist in their transportation." The others in the Alpine region, for their part, did not envision the effects as extending beyond where glaciers were still visible to them. Cryptic. WYSIATI?

The scope of vision changed with young Swiss geologist Louis Agassiz. Borrowing an idea from a German friend, Agassiz envisioned in grand scale a former *Eiszeit* or, translated from the German, an "ice age." Agassiz's thoughts extended far beyond the Alps, finding a possible cause for the extinction of mammoths in Siberia and other fossil mammals no longer found in Europe. In 1840 Agassiz published his essay, not only on the action of glaciers but also on the evidence for an Ice Age when glaciers once spread from the Arctic all the way down to the Mediterranean. In a tour of Britain, he found relevant evidence near Ben Nevis. By the 1860s the once unimagined was loosely accepted by the worldwide geological community.[43]

A wholly different surprise awaited chemists studying energy processing in the cell in the mid-20th century. It's not difficult to conceive of the energy transfer as the chemists did: one molecule releasing its energy to another in a series of enzymatic reactions, handing off the energy like a baton in a relay race. The chemists wanted to identify the last set of high-energy intermediates that carried the energy to the final energy molecule, ATP. A number of prominent biochemists focused their attention on them, anticipating that whoever could find them was probably guaranteed a Nobel Prize. From 1953 to 1970, using different extraction techniques, sixteen different alternatives would be proposed. Not just offered as theoretical possibilities, but as experimentally isolated samples. But all sixteen were errors. Indeed, no one would ever find the chemical intermediates. They do not exist.[44]

The alternative was cryptic indeed, at least for most chemists. It involved another form of energy altogether. One cell function that requires energy is moving molecules across the cell membrane. But most cell chemists viewed these gradients as ways to use energy, not possible sources of energy themselves. One biochemist working in this area, Peter Mitchell, puzzled philosophically over how these *chemical* reactions were associated with the *physical* processes of movement. For him, the framework of conventional chemical equation linking reactants to products was incomplete. There was also, simultaneously, a spatial dimension: molecules crossing the membrane. He called the hybrid notion "chemiosmotic." Clued by some experimental findings from one of his students, he applied this concept to these ATP reactions.

It was a complete gestalt switch. Chemists have long been frustrated, trying desperately to study the reactions in solutions, freeing the enzymes from the mitochondrial membrane to study them in isolation. Mitchell was now claiming that this was not possible. The membrane itself was an essential part of the process. It was not an extraneous feature to be removed. The "high-energy intermediates," he proposed, were an electrochemical gradient across the mitochondrial membrane. There were no intermediate molecules to isolate or identify. No wonder no one could isolate them. Mitchell published his chemiosmotic model in 1961, and his revolutionary insight was finally recognized by a Nobel Prize in 1978.[45]

As noted earlier, all these cases of cryptic alternatives pose a deep problem for science. In a historical context, how can one ever address the limits of current knowledge, the limits of technology, or the limits of our imagination? It may seem "unfair" to characterize this as an error type. How can a scientist accommodate an unknown? How can reasoning that once seemed fully justified be called an error? Still, knowing about their possibility might inform our epistemic postures. For example, might we be able to articulate more fully the edges of our knowledge at any given time? Or identify where qualifications and postures of uncertainty may be warranted? (See section on "Error probes" in Chapter 8.)

The discovery of new alternatives, once cryptic errors, echoed repeatedly throughout history, might precipitate a deep pessimism about science. Will we ever—*can* we ever—develop a theory that is not susceptible to being upended in the future by some currently unidentified alternative? The topic has kept a handful of philosophers of science busy in apparently unending debate. It's one thing for concepts to keep changing. It is quite another, as Kyle Stanford notes, for the prospect of unconceived alternatives to leave scientific knowledge perpetually open, unresolved, and vulnerable to error. Yet we may also conceive cryptic alternatives through its converse. Namely, what alternatives are conceivable and have been addressed? Which have been deemed inadequate? What has been ruled out, at least? Paradoxically, perhaps, this approach indicates a possible role for knowledge of what is not the case, rather than what is. It begins to hint at a "positive" role for negative knowledge (Chapter 8).[46]

Epistemic hubris

An inventory of conceptual error types in science is certainly not limited to the errors discussed here. My primary intent has been, rather, to convey the scope of the problem and to provide the framework of successive mappings as a way to organize the various error types. Again, a guiding principle in error analysis is to consider every step in how we justify the (re)mapping from observations to final claim. That includes basic logical reasoning (of various types), calculations, combining of data, conceptual transformations, model choices, and interpretive frames.

One last conceptual error may be worth noting. Perhaps the most important? Namely, the meta-error of assuming that these various errors apply only to other scientists, not to oneself. This final error is extraordinarily widespread, as indicated in the scientists' comments in the opening of Chapter 2. One might call it, perhaps, *epistemic hubris*: the belief that through a dedication to science and heartfelt commitment to rigorous logic, methodology, and critical thinking, one will escape conceptual errors. History (and even contemporary ethnography) says otherwise.

This final point is easily registered and then forgotten. It bears repeating. *No one is exempt from the effect of these and other cognitive flaws.* Overgeneralization, faulty assumptions, confirmation bias, perspectives permeated with cultural context, failure to address alternatives, and limited creativity: all foster conceptual blind spots. No one is wholly immune. The error here is not in making an error, but in the hubris of imagining that it cannot occur beyond the fringes of one's awareness. That is, some scientists may be more open (and honest) in engaging alternative perspectives and acknowledging those errors—and setting about remedying them. (For more on this strategy, see sections on "Merton's norms" in Chapter 4 and on "Social checks and balances" and "Complementary perspectives" in Chapter 7). Here, the inevitability of conceptual error at the level of individual reasoning provides a potent context for considering the social structure of scientific discourse, addressed in the next chapter.

4

Social-Level Errors

Social-level (discoursive) errors • Merton's norms • communal confirmation bias • communal cultural bias • lapses in communication • peer-reviewed publication • publication bias • credibility bias and fraud • conflict of interest • critical consensus

A keen reader might regard the cases discussed in the previous two chapters on observational and conceptual error types—especially those of biases—as errors of individual scientists, but perhaps not of science. Or at least not of science writ large. That is, not every scientist necessarily reasoned as the particular individuals did, nor subscribed to their errors. What truly matters, one may well contend, is not the *individual* claims, but the knowledge widely accepted across a diverse scientific *community*. Indeed, scientists can, at times, be very effective at finding each other's mistakes and exposing their biases. Properly construed then, science has an important and ineliminable *social* dimension. As the Editors of *Science* magazine recently noted, "Science is a social process. Discoveries—even by supposed 'lone' geniuses—do not become knowledge until the findings are shared with the scientific community, to be vetted, challenged, and expanded upon."[1]

Even so, there is no clear or widely accepted standard articulation of the epistemic processes of science at the social level. Scientists are certainly familiar with the social customs for validating claims through open discourse with other scientists. But rarely, if ever, are these social practices taught explicitly as part of scientific *methodology*. Yet if the social process of review by peers "fails," we may justly characterize the relevant factor as another *source of error* (in a sense directly comparable to the cases discussed in the last two chapters). These constitute *social-level (or discoursive) error types* (see Figure 2.2).

The social structure of modern science has been important at least since the founding of the Royal Society in England and other scientific academies in the 1600s. Scientists began presenting, discussing, and debating each

Toward a Philosophy of Error in Science. Douglas Allchin, Oxford University Press. © Douglas Allchin (2026).
DOI: 10.1093/9780197827703.003.0004

other's findings in a shared, institutionalized setting. They convened regularly and published reports under a formal review process. They documented their discussions and announced discoveries in proceedings, which contributed to a collective memory. They established standards of testimony, and of presenting claims that have been further developed over the centuries. The function of such societies (besides promoting the patronage of the scientific enterprise) was to install vetting by peers as a fundamental element in establishing trustworthy scientific knowledge. "Only after the originality and consequences of his work have been attested by significant others," sociologist Robert Merton noted, "can the scientist feel reasonably confident about it." Namely, a consensus among informed professionals began to displace individual claims as the standard hallmark of reliable science.[2]

Publishing research results and arguments is thus not sufficient. Once empirical phenomena have been explored and data collected and crafted into workable concepts, the epistemic work shifts from problem-posing and problem-solving to persuading peers (Figure 2.1). The implicit goal is to establish concurrence and develop an actionable consensus. However, scientists may disagree. They may interpret the same evidence from different perspectives or cognitive standpoints. With diverse forms of expertise, some may recognize sources of error that others have not noticed. With different background knowledge, they may be aware of data not addressed by others. Ideological or cultural biases may be exposed. Controversies may flare. The uncertainty at the social level may lead to variations of the contested experiments or to seeking further evidence that can resolve ambiguities and uncertainties. The process of reconciling different views almost always involves a certain degree of "negotiating"—in both senses of the word: first, negotiating a pathway to further relevant evidence and to more secure knowledge about the world and, second, negotiating through dialogue and brokering evidence to reach conclusions agreed on by all. Contrary to the popular mythos of the "Scientific Method" as the primary guarantor of trustworthy science, the social practices of science are probably the chief epistemic mechanism for filtering out error and ensuring reliable knowledge.[3]

Epistemically, we trust that countervailing prejudices will expose and cancel each other. That is, we expect the scientific community to function at this new, social level as an implicit system of checks and balances. Through active discourse, the collective vets claims and generates more robust scientific knowledge. Sandra Harding called the result "strong objectivity." That is, while individuals are susceptible to cognitive biases and blind spots, groups are less so. "Our truth," ecologist Richard Levins once observed, "is the intersection of independent lies." Indeed, several scholars now claim that this error-monitoring and error-correction process is primarily what distinguishes science from non-science and warrants its public credibility.[4]

Several social structures and social practices guide the epistemic discourse at this level. These processes contribute to assessing and qualifying individual claims. First, most sources of funding involve review of grant applications, assessing the expertise or competence of the investigator before a project even gets support. Second, papers are submitted to journals as nominal gatekeepers for publication. Editors typically refer the paper to several peer experts, who provide critical comments and recommendations on the value of publication. Editors then decide whether the submission meets the publisher's standards and therefore merits publication. Peer review continues after publication, however. Scientists convene at conferences, present their ideas, and discuss their merits and deficits. They correspond. Nowadays, they may post online. They critique. They rebut. They may conduct additional research and publish further evidence. On some occasions, many published papers are considered together and critiqued in a synthetic review article. When the science is important for policy, professional societies may convene consensus panels that integrate findings and sort the reliable from unreliable findings. These are the *social mechanisms* for filtering out errors, taming overly extravagant claims, and flagging misleading biases. These constitute the social, or discoursive, methodology of science.[5]

But, like experimental methods and conceptual reasoning, the social level processes do not always function perfectly. The system of checks and balances may sometimes falter. Thus, the inventories of observational and conceptual error types (last two chapters) do not exhaust a complete consideration of the sources of error in science. Just as the social level may help remedy some errors, so it may fail on other occasions. Identifying the various social-level error types thus parallels articulating the social processes of science.

As one explore errors at the discursive level of mapping, correlates of the problems found at the observational and conceptual levels—sample size, incompleteness, misframing, miscalibration, biases, and the limit of heuristics—all reappear. They can lead to faulty theories being widely accepted or to effective theories being rejected, unduly peripheralized, or forgotten. Rather than correct any error, the social interactions may perversely amplify and entrench the errors. This chapter describes these error types and offers some illustrations.

Merton's norms

To characterize and categorize various error types at the social level, one may draw on Robert Merton's sociological analysis of science dating from the late 1930s. Merton was concerned about "the growth of certified knowledge." What *social* structure or features of *institutional organization* would foster that epistemic goal? Merton postulated four norms, or idealized practices, toward achieving that end. These guiding principles have been echoed more recently by philosophers of science working on social epistemology, or how social dynamics contribute to (or work against) the development of scientific knowledge.[6]

The final norm—but perhaps the most central—was *organized skepticism.* Merton identified a role for "institutional vigilance"—an "unending exchange of critical judgment," as described briefly above. It would be expressed as "vigorous but cognitively disciplined mutual checking and rechecking." Namely, science (ideally) "affords both commitment and reward for finding where others have erred or have stepped before tracking down the implications of their results or have passed over in their work what there is to be seen by the fresh eye of others." The norm implicitly acknowledged that no single scientist would be perfect. Reciprocal critique offers a methodological check, much as a scientist uses calibration or controlled experiments to help ensure that their observations are reliable. The social system of science should, ideally, help identify and remedy individual errors. But perhaps this is not always the case in observed practice? [7]

Merton's other three institutional norms indirectly support or extend his epistemic principle of organized skepticism, or mutual criticism. First, Merton posited an ethos of *communalism.* Essentially, he viewed science as echoing the Three Musketeers' famous motto, "All for one and one for all." For scientific knowledge to grow, it must be conceived and pursued as a collective effort. Results must be public, so that others may build on them. Notably, this implies open sharing through an effective communication network. One can easily imagine possible problems: language barriers, lack of access to publications or to proprietary journals, exclusive "insider" correspondence networks, weak search engines, incomplete keyword indexing, and so on. In recent years, there has been renewed concern about this particular ethos of science, epitomized by the Open Science movement. Proponents advocate open access to data and analysis software, and even to specialized reagents, cell cultures, and/or other key materials. But it begins with describing and sharing one's work. Ultimately, one cannot take communication for granted. How might "information islands" contribute to erroneous claims by others? Communalism applies equally to the context of critical discourse. For others to notice errors or detect bias, the relevant evidence and arguments for claims must be public and transparent. Quite simply put, effective communication and exchange is essential to enabling a community of scientists to find and remedy errors. But one can easily imagine lapses, or social-level errors.

Next, Merton described a norm of *universalism*. Here, he echoed a core sentiment in America's *Declaration of Independence*: viewing all citizens as created equal. Namely, all scientists should have equal opportunity and respect in the public discourse. None should be inherently privileged or entitled to special persuasive influence. For example, when a paper is submitted for publication, editors typically blind the author's identity before sending it out for review. Just as blinding is used to control for observer bias in recording or analyzing data, the editor hopes to ensure that the reviewer's assessment will not be biased by their prior opinion of the author, favorable or not.

Universalism is rarely achieved in practice. Perhaps appropriately so? Researchers (and their labs) earn status and respect through the quality or rigor of their work over time. That is, scientists informally rank the credibility of their peers: an indirect measure of the trustworthiness of their claims, based on their track record. Scientists "calibrate" their colleagues' testimony against their own experience and reports by trusted others. This can be an important epistemic heuristic. In an environment overwhelmed by information, the credibility of the author is a proxy variable for the trustworthiness of their claims. You can trust the quality of their evidence and their interpretation of the data. The heuristic productively allows individuals to sort through the literature more efficiently and to avoid the tedious work of reproducing the results for oneself.[8]

Still, credibility judgments, like any heuristic, may fail. This may open the way to errors, too. Scientists thus need to be prepared to address those occasions where the credibility heuristic breaks down. As detailed more fully below, fraud may be characterized as a type of credibility error. That is, in the spirit of universalism, all individuals with relevant degrees and institutional positions are generally regarded, by default, as basically competent and honest. That is an extension of trust based on an assumption, generally warranted. As the numerous historical cases of fraud attest, however, sometimes that trust is misplaced. Occasionally, grossly misplaced. The level of trustworthiness, or credibility, needs to be recalibrated. Accordingly, one can conceptually separate fraud into two dimensions: the epistemic dimension of identifying a source of error at the social level (unjustified credibility), and the moral or professional dimension (violating responsibilities to peers).[9]

Finally, we may consider Merton's norm of *disinterestedness*. This was, perhaps, Merton's nod to objectivity. That is, for him "the primary goal, the furtherance of knowledge, is coupled with a disregard of the consequences that lie outside the area of immediate interest." Ideally, a scientist should have no stake, or "interest," in any particular outcome. In terms of the concerns most actively discussed today, scientists should be free of *conflicts of interest*. Those interests may come in many forms. For Merton, they included nationality, sex, age, race, ethnicity, and religion. To these we may add, more generally: power, profit, and privilege. That is, in addition to the cognitive influence of confirmation bias, scientists may experience the motivational influence of contexts where they may benefit politically, ideologically, or commercially. These cognitive factors, again, may open the door for errors to appear and persist. As Merton readily acknowledged, "scientific research is not conducted in a social vacuum." Regulating any adverse influence of interests—say, by drawing on remote parties to adjudicate disputes—thus becomes a function of the social structure of science. Given how human minds function, it seems impossible to meet the norm of disinterestedness fully. Interests will always be present. Still, one can hope to attend to these interests at the social level, and demand correspondingly more rigor. Objectivity emerges intersubjectively, by refusing to honor the prejudicial views of any particular interest.[10]

Merton's norms, also described as "the ethos of science" (Figure 4.1), have occasioned contention among philosophers and sociologists of science. Are these descriptions of how scientists actually behave, or are they merely norms that frame how scientists *should* (ideally) behave? Merton was perhaps ambiguous. He explicitly labeled them as "the normative structure of science." But he also implied that he derived them from his observations

Merton's norms / Epistemic function	*Error types*
organized skepticism / regulative interaction of diverse critical perspectives	• communal confirmation bias • communal cultural bias
communalism / open sharing and open critique	• incomplete communication • language barriers, limited journal access, poor indexing • publication biases
universalism / equal discoursive opportunity	• credibility bias • fraud • Matthew effect, aura/shadow effects • selective access to scientific profession & limited diversity in critical discourse
disinterestedness / independence from the exercise of political or economic power	• conflict of interest • biased peer review or funding

Figure 4.1 Merton's norms, their epistemic functions, and corresponding error types.

of the practice of science and the values that were expressed in the public discourse and actions of scientists. He regarded the ethos as guiding actual practice in a way that differentiated science from other social institutions. History, however, plainly demonstrates repeated "violations" of the norms. For the purposes here, we need not debate whether science is more or less than some imagined ideal. Our concern is epistemic. Do the social protocols promote reliability in scientific knowledge, and do particular implied deficits tend to precipitate or perpetuate errors (as defined in Chapter 1)? In this way, Merton's norms may guide an analysis of social-level error types.

Communal confirmation bias

In Merton's view, science exhibits organized skepticism because scientists perpetually challenge assumptions. They are imbued (he believed) with "a latent questioning of certain bases of established routine, authority, vested procedures and the realm of the 'sacred' generally." For them, "nothing is sacred." For Merton, scientists not only should, but do, embody a healthy spirit of iconoclasm.[11]

And yet (most scientists will tell you) new ideas that challenge the theoretical status quo in science are generally not warmly welcomed. In the section on "Theory-Laden Judgment and Confirmation Bias" (Chapter 3), we saw how individual claims can be shaped by theoretical contexts and individual backgrounds or preferences. Thus new ideas, not the old ones, typically receive the harsh skeptical treatment. A convenient rationale would seem close at hand, perhaps?: "Better safe than sorry." Of course, one can easily imagine a whole community of researchers sharing similar conceptual blind spots. (Consider a collection of exclusively male researchers in a patriarchal society, for example?) The sampling of relevant critical perspectives is incomplete. Organized skepticism fails. When a *group* of scientists exhibit the same theoretical biases, then, they can exhibit a distinctive social-level error type: *communal confirmation bias.*

Communal confirmation bias can be observed when scientists seem to act in concert to condemn a new idea, with very little (or only casual) reference to the evidence available. They may also fail to acknowledge or fully engage the evidence presented for the new alternative, preferring instead to simply promulgate the success of the prior theory, rendering it as apparently unquestionable. A stunning (and at the same time, troubling) example may be found in the reception to J. Harlan Bretz's interpretation of the physical geography of the Scablands in southeast Washington state in 1922.

The area exhibits many striking geological formations: a vast plateau scarred by a network of deep channels, "carved into mazes of buttes and canyons." As a junior geologist, Bretz spent several summers documenting and mapping the features that could signal what processes had produced such a remarkable, austere landscape. Bretz was perhaps as surprised as anyone to discover that many features bore witness to the violent action of an incredible, unprecedented torrent of water. Familiar features in rivers, but here on a grand scale. Neither the gradual, sustained action of streams the size of those still present, nor glaciers, could account for the wide waterfall ledges, the deep piles of rock debris at the foot of sheer cliffs, or the reticulated nature of the gorges, splitting and recombining, and so on (some 21 unique features, in all). Bretz certainly did not embark on the project with some favored theory or preconception that shaped his observations. And he mindfully followed a principle of multiple working hypotheses. Bretz was repeatedly "amazed" by the stark, unexpected evidence he encountered. After a second year of fieldwork, he published his voluminous findings and conclusions. There, he carefully addressed the plausible alternative interpretations and how, with the available evidence, "all other hypotheses meet fatal objections." The Scablands had once experienced a tremendous flood.

Bretz's claims did not sit well with established geology. Bretz was proposing a catastrophic view of geological events. That perspective had been popular in the 18th century. But the uniformitarian principles of Charles Lyell in the mid-1800s had brought an end to the age of imagining cataclysmic upheavals, with their global floods and volcanic turmoil. The only legitimate benchmark for interpreting events in the past was the forces visible today. What Bretz proposed in 1923 was unprecedented. The initial response to his claims was mostly a chilly silence. However, in 1927, after several more seasons of extending his observations, he was invited to present his ideas before an august gathering of the Geological Society of Washington, DC. When the formal presentation ended, the ultimate purpose of the gathering became painfully clear. A panel of the leadership of the U.S. Geological Society provided unannounced commentaries on Bretz's ideas. They all paraded the uniformitarian theme of "long periods of years" and the "orderly and long-continued process of head-end erosion." Bretz's conclusion about vast amounts of water was "impossible," "preposterous," and "wholly inadequate." Several months later, a colleague wrote sympathetically to Bretz that "anyone who wishes to discuss the question ought to visit the locality." But none of Bretz's critics had done any field observations, nor did they do so later. Their alternatives were plausible only by focusing narrowly on incomplete (cherry-picked?) fragments of the overall evidence. That is, they were content to argue by relying on their theoretical preconceptions alone. That blinded them on this occasion to an exceptional case: a vivid historical example of communal confirmation bias. The error was prejudiced dismissal of relevant evidence and sound reasoning.

In an interesting twist, a former president of the Geological Society of America had already cautioned his colleagues against "theoretical stagnation." In 1926, he had written a paper in *Science* on "The Value of Outrageous Geological Hypotheses." He complained that geologists of the day had become too complacent in their "accumulated convictions" and "preconceived opinions," which they were "too prone to regard as geologically sacred." He clearly recognized the temperament of the community, primed for communal confirmation bias. But it was not Bretz's flood that he seemed to have in mind. Rather, the "outrageous" idea was probably the continental drift theory of Alfred Wegener, whose work had recently been translated into English. Even though philosophically disposed to entertain bold new hypotheses, in practice he disparaged them.[12]

Not all forms of communal confirmation bias are visible in conspicuous debate. For example, in the 1970s, there was active debate about the levels of natural selection. While everyone could acknowledge (since Darwin) that differential survival and reproduction of individuals could lead to evolutionary change, there was less agreement about whether the process could operate at the group level. Could differential survival or proliferation of groups (say, different human tribes, primate troops, or populations) lead to genetic changes in *group* properties, in a way that was distinct from the cumulative effect of selection on individuals: group selection? Many models "did the math." All seemed to indicate that group selection was feasible, but only under extreme and unlikely conditions. Because they all took slightly different approaches, relying on slightly different sets of assumptions, the conclusion seemed quite robust. Michael Wade, however, was skeptical. His work on populations of flour beetles subjected to different environmental pressures in the lab indicated that their properties could indeed evolve (at the group level). When he analyzed a dozen prominent mathematical models, he discovered that, in fact, they all shared several common assumptions. Many "intuitive" simplifications facilitated the calculations, but they were ultimately unrealistic and restrictive. More importantly, they all tended to diminish the possible significance of group selection. That is, the modelers had adopted individual selection as a norm and inscribed it unwittingly into their computational equations. The confirmation bias was so deeply embedded that it remained hidden in a set of "innocent" mathematical assumptions. It took Wade's analysis, starting from a different standpoint, to make the cryptic bias plainly visible.[13]

Almost any scientist can regale you with a cautionary tale of an idea in their field that, while eventually celebrated, originally met disbelief and resistance. Or even ridicule. Copernicus and the heliocentric solar system. Deep time and the age of the Earth. Darwin and evolution. The Big Bang. The meteorite that wiped out the dinosaurs. "Resistance," here, describes the effect of communal confirmation bias. The new findings did not match what was already known. Or what people *thought* they knew, which was significantly incomplete, misguided, or outright wrong. When popularized, such stories often embody a mythic, moralistic structure. Generally, however, scientists defend the status quo for good reason. Their conservative posture is an epistemic defense of the evidence that already exists. Namely, we ought not to abandon hard-won knowledge willy-nilly. Of course, that provides a challenge for any new idea.

The chief problem in these episodes of "resistance" is not that confirmation bias existed. It was that the scientific community was largely homogeneous theoretically: that yielded *communal* confirmation bias. There was not enough breadth of perspective to notice any error or to allow new alternatives to gain traction. That is, the critical social mechanism for keeping individual confirmation bias in check was missing. Skepticism was rampant, without Merton's qualification that it be "organized." If interpretive bias can be a source of error, one needs to "control" for it by assessing its effect with parallel interpretations, just as one might do through controlled experiments at the observational level. Without a theoretically diverse community, the interpretive sample size is too small and possibly unrepresentative. *Communal* confirmation bias poses special problems.

All this generates a deep conundrum for science. How much theoretical diversity is appropriate? How does one balance defense of past evidence with the need for fresh perspectives? Thomas Kuhn labeled it the "essential tension" of doing science. He recognized that research functioned more efficiently if everyone shared a common "paradigm," or set of research exemplars and ways of doing things. At the same time, he acknowledged the periodic need for alternatives, which might ultimately lead to new, successor (or divergent) paradigms. Kuhn posed the problem, but did not have a ready solution. Later, philosopher Philip Kitcher addressed the same issue in his magnum opus, *The Advancement of Science*, treating the division and organization of cognitive labor with remarkable mathematical rigor. Perhaps all one can say is that an overwhelming degree of shared biases is unfruitful. It is not surprising, perhaps, that science often depends on "rebels, mavericks and heretics," as well as "outsider scientists" for scientific innovation. The problem of homogeneity and communal confirmation bias will be addressed again in the last section of this chapter.[14]

Communal cultural bias

An important variant of communal confirmation bias occurs when the community's shared ideas are not based strictly on prior scientific background, but on shared cultural perspectives, unquestioned social norms, or ideological assumptions. As noted in earlier discussion, such ideas are not wholly sequestered in the scientists' minds and may shape scientific thinking. Hence, when a group of scientists is culturally homogeneous, they may lack the critical perspectives that can expose and probe such conceptual influence: hence, *communal cultural bias.* That is, assumptions are not even perceived as assumptions. The bias goes unnoticed.

Such errors are of major concern in public contexts, where policy may be based on a scientific "consensus" that is ultimately biased. The claims may embody cultural prejudices that are then used to justify those very prejudices as scientifically sanctioned. Homogeneous scientific communities are as susceptible to bias as individuals are. Worse, perhaps, the agreement among many people (unaware of their shared blind spot) may foster an illusion that the ideas have been critically examined and passed muster. Thus, there is a meaningful difference between individual cultural bias and communal cultural bias.

For example, the notorious claims about drapetomania and dysaesthesia aethiopica among African American slaves, described in the section on "Racial Bias" (Chapter 3), were *not* widely adopted, so far as we know. On the other hand, Samuel George Morton's *Crania Americana*—equally a product of racist assumptions—was widely endorsed as exemplary science by his (Caucasian) peers. Praises came from journals around the world. James Cowles Prichard, who helped establish anthropology as a discipline in Britain, considered the work "exemplary." Phrenologist George Combe promoted it in his pamphlets for the working class. A magazine for Methodist women in Ohio quoted Morton favorably in characterizing the native inferiority of Native Americans (Indians). The important checks and balances were absent—the essence of this error type at the discoursive level. *Communal* cultural bias is substantially more threatening to the integrity of science than individual bias.[15]

Again, it is not the individual but the group that matters here. Recall the racist eugenic views of Charles Davenport, with their misleading pedigrees of pellagra, feeble-mindedness, and other conditions of poverty and destitution. It is equally important to the status of the error on this occasion that many others heartily endorsed his claims. Davenport's 1912 lecture at the American Association for the Advancement of Science was published in *JAMA*. So, too, was the discussion by attendees. There, one finds a Dr. Woods Hutchinson of New York concurring, adding his personal experience: "As Dr. Davenport says, the more carefully these defectives are studied, the more the hereditary element comes out in them. ... the criminals, prostitutes, insane, epileptics, feeble-minded and inebriates." He referred to his work with criminals in Indiana, his investigation of a notorious pair of labor activist saboteurs, and the study of feeble-minded at a school in Vineland, New Jersey. His approval was not alone. The news account recorded positive comments from Dr. L. Pierce Clark of New York, Dr. William W. Graves of St. Louis, Dr. E. E. Southard of Cambridge, Massachusetts, Dr. Tom A. Williams of Washington, DC, Dr. Warren D. Calvin of Fort Wayne, Indiana, and Dr. S. Grover Burnett of Kansas City, Missouri. They all considered their cultural views firmly grounded in scientific data. Merton's ideal of organized skepticism failed because everyone shared the same cultural bias.[16]

Similar patterns of community concurrence permeated 19th-century discussions of craniology and women's intelligence. The widely shared view was articulated by McGrigor Allan in 1869 when he addressed his fellow members of the Anthropological Society of London—all male, of course:

> The assertions and claims put forward under the term "Woman's Rights", are a challenge to anthropologists to consider the scientific question of woman's mental, moral, and physical qualities, her nature and normal condition relative to man. Nowhere, then, is the question be more appropriately and profitably discussed than in the Anthropological Society.

The answer—based on politics—seemed predetermined. The science was engaged, in a sense, to rationalize, or "naturalize," the conclusion. So the discourse flourished, without the critical perspective of women. At least until the turn of the century. The male bravado waned when a few women entered the field and started presenting evidence that exposed the unsupported assumptions and unwarranted overstatements—described more fully in Chapter 6.[17]

As in the previous section, it is worth noting the role of a homogeneous scientific community in providing an occasion for communal bias to take root. Scientific communities dominated primarily by men are ripe for communal gender bias. Scientific communities that are uniformly Caucasian are susceptible to communal racial bias. Scientific communities of well-educated social elite are subject to communal class bias. Scientific communities in one country may easily drift toward communal nationalism or ethnic bias. And so on, for various forms of bias that may shape scientific questions or scientific interpretations of observations. Namely, the nature of *who* participates in science can shape the *what* of the critical discourse. It may also apply to whoever has had an opportunity to choose the direction of research and to engage in collecting data or interpreting results. That may be an indirect dimension of Merton's epistemic norm of universalism. Equal discoursive voice depends, in part, on equal opportunity to participate in the community or discourse. The question of theoretical and cultural diversity among the practitioners of science is addressed at the end of the chapter and again in Chapter 6.

Lapses in communication

Critical discourse is essential to rooting out error. It is predicated on a system of effective communication. Fellow experts need to be aware of new results if they are to scrutinize the observational methodologies, the data reduction, the statistical analysis, the background assumptions, the interpretations of the data, theoretical claims, and so on. The scientific community certainly seems to operate on the trust that publication is fair and open and that, once published, the information will reach all relevant colleagues. But there are exceptions.

For example, Archibold Garrod worked from 1902 to 1923 on the link between individual genes and individual enzymes, or proteins. George Beadle and Edward Tatum would later win a Nobel Prize in 1958 for essentially the same discovery, but were wholly unaware of his work initially. Garrod was a physician and biochemist interested in certain "inborn errors in metabolism": biochemical deficiencies with telltale physiological effects, such as black or red urine. Garrod had tracked the pedigrees and determined that the conditions exhibited Mendelian inheritance. He shared his findings with famous geneticist William Bateson, who shared the news of the first Mendelian trait found in humans. However, Bateson was not really interested in biochemistry, and Garrod was not terribly interested in genetics. The remarkable findings fell into shadow, and failed to inform a half-century of work on the nature of genes and their expression. Beadle eventually acknowledged Garrod's work in his Nobel speech: "In this long, roundabout way, first in *Drosophila* and then in *Neurospora*, we had rediscovered what Garrod had seen so clearly so many years before. By now we knew of his work and were aware that we had added little if anything new in principle." That marked a lapse in the communication system, at least.[18]

Similar slips occurred in the history of polywater in the 1960s. The error was discovered, but for several years, it remained unknown. By 1968, Vladimir Tal'roze, working in Russia, had spectroscopically identified significant organic impurities, possibly sweat as a contaminant. But his results appeared in an obscure Russian journal, *Khimiya I Zhizń*, which was not translated. News of his finding did not reach the West until late 1970. By then, hundreds of thousands of dollars had been invested in research, including by the U.S. Department of Defense, concerned about the possible use of polywater as a doomsday weapon. A few months later, Denis Rousseau (aware of Tal'roze's findings now) published a similar analysis in *Science* (but with no reference to the original study). Rousseau's prominently placed paper was more widely received and helped turn opinion against polywater. Rousseau thereafter adopted a posture as the conscientious debunker of polywater—work that had already been done by someone else two years earlier, save for translation and communication. Effective communication is as important in conveying errors as in announcing other discoveries.[19]

Even if published, not all scientific papers are read by those who may find them relevant. For example, the Nobel Prize–winning discovery of the antibacterial properties of penicillin was not completely unprecedented. In 1871, Joseph Lister (noted for introducing antiseptic practice into surgery) had found that a mold in a sample of urine seemed to inhibit bacterial growth. In 1875, John Tyndall reported to the Royal Society in London that a species of *Penicillium* had caused some of his bacteria to burst. In 1877, Louis Pasteur and Jules Joubert observed that airborne microorganisms could inhibit the growth of anthrax bacilli in urine that had been previously sterilized. Most dramatically, perhaps, Ernest Duchesne completed a doctoral dissertation in 1897 on the evolutionary competition among microorganisms, focusing on the interaction between *E. coli* and *Penicillium glaucum*. Duchesne reported how the mold had eliminated the bacteria in culture. He had also inoculated animals with both the mold and a lethal dose of typhoid bacilli, showing that the mold prevented the animals from contracting typhoid. He urged more research, but went into the army following his degree and died of tuberculosis before ever returning to research. When Fleming encountered his now famous discarded bacterial culture in 1928, he was not the first to have observed the antibiotic effect of *Penicillium*. Should we characterize the neglect of these discoveries before Fleming as errors in the communicative practices of science? At this point, perhaps all we can do is speculate on how history may have unfolded differently if the import of these findings had been pursued earlier. But note first that Fleming himself abandoned the therapeutic potential of penicillin. It was left to Ernst Chain to pick up the trail later by encountering Fleming's publication and viewing it in a different perspective.[20]

Lapses in the communication system are not errors as we normally conceive them—specific "mistakes" that can be traced to particular individuals, with a sense of accountability. With no one to "blame," it may seem as if there is no error. Again, we must separate the epistemic issue of errors from the moral or professional dimension of accountability. Not all sources of error are "human error." Some arise simply because of how the process of science unfolds. Still, lapses in communication are sources of error—at the social level. The process of mutual criticism breaks down and fails to be effective.

Peer-reviewed publication

When does a conjecture or claim become a well-established and secure "scientific" fact? Among nonscientists, the benchmark is typically publication. Specifically, publication in a *peer-reviewed scientific journal.* One could appeal to just the quality of the evidence. But publication is widely recognized as affording an additional dimension of legitimacy. Namely, in an idealized model of science, peer review functions to help filter out bias and unreliable research. Supporters of fringe beliefs or conspiracy theories, for example, typically appeal to this standard whenever they can, in order to project an image of respectability for their marginal or dissenting views. Prepublication peer review is important. But it is far from perfect. And errors do get published.

The flaws in the process become most visible when publication decisions are, in a sense, reversed. Not everything that is published nowadays remains "published." Some papers are retracted. Many of these are initiated by authors who have discovered lapses in their own procedure (recall the case of the lab that discovered that their instrument measured diamagnetism, not the sample, but in the sample *holder*). Some retractions, however, are initiated by the Editors themselves, who essentially acknowledge the ultimately unwarranted publication. For example, in 2022, the Editor of *Science* advised readers about the fate of a 2017 paper on (ready for it?) "chiral Majorana fermion modes in a quantum anomalous Hall insulator-semiconductor structure." Some readers had tried and failed to reproduce the findings. They had then "requested the raw data files from the authors, which they provided. Subsequently, the provenance of the raw data files and published data revealed serious irregularities and discrepancies." The published results were now highly problematic—and would surely not have been published originally if this information had been known at the time (for more on errors in trust, see discussion in the section on "Credibility bias, credibility gaps & fraud" to follow). "These issues have caused the editors of *Science* to lose all confidence in the conclusions of the paper, and we are therefore proceeding with an Editorial Retraction." That is, the paper was "unpublished"—an indirect admission that the original publication of the paper was (now) an error. Unfortunately, perhaps, that process took five years. Still, one can imagine how, historically, such problems could fester indefinitely before retractions became standard practice. Now, online repositories can withdraw a paper or post an accompanying advisory disclaimer. Here, the error was not just about the quality of the evidence or the argument. Retractions are reserved for more fundamental problems, usually about the professional responsibility in presenting competent and honest claims.[21]

In some cases, such as the one just described, editors and reviewers do not have access to all the relevant information to make a fully informed judgment. Indeed, this notably limits what the review process can do, even optimally. In other cases, however, "mispublication" decisions are more specious. For example, vaccine scientists were caught off guard in 2021 when the highly reputable journal *Vaccines* published a paper with the startling claim that for every three deaths prevented by COVID-19 vaccination, two deaths would result from the vaccination itself. That would profoundly affect public policy, if true. The claim quickly found its way into anti-vax activist networks and social media, including 14,000 tweets in the first few days. The paper was deeply flawed, however. For example, it used an open reporting surveillance database to try to measure (inappropriately) causal influence. It also made crude calculations based on inappropriate statistics from another national group entirely. These were errors no competent scientist would make. Indeed, the three authors of the paper had no experience in virology, epidemiology, or vaccine science. The three reviewers of the paper all failed to comment on the glaring errors. (The relevance of their expertise has now been called into question, too.) The publication triggered a storm of criticism. Six participating editors resigned over the sloppy peer-review process. The paper was eventually retracted—eight days after publication. Not for its controversial claims, but for its invalid methodology. Basically, for incompetence. Namely, it should never have passed an effective peer review. But by the time of the retraction, the paper had been accessed online over 380,000 times. For many observers, the pre-publication peer-review system failed on this occasion, allowing an egregious error to cascade into public discourse with the imprimatur of science. Such occurrences, though infrequent, do occur.[22]

Formal retractions are rare: perhaps less than 1 in every 5,000 papers. And, historically, they seem to be concentrated in a few bad actors: over ¼ of the documented retractions stem from less than 2% of the errant authors; and 40% of the retractions come from just one publisher. Still, the effect can be substantial. That includes misinforming other scientists, not just public consumers of science. According to RetractionWatch, which has been monitoring cases since 2011, the "top 10" culprits have been cited by other papers over 800 times each, and each many times *after* the original paper was officially retracted. The most highly cited retracted paper, a paper on the Mediterranean diet, has been cited over 2,855 times. The paper has been revised and republished. The new version now acknowledges the original problems with the randomization of study subjects; hence, it documents only a correlation and not more powerful evidence of direct causation between diet and health. Some studies seem to indicate that retractions are increasing. That is probably because of more aggressive oversight and policies in purging publication errors. At the same time, the overall *rate* of retractions seems to be waning. That seems to reflect that journals are adopting new review policies, tightening controls, and limiting ultimately faulty publication decisions. The social practices of science seem to be evolving rapidly to regulate this source of error more effectively.[23]

Errors in peer review and publication, one should note, can also go the other way. That is, while some major errors slip through to publication, in other instances deserving papers may *not* get published. Such cases are difficult to document historically, of course. However, one notable example is Lawrence Morley's work on ocean floor spreading. It is worth describing in some detail, to articulate the social process of peer review and how (or when) it may fail.

Morley was interpreting a magnetic "zebra pattern" in the ocean floor that had been published in 1961. His bold suggestion, threading together three disparate lines of research, was that the Earth's crust was breaking apart, and as it slowly separated from the seam, it left a magnetic record in the newly formed rock of reversals in Earth's magnetic field. For the non-geologist, that may seem a highly technical and speculative hypothesis worthy of little note. However, this idea eventually became a fulcrum in vindicating Alfred Wegener's concept of continental drift and in transforming it into the more sophisticated modern theory of plate tectonics. Historically, at least, an enormously important idea.

Morley submitted a brief letter to the journal *Nature* in February 1963. It was rejected, with a note that there was not room to print it. Morley then submitted a more complete paper to another journal. It, too, was rejected. The (anonymous) reviewer's comment was, "His idea is an interesting one—I suppose—but it seems more appropriate over martinis, say, than in the *Journal of Geophysical Research*." The irony, of course, is that in September

Nature published a different paper with the very same idea—a paper that is now regarded as a classic in the history of Earth Science. Based on recollections and circumstance, it is probable that the Editor's contrasting decisions reflected a highly informal review system that relied heavily on personal connections (and a prejudicial favor for England's two most prestigious universities—Morley was Canadian, and the accepted paper from Cambridge). It may also have been influenced by a word-of-mouth report that the second paper was forthcoming. If so, Merton's norm of universalism (equal opportunity) was surely not at work here. As early as 1974, one geologist noted that Morley's work was "probably the most significant paper in the earth science ever to be denied publication." One can debate endlessly whether the rejections of Morley's papers were justified or not at the time—a counterfactual historical question that is immensely problematic and largely unfruitful. However, geologists have come to recognize Morley's unpublished contribution as worthy, and now refer to the core idea as the Vine-Matthews-*Morley* hypothesis, honoring all three relevant authors. In retrospect, we may say not publishing it was an error at the social level of science.

In a poignant contrast to retractions, Morley's once-rejected letter to *Nature* was eventually published. It appeared in full in *Saturday Review* in 1967 as part of an article on Morley's underappreciated role. It was then *republished* in 1982 by historian William Glen in his authoritative history of the period. Apparently, it was important, after all. Setting aside the question of credit and priority, the case serves as another compelling example of how peer review and editorial decision making in the publication system are far from the imagined ideal as an accurate filter of error.[24]

The question of the effectiveness of peer review is significant in part because the process is enshrined in science folklore as a bulwark against error. It is commonly cited as integral to the "self-correcting" mechanism of science (see also Chapter 8). Yet the process seems limited in what it can do, versus what it is called upon to do in popular rhetoric.[25]

Publication bias

As noted above, critical discourse depends on research results being shared. But not all research is published. The public record may thus be incomplete and unrepresentative. Errors may emerge, just as they do when field or laboratory observations are based on small or biased samples (section on "Observational Errors" in Chapter 2. That is, the pattern of what is published and what is not may itself exhibit a bias that misleads scientists.

One pernicious problem is the tendency of journals to favor publishing research based on "positive" results. The journal wants to be known as a source of actionable findings, work that others can build on or that will change the course of ongoing research. The assumption is that few readers want to read about (or perhaps even need to know about) negative results, viewed as "failures." Such papers are often rejected for publication. Anticipating such rejection, some investigators do not even bother to submit some papers for consideration. Consequently, authors may discard their papers outright or stash the results unceremoniously in the back of some filing cabinet: the now notorious "file-drawer problem." As a result, the body of published results is skewed. One study estimated that as much as 90% of the negative results do not reach public awareness. Another analysis focused on a set of studies on the efficacy of antidepressants, all registered with the U.S. Food and Drug Administration (FDA) as part of the drug approval process. Only 69% were published publicly, almost all with positive results. According to the published literature, 94% indicated the drug was effective. According to those in the FDA record, however, only 51% were positive. In addition, the published studies indicated a 32% higher effect size. Namely, the published literature misconveyed and overstated the efficacy of the drug. Such discrepancies have been found in other studies comparing preregistered research with ultimate publications. Negative results frequently get lost.[26]

Another bias is toward publishing statistically significant findings. The threshold for deeming the probability of sampling error unacceptable is a convention (see, e.g., the case of the oops-Leon particle; "Small sample size" in Chapter 2). Still, results that can be shown to meet the arbitrary standard of $p \leq 0.05$ are generally three times more likely to be published than those that do not. Such biases can wreak havoc with meta-analyses—increasingly common—which aim to synthesize results from multiple studies. The missing evidence—although relevant—cannot be factored in. Namely, the published literature itself is a biased, unrepresentative sample. Hoping to be more thorough, reviewers must scramble to find the unpublished results, if at all available from other sources.[27]

In a similar way, perhaps, journals place a premium on novelty. Of course, it is vital for a community of researchers to learn about challenges to or changes in established knowledge. At the same time, it can be equally important to check earlier results, perhaps with larger samples, or in under-explored domains. The net effect of the novelty publication bias is to discourage just this type of basic, but essential work in science.

Science inevitably grows through trial and error. Memory of failed efforts can be important in shaping subsequent research. Namely, others should not need to repeat earlier missteps. Negative results, even null findings—where one fails to find evidence for a proposed (yet presumably plausible) hypothesis—can be important benchmarks at the collective level. Publication bias leaves such communal work underinformed. A focus on negative results may, however, remind researchers of the importance of their research design. Namely, appropriately planned research should yield results that are informative, whether they are positive or not. The fear should be, instead, that results may be ambiguous, unclear, or simply uninterpretable. For example, investigators need to be cognizant of measurement precision and the statistical power of their planned sampling method. That is, the sample needs to be large enough to discriminate a "signal" amid the statistical "noise," assuming a pattern or distinction is there to be found. Yes, even "negative" results can be significant, if they have been framed properly. All results should be meaningful, dissolving the role of much publication bias.

Credibility bias, credibility gaps, and fraud

Skepticism is often touted as a hallmark of science. Merton certainly gave it prominence in conceptualizing the norm of "organized skepticism." Skepticism is also found frequently in the rhetoric used by science advocates to denounce pseudoscience and extol the virtues of scientific rationality. Quite the opposite is true, however: science depends on *trust.* If science is to make progress, scientists cannot forever be questioning each other's discoveries, reproducing each other's experiments, or doubting their methods. But what warrants trust? What guarantees against the possibility that someone has not succumbed to one of the many errors catalogued in the last two chapters? That is a philosophical question—and a practical challenge.

Trust was a major concern for the Royal Society in the emergence of modern science in the late 17th century. Members aimed to share observations and experimental findings and thereby to profit from each other's work. As superbly described by Steve Shapin, trust was essential. But given the unique circumstances, trust could also be expected. Members were drawn from the social elite. They were all *gentlemen*. The cultural norms and accountability with their social network meant that they were bound to exhibit the utmost integrity.[28]

Nowadays, things are not so simple. Judgments of trust are typically based on a researcher's track record. That is, knowing past performance, one expects a scientist to continue to exhibit the same level of competence. It is a simple form of inductive reasoning, applied to a researcher's demonstrated acumen and abilities. Accordingly, each researcher develops an impressionistic ranking from their peers: their *credibility*. Other elements may factor indirectly into the informal judgment: the scientist's educational background, mentorship, affiliations with prestigious institutions, collaborations with other credible scientists, awards, prizes, leadership positions, academic honors, or personal conversations. All these features function as indicators for whether to trust the reliability of any newly published work—based on authorship.

In day-to-day practice, the credibility of the *researcher* is often a proxy for the credibility of their *claim*, its argument, and the evidence. With the overwhelming amount of research being published, who has time to read all those papers in detail or check the results in one's own lab? It's less work just to accept the claims based on the author's credibility. Credibility thus functions as a heuristic, paralleling the use of surrogate variables in observing natural phenomena.[29] The efficiencies of the credibility shortcut can matter. In the 1960s, two labs were racing to discover the structure of a hormone, thyrotropin releasing factor (or TRF). One was led by Roger Guillemin and the other by Andrew Schally. Schally decided to trust Guillemin's results. Guillemin, on the other hand, insisted on repeating and confirming every result announced by Schally's lab. Ultimately, Schally's trust allowed him to make the discovery first. In practice, prudent use of credibility judgments can matter as much as the careful and detailed assessments of arguments.[30]

Credibility judgments can change. For example, in 2020, a team of researchers at the University of Rochester published a remarkable claim about synthesizing a hydride substance apparently capable of superconductivity at a "high" temperature of −23°C and at high pressures. However, after failed efforts by many other competent labs to attain similar results—and with the original lab declining to share materials and detailed procedures—the journal *Nature* retracted the paper in late 2022, in a sense disqualifying its claims (as noted above in the discussion of retractions). In 2023, the same lab published a new claim—even more remarkable—about another substance that could superconduct at room temperature and at modest atmospheric pressures. Such a discovery would have enormous implications for energy technology—if true. However, the credibility of the group was now tarnished. The paper went through an unusual five rounds of peer review before publication. Even so, many potential readers were skeptical. One colleague noted, "I think they will have to do some real work and be really open for people to believe it." But eyeing the potential commercial value, the lab was reluctant to share details. Namely, the lab did not accord with Merton's norm of communalism. Another critic was blunt. "I doubt it, because I don't trust these authors." Evaluation of credibility can sometimes trump evaluation of the evidence.[31]

Like all heuristics, however, credibility judgments can, on occasions, fail. The constellation of experiences that allow a scientist great insight on one occasion may not be equally effective in deciphering different phenomena. The investigator may branch into another unfamiliar field and not have sufficient background. Or it may be a matter of luck.

In a sense, heuristic reliance on credibility is akin to confirmation bias. Here, however, the persistent precedent is not a concept or theory, but the imagined scope of someone's scientific capabilities. Our guide to the sociology of science, Robert Merton, was fascinated with peer assessments and observed that those who did well tended to earn favor: receiving more grants and so on. He called it the Matthew effect, named after the Biblical passage: "For to every one who has will more be given." This effect has now been documented in a large study on peer review. Hundreds of researchers were asked to evaluate the same test manuscript. In some cases, the sole author was obscure. In other cases, it was a Nobel laureate. The former scored a 10% "acceptance" rate for publication. The latter, 59%—a sixfold difference. This is *credibility bias*, a social-level source of error.[32]

Credibility bias can be found throughout history. Legendary scientists, with well-deserved reputations, sometimes made claims that ultimately proved wrong. Their legendary status, however, made it more difficult to challenge and correct those errors. For example, Linus Pauling won a Nobel Prize in Chemistry for deciphering the alpha helix structure of proteins. Hence, many unhesitatingly believed his claims for vitamin C as a preventative cure to the common cold. Here, however, his research was meager and he was famously wrong. Yet his credibility lingers even now and the effect of his advocacy can still be observed, despite the updated science and strong consensus against Pauling's claims. Credible in many instances, Pauling was not credible in all.[33] Credibility bias was also integral to the debate over the age of the Earth in the late 1800s. Physicists and geologists vied for authority, using their alternative assumptions. But it was primarily the enormous prestige of William Thomson (Lord Kelvin) that allowed him to denounce anyone who did not adhere to the "exact" science of thermodynamics. For decades, his fame shadowed over the geologists' claims, even though his assumptions were wrong and his estimate of the age far too low.

Accounts of "unsung heroes"—scientists neglected or maligned by their historic peers, but who were vindicated in the long run—are stock anecdotal material in professional folklore. They often provide further examples of credibility errors. I refer the interested reader to those stories, shared widely elsewhere: the rejection of Alfred Wegener's continental drift (in the United States, at least); Ignaz Semmelweis' advocacy for handwashing by doctors; Gregor Mendel, his pea plants, and the basic laws of inheritance; Barbara McClintock's early insights on "jumping genes" in maize; Oswald Avery's discovery of DNA as the hereditary molecule in bacteria; Svante Arrhenius' prescient work in 1896 on the greenhouse effect and global warming; Marie Tharp's map of the mid-Atlantic rift and its meaning for continental drift; Jean-Baptiste Lamarck's ideas on adaptive structures and evolutionary transformation; and more.[34]

Judgments of credibility may also be influenced by non-epistemic factors. Psychologically, our minds seem susceptible to congenial personalities, forceful dispositions, demeanor, appearance and style, and even unrelated contextual emotional factors. These other factors, in practice, also contribute to appraisals of trustworthiness, although not justifiably so. Accordingly, researchers can sometimes develop reputations that outstrip their actual credibility. The credibility is not really "earned." The difference between "reputation" and "credentials" may be called a *credibility gap*, another social-level source of error. For example, biochemist Albert Szent-Gyorgyi was described by one colleague as "the most charming scientist in the world." He was an eloquent speaker and an inspirational presenter. Along with a Nobel Prize from early in his career, that personal charm helped earn him many grants. But his research consistently failed to measure up to his rhetoric and one funder after another abandoned their support for him. The exercise of trust based on impressions of credibility was, ultimately, ill-founded.[35]

Credibility gaps provide opportunities for fraud to flourish. People (in general) may lie. They may cheat. Scientists are no different from other humans. So, one should not be terribly surprised to encounter fraud in science as well. Commentators generally portray fraud as a travesty in science: an affront to the quest for genuine knowledge. Whole books parade a panoply of egregious cases from history and bemoan the lack of integrity. Much verbiage is devoted to moral judgment and to castigating the perpetrators as villains (see "Error naturalized" in Chapter 1). However, fraud succeeds in part because *other* scientists *trusted the credibility* of the author. At least, initially. Of course, as noted above, communicative trust is integral to how science functions. So it seems hard to avoid fraud if science is to progress: a puzzling dilemma. Here, then, I discuss fraud in the context of the conventional social practices of science. The social system is intended (in Merton's ideal) to bring errors to light. But when errors are deliberately hidden, it requires extra work to find them and to isolate the source of error in fabricated data or deceptive reporting.[36]

Jan Hendrik Schön, Wunderkind?

As a more extended example, consider the case of once acclaimed physicist Jan Hendrik Schön. The scale of the case may be measured by its profound outcome: eight papers were retracted from *Science*, seven from *Nature*, 13 from other journals. Three awards were rescinded and the prize money returned. And a doctoral degree was revoked. One may say that the social system of science ultimately worked, exposing the fraud. On the other hand, credibility judgments eclipsed deception on numerous occasions—for years—prior to that remedy. In essence, the system of trust failed to discriminate between those worthy of trust and this one person who was not.

Schön burst on the scene in late 1999 as an emerging talent at the prestigious Bell Laboratories. He had been recruited with the hope that he would help sustain the lab's famed history in innovative research on transistors. Schön would investigate the potential of organic crystals to replace silicon in electronics, work relevant to his PhD research in Germany. He did not disappoint his managers. At the end of a year, he had published positive results—amazing results. The field was abuzz with the potential for developing cheap semiconductors and, simultaneously, for pushing the boundaries of theoretical physics. The discoveries continued (apparently) over the next two and a half years. Each seemed to conquer an ongoing experimental challenge and to vindicate various theoretical predictions. Superconductivity under practical conditions, field effects, organic lasers, narrow electron channels, ambipolarity, single electron gating, self-assembling monolayers. Schön seemed to exhibit extraordinary skills managing materials in the lab under very challenging technical demands. On top of that, he was amiable, polite, self-effacing, and yet self-assured: overall, an exceptionally agreeable personality. Schön's credibility skyrocketed. News articles hailed him as a wunderkind, a scientific prodigy.

But it was all well-crafted lies. Schön had fabricated his results, working backwards from targeted results to appropriate measurement figures and adapting and mislabeling real data out of context. Surely his acts were professionally reprehensible. But his errors fooled colleagues and slipped through to publication in high-prestige journals. Why did the social system of science not catch the errors? Here, credibility judgments eclipsed the presumed safeguards.

Schön developed masterful techniques for weaving through the publication process. First, he learned the rhetoric of papers that would impress editors and reviewers, but not attract too much attention to the absence of relevant details. For example, present compelling data. Use figures to tell a story. Resonate with and confirm earlier theoretical speculations. Be incomplete, yet humbly acknowledge one's limits—while enthusiastically inviting further research. Namely, tell the readers just what they are eager to hear.

Even so, Schön's papers did not all sail through peer review without question. Reviewers frequently cited unacceptable lapses of technical information, even if they endorsed the context of the paper as a whole. In retrospect, some regret not having expressed those reservations strongly enough. Here, too, Schön was adept. Rather than bluster defensively, he openly admitted to the cited problems. That is, his response suggested that he had no self-serving motive, nor anything to hide: a posture that (ironically) tended to enhance his credibility. Sometimes he added the requested information, but inevitably he artfully dodged providing enough details that would betray his wholesale fabrications. Editors seem to have relied partly on the credibility of Bell Labs, assuming that papers had earlier passed rigorous internal reviews. They had not. Because the missing details had possible implications for securing patents and guiding further research, certain omissions were tolerated: a promissory note, to be fulfilled in due time. The journals were often more interested in communicating the groundbreaking theoretical developments, and assumed (based on Schön's position at Bell

Labs) the evidence embodied basic competence. Later, they relied in part on his publication track record and awards as an indicator of his credibility, unaware that it was all built on sand.

Efforts at replicating Schön's supposed achievements—the first step in trying to build on them—also encountered problems. Schön's publications supposedly reported how to produce materials with exceptional performance. However, no one was ever able to produce those materials or the effects for themselves. The "recipes" were incomplete. For example, it seemed impossible to produce his uniform layer of aluminum oxide on a crystal, even though that's what the data seemed to indicate. Or to lay down such a thin gold layer on the electrode, as Schön claimed. Or how could one get such a symmetrical distribution of conductance values, statistically improbable? In another case, based on the designated channel width, the possibility that measurements reflected an artifact seemed all too high. Or the measured currents seemed to belie the basic thermodynamics of electron behavior. Or the graphs indicated a form of superconductivity inconsistent with the oxygenated structure of the material as reported. Or the copper coating consistently oxidized, preventing any gating effect, as supposedly observed. And so on. All seemingly "minor," yet inevitably critical variables. Many colleagues, both within Bell Labs and elsewhere, consulted Schön himself, trying to resolve the technical ambiguities. Once again, Schön was modest, typically admitting his shortcomings and promising to address them. In retrospect, we can see that Schön primarily sought social acceptance. He was not motivated by ambition, fame, or money. His desire for peer approval apparently led, in many instances, to yet new fabrications, explicitly tailored to please the would-be critic. Alas, these only resolved the problem on paper, not the realities in the lab. With time, others in the field began to doubt Schön's claims as unfounded. But with Schön's widespread celebrity status, no one was in a position to challenge him. It was just easier to disregard his work. So the disparate negative appraisals never coalesced into something substantive. By the time the nature of the fraud was ultimately exposed, much of Schön's work had already become marginalized.[37]

One younger colleague, whose tenure at Bell Labs overlapped with Schön's, later reflected on the power of Schön's credibility. Like others, he had tried and failed to reproduce some of Schön's results. In sharing his doubts about Schön, he met with a typical response: "But he's a genius." Namely, reputation trumped reality. Schön's inflated credibility stymied the very process that might have led more promptly to exposing his fraud as a source of error here. The credibility was mismeasured, and the familiar heuristic of trust failed.[38]

•

Many people who comment on the scientific method regard the source of error in fraud as lying. They portray it as a threat to the (moral) integrity of science. As illustrated here, I view the problem quite differently. Fraud is a lapse in the system of trust and communication in science. It is a social-level error, fundamentally similar to other misjudgments of credibility—and corrected accordingly. We will encounter cases of fraud again when we examine how errors are discovered and addressed (Chapters 5 and 6), where we can consider further how the epistemic versus moral dimensions interact in science.

Conflict of interest

Recall, finally, Merton's norm of disinterestedness. "Interests"—in the form of profit, power, or privilege—can subtly bias individual thinking at the conceptual level. At the social level, they can also signal misleading arguments or persuasive tactics. Corporate and ideological interests are well aware of the rhetorical and political authority of science. They repeatedly endeavor to "bend" science toward influencing policy, avoiding regulatory actions, or advancing their aims in other ways. In recent years, fraud by rogue individuals has raised substantial concern among scientists. In my view, a far greater danger is posed by the proliferation of disinformation: misrepresentations and other deliberately misleading claims in the scientific literature aimed at advancing certain interests, sometimes quite surreptitiously. The methods vary. Watchdogs have now documented efforts by industry-sponsored researchers to: (1) bury negative evidence—a form of publication bias; (2) cut off a study early to capitalize on favorable but incomplete results—statistical bias; (3) submit ghostwritten papers for publication, using an academic name or institution to disguise the source of the study—credibility fraud; or (4) create whole journals with lax publication standards to bypass peer review—skirting the gatekeeping function of publication. All are variants of social-level errors discussed above. They all pollute the scientific literature. Our aim should be to develop sanctions to discourage such practices and tools to limit their impact.[39]

For example, in one of the more disturbing cases, from 2003 to 2009 the Editor of the journal *Neurosurgery* approved the publication of 16 papers on brain concussion injuries caused by playing professional football, all based on research sponsored by the National Football League. Each paper denied or trivialized the problem of chronic traumatic encephalopathy (or CTE). The Editor's decisions to publish them were sometimes in direct contrast to the consensus of the reviewers' reports and the sports section-editor. Who was the Editor? A consultant to a professional football team and, as described by a colleague, a "sports guy wannabe." (He was allowed to stand with the team on the sidelines during the games.) At least the duplicity became apparent to many concussion researchers, who dubbed *Neurosurgery* "the Official Medical Journal of the National Football League." All the work has since been discredited, along with its cherry-picking of data and faux-testing of football helmets. But the imprimatur of scientific publication was leveraged to help delay any substantive action on the health risks, which became clearer as time went on. Conflict of interest had thoroughly corrupted the very system of peer review, which normally aims to help deny bad science the privilege of publication.[40]

Conflicts of interest arise when private interests rise to eclipse the public interest in reliable knowledge. Disinformation emerges in many ways. Investigators may frame research questions to exclude relevant (but damning) variables. They may use small samples and repeat experiments hoping to capture statistical exceptions in unrepresentative (but "favorable") samples. They may gerrymander statistics or avoid addressing alternative explanations. All yield misleading conclusions. Yet they can nevertheless be presented as consistent with "the evidence." Disinformants seem to subscribe to a view that if you have not falsified or fabricated data outright, it cannot be deemed misconduct or malfeasance. The claims certainly appear legitimate—on the surface. The epistemic sleight of hand can easily escape detection. Yet the errors continue to circulate, to benefit some at a cost to many. Scientific readers who are at least informed of conflicts of interest (who sponsored or designed a particular study) are in a position to scrutinize the arguments and data more closely. They are alerted to recalibrate their level of trust and to probe the questionable credibility. Accordingly, many journals have instituted policies that require authors to disclose any potential conflict of interest. However, the record of heeding those policies appears mixed. For others, unraveling any deception involves unfamiliar and onerous terrain. Managing conflicts of interest and their disclosures may be the next great challenge in regulating error in science at the social level.

Conflict of interest plagues, in particular, research on diet, often funded by food, beverage, and agricultural industries. For example, a 2015 paper on added sugars in the diet published in *Annals of Internal Medicine* was widely criticized for its shoddy work (including by an editorial in the very same issue). Even so, the status of the paper—published in a respected peer-reviewed journal—was quickly used in media efforts to discredit dietary guidelines by the World Health Organization and U.S. and British health agencies. Conflicts of interest in the funding and authors had been declared—although it is not clear that the editors heeded them, or that the peer reviewers were alerted to them (in the final publication, they appeared in an obscure online appendix). The publication eclipsed the timely effect of any subsequent peer criticism.[41]

In 2019, the very same journal published another controversial paper reviewing the healthiness of eating red meat, concluding that "adults continue current unprocessed red meat consumption." The lead author of the controversial review on meat, Bradley C. Johnston, was, tellingly, also a co-author on the 2015 article about added sugar. Another co-author was Dean at a university in Texas (where Johnston would soon join the faculty), which received substantial funding from the beef industry and hosted a newly endowed "Beef Cattle Academy." Although the diet recommendations were labeled "weak" and based on "low-certainty evidence," the journal prepared a press release dramatically titled: "New guidelines: No need to reduce red or processed meat for good health." Such claims were echoed in news headlines in the *New York Times* and elsewhere: "Eat less red meat, scientists said. Now some believe that was bad advice." Of course, ties to industry do not necessarily negate a study's findings. But they should set off alarms for publishers to engage in more rigorous review of the methods and assumptions. That was not done here. Not surprisingly, perhaps, the results were soon debunked by a subsequent study. But not before all the misinformed publicity.[42]

The role of interests in scientific research, along with other forms of bias, seems inevitable. Inescapable, perhaps. Accordingly, science relies (as Merton noted) on its social dimension—its implicit system of mutual critique—to keep any resultant errors in check. That becomes difficult if interests are not disclosed or, worse, are mindfully hidden. That applies to political and ideological interests as well as to financial or commercial ones. Detecting and remedying error becomes even more difficult if authors engage in subterfuge and craftily disguise their interests. Errors will also fester and spread beyond the boundaries of the scientific community if editors (and/or publishers) do not apply appropriate levels of critical review. The stronger the interests (or the higher the political stakes of the claims), the deeper and the more probing the needed review. The first step, however, may be in recognizing that failure to manage interests effectively constitutes a social-level error type. And one of growing importance.

Critical consensus

Science is about "the evidence," surely. And sound reasoning. But, as cases in this and the previous two chapters illustrate, it is also about *interpreting* the evidence. And it is about having sufficiently *complete* evidence. A single scientific experimental finding is not enough. A single line of reasoning rarely offers enough assurance. Results may be discordant. Scientists may disagree. (Sometimes vehemently.) An integral part of science, then, is its social practices. Differences in interpretation must be resolved. Disparate bits of evidence must be reconciled and pieced together into a coherent whole. Biases and errors appear, yes, but through ongoing work and active mutual critique, they are identified and filtered out—at least when the system of discourse functions effectively. This chapter has surveyed the range of those discoursive practices, which embody Merton's vision of socially organized skepticism: the "scientific method" *at the social level.* Where might lapses at this level allow errors or biases to persist? An analysis of the spectrum of social-level error types fosters a fuller and clearer converse understanding of how science works—here, at the social (or discoursive) level.

Such considerations underscore the importance of a critical consensus. Reliable scientific knowledge is, indeed, more than just a simple appeal to the evidence. One needs robustness. One needs concurrence of views. Reaching consensus is not always easy. Debates may linger for years—or decades—as scientists pursue the critical evidence that will help persuade colleagues and resolve disputes. Science takes time. The cost of reliability may be patience and persistence? So: consensus may be the ultimate hallmark of reliable knowledge in science, but in many cases, such consensus is hard won. And it is rarely immediate.[43]

Consensus is also generally hard to document. There are no formal polls that monitor scientists' acceptance of particular claims. There are no opinion surveys. It is mostly a diffuse sense of agreement. It is often gauged by observing the concurrence of just the leading experts in a particular field. Or it may be a recognition that dissent, once prevalent, has waned. Publication based on certain ideas simply ceases, imperceptibly.[44]

Any effort to ascertain consensus would be problematic from the start. Scientific theories are complex, and it is often difficult enough just to clarify what a particular theory states or entails. Degrees of endorsement may vary. For some, support may be contingent on various qualifications. Also, whose views matter? Not everyone is a relevant expert. Who is in the field, and who is an outsider? Also, what counts as consensus? 90% agreement? 99%? A simple majority? An explicitly negotiated statement? Consensus is, unfortunately perhaps, not easily defined.[45]

On special occasions, however, scientists may be assembled to weigh in on scientific claims. The National Academies of Science in the U.S. is regularly commissioned to assemble such advisory committees on topics directly relevant to legislation or policy, such as a panel to review the evidence on EMF cancer risks. The National Institutes of Health also convene "consensus panels" to clarify the expert perspective on important medical issues (such as the bacterial cause of ulcers, or acupuncture). When controversies about the status of Pluto raised public concern, in 2006, the International Astronomical Union took the highly unusual step of asking its members to vote on what would be accepted as the definition of a planet ("What counts as error" in Chapter 1). In times of public controversy, professional institutions may issue position statements. But these cases are notable because they are exceptions, not the rule. Consensus is rarely codified. One can only say that by the time something appears in school textbooks, it is generally so unproblematic and benign as to represent consensus. That, in a sense, is an indicator that the discoursive practices of science have reached a resolution and closure.[46]

One may say, however, that to be stable, a consensus must be "deep" and probative, and reflect critical discourse among a diverse set of relevant experts. That is, a consensus based on minimal research is surely vulnerable to the later discovery of additional evidence, with its potential to shift interpretations significantly (see "Unaddressed alternatives" in Chapter 3). Also, has the relevant topic been sufficiently probed? What is known ultimately reflects what questions have been asked, and what investigations have been funded or provided with appropriate resources. So, the nature of who originally set the research agenda may, paradoxically perhaps, matter to consensus (as noted in the section "Communal cultural bias" above). In addition, organized skepticism only functions effectively when there is a sufficient sampling of relevant interpretive perspectives (free from communal confirmation bias). Here, too, the composition of the scientific community may matter. Is it appropriately diverse (especially with respect to topics bearing on politics)? The scientific norm of universalism (equal voice) is not based on ethics: that science should be equitable and *morally fair* to all genders, races, cultures, and so on (that compelling question is addressed by wholly different arguments). Rather, it is an *epistemic* norm. Namely, reliable knowledge cannot rest on the interpretations from a narrow set of perspectives or special interests. Mere agreement, even universal accord, does not suffice. Epistemically effective critical discourse involves input from a wide spectrum of cognitive standpoints. That is what constitutes a *critical consensus.*[47]

A focus on consensus is a modest reminder that not all errors in science are attributable to individuals or to their supposed blunders, delusions, or missteps. Some errors arise from institutional structural problems, incentive systems, the homogeneity of the scientific community, political relationships, publication policies, and so on: the context in which scientific discourse is conducted. Accordingly, one may well afford the social practices of science more epistemic significance. Perhaps one may see them as a form of distillation, separating enduring, reliable knowledge from the epistemic noise of error and bias. As expressed by Danish mathematician and poet Piet Hien: "*The road to wisdom?—Well, it's plain / and simple to express: / Err / and err / and err again / but less / and less / and less.*"[48]

5
From Incongruence to Error

Isolating error as epistemic work • incongruences • isolating errors • testing errors • differentiating domains • the puzzle of unlearning

It is not enough to say that scientists err or that science is "tentative" (Chapter 1). Nor is it sufficient to classify various types of errors (Chapters 2–4). We need to know how scientists find errors and fix them. That is especially true if the processes of pre-publication peer review and replication, standardly invoked as stopgaps, are inadequate (see Chapters 4 and 6). Errors, alas, do not announce themselves. That is, in essence, *why* they are errors (Chapter 1). What appears to be one thing turns out, upon further inspection, to be something else entirely. Yet errors often are found and remedied. How, then, does one become aware of errors, or even of the hint of a possible error that calls for further investigation? In this chapter, I explore how.[1]

Nor do errors magically correct themselves (Chapter 4; see also section on "Isn't science 'self-correcting'?" in Chapter 8).[2] As orientation, then, I begin by reviewing the historical details of a handful of cases where error *was* corrected. This highlights the *epistemic work* involved. Namely, error correction involves collecting more evidence—*deeper* evidence—or reconfiguring concepts into new patterns. Articulating an error is ultimately a discovery, of sorts, not the meaningless residue of some process that gradually distills the truth.

Next, I consider the general "natural history" of an error. It is rarely a case of reviewing individual claims, or brute checking for methodological lapses (Chapter 1). Rather, the process seems to begin typically with an unexpected *incongruence*: where *separate* lines of research converge and reveal a clash of justifications. What once was resolved seems now uncertain. Such discrepancies may occasion further research, where the mismatch itself now becomes the focus of inquiry. The aim is to ferret out the misstep(s) in one (or both) of the presumed justifications: *isolating the error*. That may include "testing" the proposed error and confirming the remedy or solution

Toward a Philosophy of Error in Science. Douglas Allchin, Oxford University Press. © Douglas Allchin (2026).
DOI: 10.1093/9780197827703.003.0005

with more confidence. Finally, I comment on the puzzle of "unlearning"—how evidence may be reconfigured in a way that not only leads to new conclusions but also requires us to jettison earlier ones.

Isolating error as epistemic work

First, let us address the unstated impression that error is what is discarded after we have found the truth. It is not typically regarded as "knowledge." Rather, we tend to dismiss error as the *negative residue* of producing *positive* knowledge. For example, John Losee titled his book on falsification, "theories on the scrap heap." John Grant titled his history of discredited ideas as "discarded science."[3]

A concerted study of scientists addressing error will show that finding and remedying error requires *work*. *Epistemic* work. Additional evidence is needed to ascertain an error as an error. To discover an error is thus a form of knowledge, even if one might construe it as "negative knowledge" (more on that in Chapter 8). In this section, I present three examples as a preview to characterizing the process more generally in the sections to follow.

The phantom of the OPERA

When experimental apparatus does not function properly, it may produce phantom results. One may imagine that with a cursory check, any source of error in the instrument will rise to the surface and announce itself. Not so. Here is one such case.

In late 2011, international news headlines announced that scientists had discovered neutrinos traveling faster than the speed of light, upsetting one of the basic assumptions of modern physics. One team of researchers was producing neutrinos at a particle accelerator in Switzerland while another observed their appearance at a lab in Italy over 700 km away, using an instrument called the Oscillation Project with Emulsion-tRacking Apparatus, or OPERA. When the time was taken into account, the neutrinos seemed to travel faster than was thought possible: faster than light, an anomaly for the theory of relativity. The claim was presented with some caution, of course. But the investigative collaboration nonetheless published the results, and openly invited others to find any errors.[4]

Many factors would be checked over the next several months, each reflecting a different error type. First, how were the neutrinos generated? Was their energy level responsible, perhaps? The team had already encountered that variable months earlier, and run the experiment again with a different energy level, with the same result.

Was the particle beam properly aligned? Rechecking confirmed: yes, it was.

Also, the neutrinos were produced in large bursts. Perhaps the detection of a neutrino's arrival in Italy was being confused with the release of a different neutrino far away in Switzerland? In the following month, the team refined the procedure, cleverly removing that possibility, again with similar results.

Meanwhile, other workers were checking to see if the Earth's rotation could have subtly, but significantly, influenced the effective distance that the neutrino traveled.

Yet others were redoing the mathematical analyses with blind coding techniques, so that there could be no subconscious influence on the calculations.

Velocity is determined by time and distance, so the distance measurements were rechecked. They were confident that the uncertainty (imprecision) was no more than 2 cm in the total 730 km.

That left time, a major concern from the outset. Accurate measurements required synchronization between the two locales. That was achieved in part through GPS satellites. Were they functioning properly? There were atomic clocks at each location used to help correct local drift from the signals. These had to be checked.

The system had other complications. Time signals also had to be conveyed along fiber-optic cables. How long was each cable, adding how much time? There were also cable junctions. Each connection introduced a short latency as the signal was propagated, and even these short delays needed to be measured and taken into account in the final value. This had all been done five years earlier when the station was set up. But now every element in the time signal pathway needed to be rechecked.

About 10 weeks after the initial announcement, a cable connector was found that was not fully screwed on. It was tightened, of course. But was it responsible? The experiment needed to be redone—a replication, of sorts, but now with an apparent tweak.

All this while, physicists were speculating on the theoretical possibilities, including revisions that might accommodate the findings, if true. arXiv, the online depository for manuscripts-in-progress, received 197 postings.

Six months after the original announcement, the experiment was redone, now with the cable tightened. This time, three additional research stations were simultaneously involved in independent detections. None found faster-than-light results. The error seemed localized: a misaligned cable. The extra time lag introduced by the loose connection seemed to match the original error.

Meanwhile, a nearby lab had (fortuitously) been able to compare their historical time measurements with those of the original lab. The problematic time lag seemed to have originated in 2008 and continued until the loose cable was tightened in late 2011. That was reassuring supplemental evidence. To be sure, analysis of a (now deliberately) loose cable was done, to confirm that the magnitude of the time delay indeed matched the change in the original anomalous result.

Labs in Japan and the United States were installing more precise timing equipment themselves, and a few months later, they reported that their efforts at replication had failed, too. But by then, no one was really surprised. At least three subsequent research projects were conducted over the next two years to lay the spurious result to rest.

Isolating the error had involved casting a wide net and testing multiple elements—material, experimental, and conceptual. After six months of additional scientific work, the anomaly had become a clear and confirmed *error*. What may (should) fascinate philosophers of science here is that all these checks addressed individual elements in an enormously diffuse and multi-layered justification (see Figure 2.1). That's how complex scientific justification is.

All that work—epistemic work—was required to justify the conclusion that a loose cable was the source of error. The spurious result could now confidently be put to rest. No radical upheaval of physics, after all. This time.[5]

Discordant atomic weights for nitrogen

The next case begins with an apparent contaminant, but ends unexpectedly with the discovery of a new chemical element and an overhaul of the periodic table. So (in contrast to the last case), "minor" lab errors can unfold into major theoretical revisions. But, again, it requires substantial work to unravel such subtle signals.

The story began modestly enough, with some rather mundane, even tedious work that typifies much of what Thomas Kuhn called "normal science." Physicist John William Strutt, the third Baron of Rayleigh (typically known as Lord Rayleigh), was interested in clarifying the atomic weight of gaseous elements. He was seeking sufficient precision to assess the hypothesis, advanced decades earlier by William Prout, that all elements are basically multiples of hydrogen units. Rayleigh began in 1882 with hydrogen and oxygen. Then he moved on to nitrogen. "Years devoted to weary weighings," *Nature* magazine reported later. But this time, there was a difference. Literally, a difference. In the interest of utmost reliability, Rayleigh had prepared his nitrogen samples by two different methods (one by releasing nitrogen through a chemical reaction, the other by isolating it from the air). Their weight measurements did not match. By one-tenth of 1%. Hardly enough to disturb most investigators. But Rayleigh's chief aim had been precision, and this difference was large enough to matter. Something was wrong. But what was it? Was it sloppy procedure? Rayleigh redid everything. Same result. He was stumped—out of immediate alternatives. Puzzled by the discrepancy, which he now "regarded only with disgust and impatience," he "published a letter in *Nature* inviting criticisms from chemists who might be interested in such questions." No major insights were forthcoming. Again, a problem with no conceivable answer.

Rayleigh then began to explore the available alternatives more thoroughly. He began with perhaps the easiest to check (and perhaps also the most innocuous error). Was it a basic material error based on a contaminant ("Material errors" in Chapter 2)? Rayleigh took extra care against impurities, especially from hydrogen, and was persuaded there were none. So, was it an artifact of the preparation methods ("Instrumental errors" in Chapter 2)? By repeating his methods with slight variants, Rayleigh confirmed and sharpened the discrepancy between atmospheric and chemical approaches to producing his sample of nitrogen. The problem now escalated to more substantive alternatives. There was an unknown gas lurking somewhere. And it seemed mysteriously unreactive.

A colleague kindly referred Rayleigh to similar work by Henry Cavendish, over a century earlier. That opened a new method for Rayleigh: running sparks through the gas. This gradually converted and removed the nitrogen. Rayleigh was left with a mostly pure sample of his troublesome gas, composing less than 1% of atmospheric air. Preliminary analysis showed distinctive spectral lines and an atomic weight of about 40. It was definitely not nitrogen.

At this point, Rayleigh heard from chemist William Ramsay, who had taken a personal interest in his published query. Earlier, Ramsay had worked with nitrogen oxides and also on atomic weights, so he was well positioned to explore Rayleigh's problematic results. Ramsay reported that he had now isolated the rogue gas himself by another means and he confirmed independently the distinctive properties Rayleigh had found. The error in Rayleigh's discordant nitrogen data was now resolved by a previously cryptic alternative: a new component of air, which they called argon. The couple announced their exciting discovery briefly in August 1894.

But what exactly was argon, chemically? It seemed to be a new element. Of course, in the last decades of the 19th century, such a discovery was not that unusual. But this prospective element did not fit in the periodic table, as then understood. It was unreactive, and could not be grouped with the chemical properties of any other group of elements. The atomic weight placed it uncomfortably between two highly reactive elements, potassium and calcium, whose properties each already fit comfortably with other elements in their groups, leaving no empty slot between them. If argon was an element, something about the periodic table would have to "give." An anomaly. An effective alternative was now even more cryptic.

Rayleigh and Ramsay redoubled their efforts in determining the properties of argon, hoping to reveal more clues. They engaged William Crookes, who applied his respected expertise on spectroscopy and new elements. Argon was distinctive and unlike any known element. They enlisted Karol Olszewski to determine melting, freezing, and critical temperatures. Most importantly, the phase transitions were sharp, indicating only a single kind of atom—namely, an element. Finally, Ramsay measured the ratio of specific heats. Based on the kinetic theory of gases, the value indicated, again, an uncompounded atom—again, indicative of an element. Evidence favored argon's elemental status.

Rayleigh and Ramsay had waited cautiously until the additional details were determined before making any more public claims. Now, five months later, they presented their findings publicly before the Royal Society, to an extraordinary and eager crowd of over 800 people. The methods for isolating argon and validating it as a distinct atmospheric gas hardly mattered now. The presentation focused mostly on addressing in turn all the alternatives to its elemental status (much in the way that Suppe[6] has described the argumentative structure of the modern scientific paper). The arguments were certainly not lost on the person who reported the occasion in *Nature* magazine. However, the scientific reception was mixed. As Rayleigh noted later, "The result is, no doubt, very awkward. Indeed I have seen some indications that the anomalous properties of argon are brought as a kind of accusation against us." Their findings had only amplified the conceptual tensions.

In times of scientific crisis, Kuhn posited, alternative theories proliferate. And so they did here. Troubled by the specific heat claims, some chemists declared that the underlying kinetic theory was mistaken. That is, one could not conclude that argon was a simple (monoatomic) element. Some accepted that argon had only one atom, but was composed of known atoms in an atypical structure—perhaps three atoms of nitrogen, or N_3—hence, *not* a new element. These options would protect the cherished structure of the periodic table, which some regarded as inviolable.

Others began to speculate on adjusting the periodic table. For example, Ramsay himself suggested the following year that a new group be added after the halides (fluorine, chlorine, bromine, and iodine). Others conceptualized that elements with no valence made sense, while placing them in various locations in the table. But these new alternatives still did not solve the fundamental problem of the sequential arrangement by atomic weights. The community's dilemma was articulated by Italian chemist Raffaello Nasini: "In the present position of the question, we must abandon either the conclusion universally deduced from the kinetic theory of gases or the periodic system." It seemed a no-win situation. Unfortunately, even with plentiful theoretical alternatives, few proposed practical means to confirm them empirically.

Ramsay, guided by the theoretical possibilities, began immediately to search for other inert gases that might be related to argon. Months later, he isolated helium. Neon followed in 1898, and krypton and xenon in the years following. All with atomic weights that indicated where they belonged. The uncomfortable exceptions had now formed a new pattern.

By 1901, the once cryptic group of elements had been accepted, inserting an additional column into the periodic table that had once seemed complete. The periodic table was dramatically transformed, even while its basic structure was essentially preserved. That achievement was marked by a Nobel Prize to Rayleigh and Ramsay in 1904.

Rayleigh's apparently modest discordance of measurements on nitrogen in 1892 had revealed a series of errors that escalated from (1) a hidden component of the atmosphere, to (2) an unexpected new element, to (3) a whole group (column) of elements missing from the periodic table. Even a small incongruence can open the way to significant errors and major conceptual change.[7]

Element 118

Epistemic work in tracing errors includes social-level errors, as much as observational or conceptual errors. Credibility bias or communal cultural bias can occupy our blind spots, and they need to be discovered as much as faulty instruments or cryptic theoretical alternatives. Unfortunately, it need not even be clear at first that the error is at the social level: that is part of the epistemic work in isolating sources of error among the grand inventory of error types (see Figure 2.2).

Consider the purported discovery of elements 116 and 118 in 1999. These "superheavy" elements exist only by fusing other atomic nuclei in a cyclotron. The major challenge is to find a pair of atoms that fuse, rather than fragment in the collision. So, when a team at Lawrence Berkeley National Laboratory (LBNL)—a lab with a distinguished history of creating such elements—announced the result of bombarding lead with krypton, it was considered a triumph. A lab in Germany, now clued to the unorthodox but apparently special pairing, immediately tried it themselves. No luck. Over the next few months, labs in France and Japan also reported failed replication efforts. Same combination of atoms, different results. An observational discordance: a new, unexplained incongruence.

The burden of proof shifted back to LBNL. The team had already repeated their own results, prior to publishing, to ensure that they had not experienced an experimental fluke. They had already run through the list of likely sources of error, with no obvious candidate for raising a doubt. Now, it seemed, they would have to do it all again. So they did. But this time, even LBNL could not get the same result.

Now what? That sort of lapse tends to trigger a referral to an independent investigator. The assumption is that a "fresh pair of eyes" might be able to spot an error, where customary practices, habits, and personal interests might make it difficult to notice them. The outside team ruled out several familiar technical problems: yes, the beam was aligned; yes, the detectors seemed to function properly; yes, the data acquisition and statistics seemed okay. The most likely cause, they speculated, was the magnet settings—but there was no test to demonstrate that they were responsible. So, repeat again. With a few upgrades in detection equipment, the LBNL team tried once more—now, the fourth time, almost two years later. Fortunately (or so it seemed), the data indicated once again a signature decay-chain of element 118. Discordance attenuated, perhaps?

By now, the expert who had analyzed the data was not the only one who could run the computer program. In the spirit of due diligence, a postdoc (Don Peterson) developed his own variant and ran a separate analysis. He found no decay-chain in the data. At last! A replication-with-a-twist seemed to isolate the differing results to a single factor: the analysis software.

Good news, bad news? This new discordance triggered a second review committee. Meanwhile, a graduate student (Joshua Patin) learned to use the original computer program. His analysis found all the relevant observational "events," but not in a temporal sequence that would indicate a continuous chain of successive decay. That sent the team back to the raw data file. It seemed lost, but was then recovered from a back-up storage. The whole data analysis, they found, had somehow bypassed the program. With the questionable validity of the data analysis, the team returned to the results from two years earlier. Records were missing, requiring more work to retrieve the original raw data. Once reanalyzed (with the original program, but by a different person), the evidence from 1999 for elements 116 and 118 dissolved, as well. No positive evidence remained. Retraction notices were sent out.

A few months later, the official review committee concurred. The data had been manipulated. All the attention then shifted to Victor Ninov, the expert whose golden touch with data analysis had once seemed to rescue LBNL's reputation by helping them discover these new superheavy elements. But he denied any wrongdoing. Hence, a third review committee.

As concerns escalated, so too did the localization efforts. Had the alterations simply emerged spontaneously from an admittedly finicky computer program? Plausibly. But in this bout of error research, historical logs that monitored computer use provided unequivocal evidence of editing the data file over the course of a few hours. Ninov was the only variable remaining, the only explanation for the error.

That led to a *fourth* committee. Their final, more holistic assessment holds some important lessons for the student of errors in science. Ninov was dismissed. That was largely an *institutional* and a *moral* judgment, based on professional misconduct. But the committee had equally stern words for everyone else—for not catching the error earlier, and for not having a second person check the data analysis. Namely, *epistemically*, they seemed to attribute the source of the error to the team's misplaced trust in Ninov, not to Ninov's own falsified data. Of course, Ninov had been hired as the expert. No one saw the need to question or check his work. But in the committee's view, relying on his reputation was not sufficient. So, one might equally say the source of error was a lapse in the heuristic of credibility—a social-level error.

As noted in Chapter 4, scientists regularly use past achievements and credentials as a guide to interpreting subsequent performance—an almost necessary shortcut at the discursive level that obviates redoing everyone else's work yourself. But such credibility judgments are inferences. They are, as this episode poignantly illustrates, part of the implicit justification. And they may fail. If science is to avoid the occasional hiccup, the relevant checks must be institutionalized, or a system of error correction adopted to accommodate them (see Chapter 8). Finding error—and guarding against error—requires epistemic work.[8]

•

In summary, as illustrated in this handful of cases—the erstwhile faster-than-light neutrino, the discovery of argon (and then the noble gases) in discordant atomic weights, and the fraud behind element 118 in 1999—errors may certainly get corrected. But they are not just the residue of "failed" or unsuccessful replications. They are conclusions themselves. In each case, targeted and creative research was needed to gather relevant missing evidence to resolve an uncertainty and to reach the unexpected conclusions. That is, contrary to the folkloric view that "the truth will out" on its own, simply with some unspecified passage of time, rooting out error requires scientific work. And that is equally true, as these three cases show, for observational, conceptual, and social-level errors.

Incongruences

Having clarified that remedying error involves epistemic work, we might ask what that work entails? What is the pathway from an unrecognized error to its remedy?

First, one may acknowledge that everyn error *begins* as established "fact." An error is a blind spot not yet realized to be such. The original claims must clearly undergo a transition. But further information is needed to open that awareness. What prompts a scientist to doubt an already accepted claim? Or to consider, or even to look for, evidence that would justify a contrary conclusion? How does one recognize that an error might even exist?

The first step is simply noticing that something that should be ordinary is somehow awry. Rarely, I think, does any scientist spontaneously "discover" an error. That is, one first encounters some indeterminate hint or clue that the original justification may not be warranted. A glimmer of doubt. Not necessarily prompted by deliberate "skepticism." Just a concrete indication that "something is wrong." Well, "something."

Correcting error is, essentially, a process of learning. It is a form of conceptual change. We ultimately overhaul our justification framework and replace it with an alternative gestalt. Learning—or "deep" learning—entails a cognitive "revolution" of sorts. In this case, the elements of the original pattern are reconfigured. The new conceptualization recontextualizes previous information, exposing how the earlier justification was incomplete and misleading. That is, even as knowledge grows, every error marks an episode of active "unlearning."

Psychologists and educational theorists (and philosophers of science, too) have much to say about the process of conceptual change. Here, I want to focus on one particular element. As conceptualized by Jean Piaget, structured learning typically begins with a *discrepant event.*

Some observation, perception, or event does not fit into one's expectations, based on prior experiences. Basically, there is a mismatch. For Piaget, this triggers a sense of *disequilibrium*—a metaphoric term indicating a destabilizing effect. There is a need to restore balance or a sense of order. Ideally, the discrepancy triggers the mind to reconstruct knowledge by incorporating the new experience along with the old: a process of *accommodation.* So, in this view, errors are cracked open by a mismatch of some kind. The mismatch introduces an element of ambiguity in what once seemed intellectually sure and emotionally secure. Basically, an interpretive conflict seeds learning.

Other cognitive scientists have commented on the importance of such mismatches. Leon Festinger called it *cognitive dissonance.* His implicit metaphor—of disharmonious music—is also apt. He likewise noted that the experience primes a process to resolve the inconsistencies or conceptual tensions—to restore harmony.

Some historians and philosophers have landed on the phrase "going amiss," without ever quite defining "amiss." Here, one may identify "amiss" as two lineages of scientific work that converge in an incongruence. It is not an error—yet. But it signals a latent error (large or small) yet to be isolated.[9]

Each discrepancy (or dissonance) reflects an intersection of two or more histories. Each conclusion may seem justified on its own. But they do not fit with each other. How does one reconcile the discord? What emerges *first,* then, is uncertainty. No error is evident yet.

Of course, the process of accommodation, resolving cognitive dissonance, or reconstruction is not necessarily easy or simple. We have already encountered confirmation bias (Chapter 3). Namely, the mind is disposed to interpret new experiences in terms of earlier ones. Accordingly, novel situations may well be interpreted in conventional terms, and thus strike no discord. The very clues that an error exists may be buried in an impression of normality, and hence go unheeded. In addition, the sense of disequilibrium is disquieting. The discomfort places us in *emotional* limbo, not just cognitive uncertainty. It disturbs one's sense of stability, or security. It is easy for the individual to retreat to pre-established conceptions. Indeed, one may even embrace them more firmly, by discounting or rejecting the legitimacy of the new information outright. The potential threat is thereby disarmed. Namely, an opportunity for prospective learning can backfire. The error may sadly become further entrenched. Still, if an error is ever to be remedied, it will inevitably begin when a mismatch is significant enough to nudge the concept from an unquestioned status into a zone of uncertainty.[10]

The role of mismatches was highlighted in the context of history and philosophy of science by Thomas Kuhn in his notion of an *anomaly*. For Kuhn, an anomaly is "the recognition that nature has somehow violated the paradigm-induced expectations that govern normal science." Namely, something has "gone wrong." Observations "cannot, despite repeated effort, be aligned with professional expectation." Anomalies thereby "subvert the existing tradition of scientific practice." Anomalies, Kuhn claimed, spur the development of new theories and new ways of doing research, much as discrepant events prompt learning in Piaget's account. Kuhn also noted, like others, the problematic tensions between interpreting evidence according to existing paradigms (or standardized models) versus as novelties whose anomalous nature might spur conceptual reorientation. In a subsequent historical analysis, Alan Lightman and Owen Gingerich examined exactly how and when anomalies begin. They noted that alternative perspectives were usually already in place, which allowed one to perceive or appreciate the anomalies as anomalous. In my own research on the ox phos debate in the 1960s–70s, I found that various participants certainly interpreted anomalies differently, commensurate with their theoretical perspective. So, the initial detection of anomalies as a critical error signal may surely be complex. Even somewhat problematic, perhaps. Still, they are key to opening awareness of an error hiding in a blind spot.[11]

I want to expand Kuhn's familiar notion of anomalies to be more inclusive. Along with his contemporaries, Kuhn was primarily interested in the relation of theory and observation and how evidence "underdetermined" theory. By contrast, a focus on error, with its large spectrum of error types (Chapters 2–4), requires a larger, more encompassing concept. Hence, I adopt a more general term: *incongruence*. This includes multiple forms of mismatches: (1) between two sets of observations, or (2) between observations and theoretical predictions or interpretations, as well as (3) conflicts at the discursive level, between different theoretical interpretations of the available evidence (see Figure 5.1). The term "incongruence" should echo the concepts of discrepant events and cognitive dissonance, but with a more detailed articulation about what is mismatched, as described in the following sections.[12]

Discordances

Incongruences come in three basic types. First, there are discrepancies in two or more sets of observations or experimental data. Following Allan Franklin's terminology, we may call these *discordances*. They tend to indicate, as a first assumption, that there is something awry in the measurements, the experimental design, or technical expertise in conducting the prescribed methods. Or that some data has been collected with some unknown bias—perhaps a confounding variable. Or there is a faulty assumption about how one instrument works. Or the observations were not properly calibrated or standardized. The error is probably *observational*. Yet one can also find cases where the conceptualization of one experiment or the interpretation of results was based on theoretical assumptions that, when reviewed with a more critical eye, turned out to be unfounded.[13]

Incongruence	Mismatched Items
discordance	observation–observation
anomaly	theory–observation
ambiguity (or interpretive ambiguity)	theory–theory

Figure 5.1 Types of incongruences.

We have already encountered numerous illustrative examples, such as the discordance of Rayleigh's atomic weights for nitrogen or the discordance of heavy element decay patterns in the search for element 118.

Conflicting data is not that uncommon in scientific research. But the discordance needs to be resolved, and potentially misleading sources of errors identified. As noted earlier, when innovative results are announced, other labs frequently endeavor to copy them as a first step to build on them. Discrepancies sometimes arise. For some, that may affirm the folkloric belief that failures to replicate expose errors. But the mismatches are only signals of possible error. They do not identify where any possible error occurs. That requires further research. Discordances, like other incongruences, are clues that alert researchers to the presence of an error. They are not the errors themselves. They are only a first step in identifying and remedying them.

Anomalies

A second type of incongruence is *anomalies*. Conventionally (since Kuhn), these have been construed as conflicts between theory and evidence, or between theory and observation. These are the most problematic in terms of localizing the error, because the source of the error may be in the observation, or it may be in the conceptual interpretation. There may be no way of knowing at first. It may seem wise to double check one's data or experimental setup first, to know that the problematic evidence is observationally sound. But this may be very expensive or time-consuming, or require substantial resources, depending on the circumstances. So, where theory does not accord with observation (or vice versa?), one is poised to peruse an open array of strategies, as used in *both* the other two "flavors" of mismatch.

Philosophers have invested considerable time reflecting on anomalies, so not much need be added here. Anomalies have been regarded, for the most part, as vehicles for theory change, most vividly exemplified, perhaps, in the work of Lindley Darden[14] on Mendelian genetics. For example, Thomas Hunt Morgan was breeding fruit flies in 1910, when an anomalous white-eyed mutant appeared and exhibited an equally anomalous sex-linked inheritance pattern. That led to reexamining theoretical assumptions, and eventually to the discovery that genes are located on chromosomes, forming linkage groups.[15] Similarly, Howard Temin encountered heritable mutations of the Rous sarcoma virus, which eventually led to the discovery of reverse transcriptase: an anomaly that precipitated a revision of the central dogma of molecular biology.[16] In much the same way, Stanley Prusiner confronted the anomaly of a nucleic-acid-free slow virus, which ultimately led to the concept of prions, another exception to the central dogma.[17]

Yet while anomalies may signal a latent error, they need not be promptly resolved. That may be true, even if the anomaly is well known and resolving it seems important. For example, Newton proposed a formula for the speed of sound in 1687. By then, there had already been several efforts to establish the value empirically—by Mersenne, Gassendi, Roberval, the Academia del Cimento, Cassini, and Boyle. Newton's predicted value was off by 15%: a vivid anomaly.

Given the monumental scale of Newton's work and his exactitude in everything else, the mismatch was not easily dismissed. Newton tried to finesse the discrepancy, referring to the "crassitude of the particles of the air" and "vapours floating in the air" (*Principia*, Book II, Proposition 49). Still, the error became a minor embarrassment in the midst of the great success of Newtonian mechanics. According to Thomas Kuhn, by the turn of the 19th century, it "had been one of the scandals of physical science for more than a century and had repeatedly though fruitlessly drawn the attention of Europe's outstanding theoretical scientists, including Euler and Lagrange." Laplace was surely aware of the embarrassing notoriety of the anomaly when he finally proposed the relevance of adiabatic compression (and subsequent expansion) in the early 1800s. But the anomaly had not itself precipitated the critical research that Laplace used in his proposed solution for Newton's formula. As illustrated in this case, we cannot expect (as perhaps Kuhn hinted) that anomalies inevitably leverage theory or paradigm change. In this case, the resolution was basically all within "normal science," but it took a long while.[18]

Nor need anomalies be resolved by conceptual changes or reinterpretation of observations. The incongruence between theory and observation may be precipitated by errors at the social level. Consider the human skull found in a gravel bed trench in Sussex, England, around 1912. It was not an anomaly at the time. Quite the opposite: the odd combination of an ape-like jaw and large human-like cranium features in the Piltdown specimen was precisely what many anthropologists were expecting to find in a "missing link" in human evolution. However, with the discovery of other human remains in the 1920s, this one fossil began to feel out of place. With more data as context, it *became* anomalous. By the late 1940s, the role of the once-central skull had actually become peripheral: not only an anomaly, but also an unaccountable error. So, it was not all that surprising when in the early 1950s the two pieces, along with some related specimens, were revealed as a forgery. Unraveling the Piltdown fraud was, ultimately, somewhat anti-climactic.[19]

Anomalies can still be found in contemporary science, too, of course. While they may indicate an error somewhere, sometimes they are treasured as a potential segue to deeper insights. For example, in spring 2022, a news headline in *Science* magazine advised readers, "Particle mass may tip the scales to new physics." It continued: "New estimate of *W* boson mass conflicts with prediction from 'standard model.'" The announcement virtually defined a paradigmatic anomaly. In this case, the "standard model" (SM) has governed the field and research for four decades, receiving significant confirmations along the way. The occasion for the dramatic news was the publication of data collected over 10 years measuring the mass of the *W* boson, the particle that mediates the "weak interaction" and is responsible for radioactive decay. The new value strayed significantly from the model's predicted value—in this case, by a remarkable seven standard deviations. That new determination means that all the other SM values do not dovetail neatly into a coherent whole. Incongruence, on a large scale. One might imagine that physicists would be upset about the "upset," knowing that they will have to revamp their hard-won theory. However, they were "very excited about the result!" An accompanying commentary noted that, by contrast, "physicists have looked for phenomena that directly challenge the SM in the hope of finding hints on what a more complete theory may look like." That is, the anomaly is a valuable tool for leveraging a deeper understanding. The violation of expectation also contains clues about how to construct a successive theoretical alternative. No one could guess at the time how the episode would unfold. However, it is instructive for us to see how a group of scientists was essentially celebrating the apparent anomaly, rather than fretting about an impending error. Anomalies, like the mass of the *W* boson, are potential gateways to uncertainty and, thence, to future errors and their corresponding remedy.[20]

A few notes may be valuable in summary. First, the anomaly is not the error. The error is identified later, after a process of "isolating" the anomaly, as described further below. Still, an anomaly typically marks a significant shift in epistemic posture or commitment: from relative confidence in a stable conclusion to a position of uncertainty (see Figure 8.1 and discussion of "Negative knowledge" in Chapter 8). In addition, an anomaly may sometimes hold clues to its interpretation or subsequent resolution. That resolution is not necessarily limited to theory change. As an incongruence, it may ultimately be traced to any of the error types. Anomalies, like other incongruences, are only the first step in the natural history of an error.

Ambiguities, or disagreements

So: observations can sometimes stymie theoretical expectations—anomalies—or different paths of research can lead to incongruent sets of observations—discordances. A third type of incongruence may occur between two or more theories, or concepts. That is, divergent paths of reasoning (perhaps rooted in different or overlapping sets of observations) may lead to conflicting conclusions. There may seem to be no easy way to reconcile them. That is, there are (at least) two ways of construing the evidence: what we may call an interpretive *ambiguity* or perhaps, more plainly, a *theoretical disagreement*. Ambiguity, as a term, helps underscore the uncertainty inherent in the incongruence, here at the level of discourse among scientists.

Philosophers of science have long reflected on these episodes, but typically under the banner of "theory competition." That is, discussions of disagreement have generally not been framed in terms of inherent error in one or both (or more) of the theories, nor oriented to understanding how to elucidate (and reduce or remedy) the errors. A focus on uncertainty in lieu of "competition" highlights the opportunity of ongoing research to resolve the uncertainty, rather than trying to issue a (possibly premature) summary judgment. Disagreement or controversy reflects science as a work in progress.

A reorientation from competition to ambiguity and error correction has profound overtones for how one approaches these episodes. Framing debate as one of "competing" ideas tends (at least in modern culture) to assume a zero-sum contest. One winner, one loser. One or the other, not both. That is, the widespread metaphor of competing theories evokes an either-or, winner-take-all framework. Which theory has the most evidence, or the best evidence, or solves the most problems, or is most progressive, and so on? The justification for endorsing or accepting one theory thereby strangely eclipses the justification for the other. Historically, at least, philosophers have invested too little thought in how to differentiate the theories, partition their domains (or scope), or characterize them as complementary, partial solutions.[21]

Yet historically, sustained scientific debates typically close with multiple theories being found justified, each in a different context. Despite some of the politically charged stories of many famous disagreements—for example, the Volta/Galvani debate, the Cuvier/Geoffroy debate; the Great Devonian Controversy, the ox phos controversy—they all ended by validating each participant in some respect. Historically, the theoretical ambiguity was an occasion to sort out the particularities of context, and thereby dissolve the apparent conflict. One typically finds that each theoretical perspective slipped into overgeneralization or overstated domain. These errors needed to be clarified, and the proper scope of each theory or concept determined more precisely. Ultimately, the outcome was to resolve disagreement, not to establish one exclusive winner. My orientation here will resonate with that view. Episodes of theoretical disagreements are occasions to *reconcile* the justifications, by isolating errors, revising concepts and circumscribing domain (or scope). The alternative to the conventional either-or, winner-take-all framework of theory competition is the *resolution of disagreement* and *differentiation.*[22]

Some may imagine that disagreement represents a weakness in the process of science—that, *ideally*, science should proceed without such incongruence or ambiguity. However, alternate conceptual perspectives help highlight deficits in the evidence, or expose conceptual blinkers. Alternative perspectives—from various disciplines, biographical backgrounds, cultures, social classes, genders, and so on—enhance collective awareness. Despite popular beliefs, debate and disagreement are a sign of healthy science. The discursive level of science, with its diversity of practitioners, is integral to highlighting, and thus remedying, error, as much as to generating creative theories—a theme addressed more fully in Chapter 6.

•

Incongruences are occasions for reorienting research. The focus pivots from collecting evidence for some particular concept to the new problem at hand: why gathering that very evidence suddenly seems to be problematic. Hans-Jorg Rheinberger has previously noted this pattern of diversion, or *displacement* in research trajectories—say, when a presumed contaminant abruptly shifts from being experimental "noise" to being the target "signal." For example, Rayleigh's work on atomic weights was *displaced* as the analysis of the unknown gas in his putative nitrogen sample took center stage. OPERA's work on neutrinos was temporarily displaced by work on time signals. LBNL's overall work on element 118 was displaced by work on data analysis programs, how data files were edited, and checking the work of expert colleagues. They all led to unexpected outcomes and to revising original conclusions.[23]

Incongruences introduce new uncertainty, wherein unproblematic background becomes foreground. The tendency to shift gears with the emergence of discordances, anomalies, or disagreement accounts, in part, for the apparent meandering or wandering character of the history of science when viewed at the level of individual practice. They open a new phase of research: characterizing and identifying the precise nature of the incongruence—namely, isolating the implicit error.

Isolating errors

To recap: At first, every error begins unnoticed—as an apparently justified "truth." That status of error can persist for many years or even decades. Then, in subsequent work, researchers may encounter an incongruence. Confidence in the claims drifts back into uncertainty. "Something" in the justification(s) is wrong. But what? What happens next in the natural history of an error?[24]

More epistemic work. The snag may be something small, something big, something central, or something relatively peripheral: any of the error types that we have already noted (Chapters 2–4). The exact inadequacy in the evidence or reasoning needs to be determined. Namely, the incongruence must be *localized*. Or, in lab jargon, they need to *isolate* the source of the error.

The process of isolating errors is not unfamiliar. For example, auto mechanics *troubleshoot* mysterious noises or oddities in the way a car performs. In the same way, a service repairman may troubleshoot a malfunctioning appliance, or an IT tech may troubleshoot a Wi-Fi network that is not working. So, too, for early computer programmers, who *debugged* their programs. When a doctor meets a patient who feels ill, the first step is *diagnosis*. What is the exact or detailed nature of the problem? Once clearly and precisely identified, a quick remedy usually follows.[25]

For scientists, isolating errors is a genuine process of discovery. It resembles how they try to isolate causes in a multivariate system or complex causal network. For example, physiologists seek to understand how an organism functions. As part of their functional analysis, they may want to localize a particular process to the relevant organ or tissue (perhaps a structure that is not yet known). Similarly, biochemists may search for a particular reaction or particular enzyme that is responsible for an observed product or effect. That step may lead them, in turn (in cases of diseases, say), to identifying the relevant gene or crafting an appropriate inhibitor drug. Physicists may seek to dissect subatomic particles to account for patterns in the properties of matter, along with observed exceptions to those patterns—why they would want to know about the mass of W-bosons, for example (in the case mentioned earlier). In all these cases, can one "reduce" the observed behavior to the actions or interactions of component parts? In isolating errors, one follows similar strategies, trying to localize the "cause" of the incongruence.

There are several conventional ways to decompose or reduce a system to its parts. First, one may envision it mechanistically. Here, the aim is to subdivide a system into its components, describe the action of each part, map the interactions of the parts, and then show that you can reproduce the behavior of the whole. For example, cell biologists centrifuge cell extracts to separate parts of the cell. Chemists use various forms of filtration, chromatography, or electrophoresis to separate different substances in a liquid or gaseous mixture. Each part can then be studied individually, and characterized for how it contributes to the whole. Functions may be localized to individual parts or to particular interactions. Decompose, then reconstitute.[26]

Alternatively, a system may be viewed procedurally. There, the investigator aims to break a bulk process successively into smaller increments, hoping to characterize each particular step. Again, one step may be identified as critical for a particular effect of the whole. Analysis and synthesis. In both these ways, scientists endeavor to localize functions or, alternatively, malfunctions.[27]

In the case of isolating errors, we are concerned with dissecting scientific justifications (not so much the systems under study). That is, an incongruence indicates the presence of a flaw in how we have reasoned independently toward the mismatched conclusions. So, to identify a source of error, one "deconstructs" the arguments into their component layers and reconsiders the warrant for each separate remapping step in the reasoning process. For example, in a conventional philosophical approach, one might reexamine a theoretical "prediction" (or deduction, or explanation) based on its logical structure. That could include revisiting every theoretical assumption, any boundary condition, the background knowledge (or auxiliary hypotheses), any ceteris paribus clause (implicit or explicit), the implied scope (or applicable domain) of the conclusions, and the quality of the data (the indeterminant mess of the all-too-familiar Duhem–Quine problem). But if the measurements or data are suspect, then all the experimental methods or parts of the setup or the observation protocols are open to question, as well. A long list, indeed. We might well track the source of error to any of them. The extensive inventory of error types is thus a prospective guide to isolating possible errors. At the same time, it is also an inverse map to how we structure our scientific justifications.

I opened the chapter with three cases of the epistemic work involved in resolving errors. They exemplified well the process of broad-based search. It may sometimes appear somewhat haphazard. Tracking errors and testing possible sources of error is highly contingent on context and the unpredictable findings. Here, I will add two more cases here for illustration—each apparently minor in scale, perhaps even mundane, yet indicative, I think, of the kind of troubleshooting that typifies much of scientific practice.

The case of the mislabeled hafnium

First, consider the case of postdoc Greg and his presumptive discovery of a new high-temperature superconducting compound. The original study at the prestigious Institute for Pure and Applied Physical Science in San Diego was a relatively routine exploration, hoping to find superconductivity in new compounds: zirconium–, hafnium–, and titanium–rubidium phosphide (ZrRuP, HfRuP, and TiRuP). Greg prepared three samples, varied the temperature of each, and measured magnetic susceptibility as an indirect indicator of their superconducting properties. All went smoothly and the research team announced the two new compounds as promising superconducting materials. Greg and his advisor were certainly aware of possible errors (rooted in the limitations of their technique). But with no explicit anomalies to prompt doubts, they proceeded confidently. They next enlisted a collaborator at Los Alamos to test the same compounds in more sophisticated ways. This would reveal further important properties, while confirming the critical temperatures (T_cs) that Greg had already determined, but via a different method (disturbances in specific heat). Alas, the collaborator reported a much lower T_c for HfRuP. Discordant results. Who had erred? How? Normally, Greg would have checked his original sample. But not enough of the specially prepared compound remained. When Greg and his advisor reviewed the original data, they noticed a subtle pattern in the X-ray diffraction spectrum that indicated impurity phases. How could that be? The manufacturer had listed the hafnium in the catalog as "99.9% pure." Greg commiserated with colleagues in the lab. Sympathetic, they suggested that he check the actual label. "May contain 2–3% zirconium," it read. Well, even a small amount of zirconium would explain the results. That was not new. Greg had to retrace his steps. He ordered purer hafnium and made new samples. He remeasured. The new T_cs now accorded with the collaborator's. The source of error had been localized: zirconium hiding in the "pure" hafnium. Greg, appropriately humbled, acknowledged his modest error in an aside at a conference the following year, retracting the original report about the promise of HfRuP. And his career went on as usual, unmarred by the error, which had been isolated and "fixed."[28]

The fugitive coupling factor of ox phos

The problem of isolating errors applies as much to fraud as to other error types (as indicated in the case of Ninov's manipulation of data in the case of element 118, above). Fraud is the Achilles heel in the normal heuristic of trust among scientists at the discoursive level. Once trust has been exercised, it is not always obvious that deception factored in, any more than a contaminated cell culture announces itself. Subsequently, researchers may find the reported results incongruent with other observations or theory. But how does one learn that the problem is fabricated data or a doctored electrophoretic gel, rather than capricious equipment, poor technical skill, or simply a statistically rare result? Fraud needs to be identified, too.

Here, we may consider an episode involving biochemist Efraim Racker from Cornell University, trying to find the elusive high-energy intermediates of ox phos in the mid-1960s (a case introduced in Chapter 2). Already there had been several promising accounts that were later discounted—the beginning of an "error cascade." Then, in 1964, a highly respected lab at the University of Wisconsin published a series of papers proposing a new high-energy coupling factor. Racker's interest was piqued. Yet he found that the claims did not jive with his own experience in the lab. So he tried to replicate the experiments—not to check for any error, but ostensibly to adjust his own work (based on the apparent success in another lab). At first, he noticed errors in the methods for measuring ATP. But even fixing that, inconsistencies persisted.[29]

It was an important set of reactions to set straight. So, Racker took the initiative. He prepared some samples using his own familiar methods and then flew halfway across the country with two colleagues to the original lab to resolve the problem. In Wisconsin, the postdoctoral fellow responsible for the original work, George Webster, applied his reported methods on them. But the experiments failed there, too. Still looking for an experimental error (and trusting the results as reported), Racker asked to consult the laboratory notebooks. The following morning, Webster admitted having fabricated all his data. The source of the error was isolated: Webster's claims were counterfeit.

Here, the error started modestly as awareness of discordant results from two labs. Tracing the source of that discord—here, motivated by the significance of the results—led to someone's lies, rather than to improper preparation of samples or incorrect conceptual assumptions. The basic trust in the original publications had unfortunately been misplaced.[30]

"Testing" errors

Isolating an error is typically not complete until one has conducted a decisive final test, confirming the error and its alternative (remedy), and clearly differentiating the interpretive frames. The original results (once interpreted as an incongruence) are compared with a parallel instance, with the key factor(s)—the presumptive source of error—"corrected." For example, OPERA tightened the loose cable and ran the experiment again. Greg remade a new sample with purer hafnium, untainted by zirconium, and measured superconducting temperature again.

This is, of course, the standard reasoning of a controlled experiment, which embodies John Stuart Mill's "method of difference":

> If an instance in which the phenomenon under investigation occurs, and an instance in which it does not occur, have every circumstance in common save one, that one occurring only in the former; the circumstance in which

alone the two instances differ, is the effect, or cause, or an indispensable part of the cause, of the phenomenon.

Controlled experiments (or observations) are a standard tool for confirming the isolation of an error, just as they are in scientific reasoning about causes.[31]

The Mesmer Commission

We may illustrate the conspicuous (and damning) use of this strategy of testing errors by the royal commission appointed to investigate the practice of "animal magnetism" in the 1780s. The claim was that a subtle "magnetic" fluid circulated in all living things and could be harnessed to cure multiple ailments. The social elite of Paris flocked to salons to benefit from it. As indicators, they convulsed. They swooned. They fainted. There was no doubt about that, at least. But was the mysterious "magnetic" fluid or tactile treatment the real cause? The members of the commission attributed the observed affects, instead, to the "influence of the imagination on the animal frame," or as we might say today, the power of suggestion or some other psychological influence. Two explanations at odds: an interpretive ambiguity.

To help localize the alleged error, the commission devised "a series of experiments on animal magnetism separately from imagination and on imagination separately from animal magnetism." That is, they aimed "to test the effect of the imagination on people who are not magnetized, but who believe themselves to be." In complementary tests, they would expose people to magnetic objects without their conscious awareness. The challenge, here, was tracking two variables, typically found together under ambiguous circumstances, and keeping their possible influences distinct. In language that resembled what Mill would write later, Antoine Lavoisier explained:

> By combining these two types of experiments one will obtain separately the effects of magnetism and those of the imagination, and, from this, one will be able to conclude what should be attributed to the one and what to the other.

Namely, compare conditions differing in one variable alone: the essence of a controlled experiment (Figure 5.2). So, equipped with not much more than

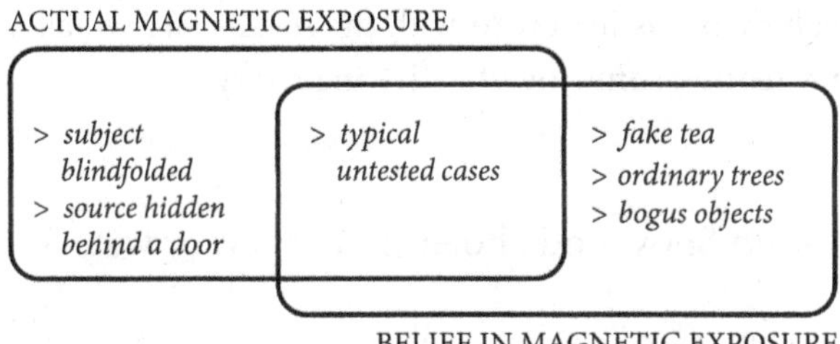

Figure 5.2 Comparison of cases used to sort causal factors in animal magnetism (or Mesmerism). Note differentiation of observational frames.

teacups, blindfolds, and nearby trees (and a practitioner who had obliged himself to follow explicit instructions)—along with the judicious exercise of deceit on unwitting patients—the commission carried out their probes at the Parisian homes of Benjamin Franklin and others.

The verdict of the commission was unanimous and unequivocal. Magnetism was effective only when coupled with cues to inspire belief. It was not effective alone, when patients were blind to its supposed presence. Conversely, "imagination" alone *was* effective (even when no "authentic" magnetic stimulus was present). The cause of the various behaviors had been isolated. Henceforth, appeal to "animal magnetism" as a physiological force was an error. Mesmer's avid disciples, in thinking about causation, had mistakenly focused on physical objects alone, not the social or psychological context.[32]

•

Localizing sources of error (or causes) is most easily understood in the context of laboratory or field experiments, where one can clearly identify a simple or very narrow causal element, such as a contaminant, malfunctioning instrument, or spurious coincidental element. But the same logic applies to alternative theoretical explanations, as in the case of animal magnetism. Namely, one can examine cases that would be commensurate with one theory, but that would simultaneously be contrary to the terms of the other. One also conducts the converse analysis, as well. But it is not always easy to establish the observational frame for just the one variable in question, where one must simultaneously exclude instances of overlap with any relevant alternative explanation. For clarity, the differences in the outcome should align unambiguously with the different theoretical explanations, and indicate which theory is in accord with the results and which seems ruled out. That may require some investigative ingenuity.

John Snow and cholera: miasmas or water?

Such was the challenge for John Snow in demonstrating the ambiguous cause of cholera, as epidemics swept through London in the 1850s. As a physician, Snow had reasoned from clinical symptoms that cholera was likely transmitted from one person to another through water contaminated from the victim's discharges. That contrasted with the widespread view that such diseases were spread via miasmas, or poisonous vapors that were produced, say, by decaying matter in a given locale. Both explained local outbreaks: another case of ambiguity, arising from conflicting interpretations of the available data. How to resolve the theoretical disagreement?

A new epidemic broke out in late 1953 in the area of Golden Square, not far from Snow's home. Over 800 people would die over six weeks. Snow was primed to suspect the local pump on Broad Street. But how could he prove that? A map showed all the cases clustered around the pump. But with pumps scattered everywhere throughout London, it was difficult to distinguish between a neighborhood that shared a pump as a source of water and an area that might contain a local source of miasmas: overlapping observational frames. Moreover, there was a more specific alternative statistical explanation by William Farr, who directed the office where all the deaths were recorded, relating elevation to the incidence of cholera. How could Snow distinguish between drinking water from one particular local pump and any arbitrary local or low-lying area where the purported airborne poison might be present?

The ideal for Snow was to isolate the cause of cholera to drinking water alone, exhibiting a unique pattern that was not consistent with some possible exposure to any miasma. Namely, could he show that one succumbed to cholera if and only if one drank from the same suspected source of contaminated drinking water? With the ambiguities afforded by geographical coincidence, that would be difficult indeed. But that became Snow's strategy. He would examine cases where one could separate the local area (in general) from the drinking water (specifically)—namely, those telltale instances where the two explanations diverged (Figure 5.3). That led him to interview individual households in order to document the fine-scale pattern of who

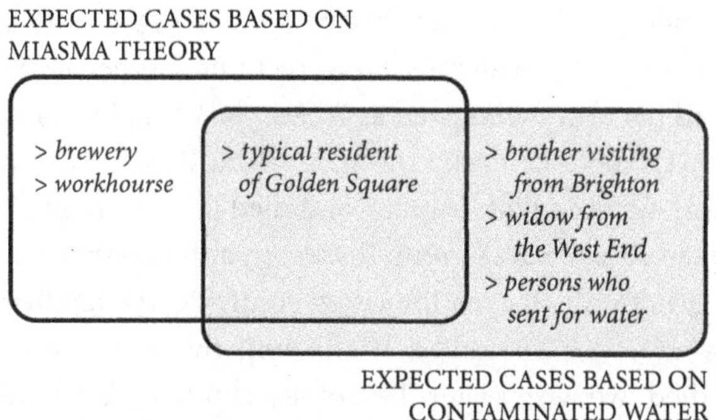

Figure 5.3 John Snow's use of observations to differentiate between alternative conceptual frames in the cases of cholera in London in the 1850s.

drank which source of water. His detailed results yield a rich appreciation of both the localization strategy and the epistemic work involved.

First, Snow had to account for "exceptions" where residents lived or worked near the Broad Street pump but did *not* get cholera. Could he establish that, despite the proximity, they had *not* drunk the water? At the local brewery, only one block from the pump, the workers escaped the epidemic—an anomaly for the miasma theory. Afforded the opportunity to drink the brewery's own malt liquor, they did so, foregoing the local water. The brewery also had their own well. A workhouse just around the corner housed 535 inmates, and was surrounded by houses where deaths occurred. Yet there were few cases of cholera—at only 1/20th the rate of the surrounding community—again, anomalous to the miasmists. However, the workhouse likewise had its own private well—obviating the need to draw from the nearby pump. All these cases were troublesome exceptions to the explanation based on geography as causally relevant. However, they did fit Snow's hypothesis based on water.

Second, Snow needed to consider the cases of cholera where the victim did *not* live near the center of the epidemic area. Presumably, they would not have accessed water from the Broad Street well: an anomaly to his own theory. Could Snow demonstrate that, despite the distance from the pump, they had nonetheless drunk water from the suspect well? These cases, perhaps, would be the most telling. Snow enumerated the details of his door-to-door investigations. In five instances, residents that lived closer to other pumps preferred the water from the Broad Street pump and sent for it specially. In three cases, students went to school in Golden Square, although they lived elsewhere. Similarly for 18 workers at a percussion-cap factory and seven at a manufactory of dentist materials. Snow recounted other stories: "the case of an officer in the army, who lived at St. John's Wood, but came to dine in Wardour Street, where he drank the water from Broad Street pump at his dinner. He was attacked with cholera, and died in a few hours." A pregnant woman went to Broad Street pump for water, although her family did not customarily drink water. Given the rising epidemic, the family relocated to Gravesend where, on the second day following, the woman was seized with cholera and died two weeks later. Two of her children also drank the water, but recovered. In one of the most dramatic cases, a gentleman from Brighton had been summoned to visit his dying brother. He "arrived after his brother's death, and did not see the body. He only stayed about twenty minutes in the house, where he took a hasty and scanty luncheon of rumpsteak, taking with it a small tumbler of brandy and water, the water being from Broad Street pump. He went to Pentonville, and was attacked with cholera on the evening of the following day, 2nd September, and died the next evening." Equally persuasive was the case of the widow of percussion-cap maker who lived in West End. She also preferred the water from Broad Street and, although she had not been to Broad Street herself, drank water that had been delivered from there and died of cholera two days later. Moreover, her niece who was visiting also partook of the water. "She returned to her residence, in a high and healthy part of Islington, was attacked with cholera, and died also. There was no cholera at the time, either at West End or in the neighborhood where the niece died."

The pattern was unmistakable. Snow had helped divide the cases neatly into relevant informative categories. His research showed uniformly that the cholera victims had drunk water from the now notorious Broad Street pump, while non-victims had not. True, there were a handful of remaining exceptions. Six individuals had drunk the water, but did not get the disease. There were other cases where information was unavailable. But overall, the cholera cases aligned strikingly with the source of water, not with a strict geographical criterion. The comparison—including resolving the details of apparent exceptions—helped isolate the cause of cholera, as well as the source of the error leading to the theoretical ambiguity. Farr's theory of miasmas (in retrospect) had been in error: justified in part by the evidence at one time, but not at a later time (once other key evidence had been collected). Elevation had been a confounding variable, hiding sewage and groundwater (also correlated with elevation) as the relevant variables. By 1866, at least, Farr had acknowledged that Snow (then deceased) had been right. The miasmists had properly recognized contagion, but their generalizations about cholera—framed by geography, local soil, foul smells, and air—ultimately proved mistaken.[33]

•

The cases of the Mesmer Commission and John Snow and cholera underscore again the epistemic work involved in isolating errors. Sources of errors produce certain concrete effects, intersecting with the researchers' main interests. To discount their relevance, their precise causal role must be mapped. That entails its own, albeit subsidiary, research. Ironically perhaps, ascertaining sources of errors (and their misleading effects), including testing them, is part of science.

Differentiating domains

Resolving errors in cases of theoretical disagreement or ambiguity entails research, as well: identifying any errors in each of the theories involved simultaneously. One major task in such cases is typically exposing and then adjusting earlier overgeneralizations (see "Overgeneralizations" in Chapter 3). The upshot is to circumscribe, or limit—or sort—the appropriate domain or scope of each concept. One thereby often *reconciles* the apparently conflicting theories by *differentiating* their domains—articulating how they function as complementary maps. That is, when their respective errors are corrected, they may "divide" the territory. As noted earlier, controversies need not be resolved by one theory wholly eclipsing the other in an either-or, winner-take-all fashion. That may lead, in particular, to a more nuanced history, despite the all-too-frequent rhetorical tendency to write triumphal narratives of winners and losers or of full-scale (revolutionary) conceptual replacement. To illustrate this form of resolving debate, I provide here an analysis of one of the most notorious episodes of conceptual upheaval, the so-called Chemical Revolution.

The Chemical "Revolution"?

In conventional histories, the episode was marked by "the overthrow of the reigning 'phlogistic' theory and its replacement by a theory based on the role of oxygen." Namely, combustion was explained for nearly a century based on phlogiston, a constitutive principle of fire, and its release from combustible materials. But in the late 1700s, Lavoisier discovered oxygen, which he showed combined with those materials and was henceforth regarded as the chief cause of burning. Historians have thus variously described the episode as the "supplanting," "substitution," "supersession," or "overthrow" of phlogistic theory: an unqualified Kuhnian gestalt-switch. Namely, Lavoisier "won," and phlogiston was justly relegated to the scrap heap of erroneous ideas: an error, if ever there was one. (Recall from Chapter 1 John Herschel's 1830 indictment of the phlogistic doctrine as a "false theory" that "impeded the progress of science" by deluding scientists with "a mist of visionary and hypothetical causes in place of true and acting principles.")[34]

Observed more closely, however, the history is more nuanced—and more philosophically complex. Here is an account shaped by a philosophy of error, rather than the widespread framework of holistic theory appraisal or theory competition.

Recall, first, that the concept of phlogiston was introduced in the late 1600s to help explain why things burn, and to unify the phenomena of combustion (of organics) and calcination (of metals), along with the reverse process, the reduction of ores to their metals. All involved the transfer of phlogiston, the thing that made metals "metal" and that could be given off as light or heat. Burning and calcination involved the loss of phlogiston; reduction, the gain of phlogiston from charcoal. In addition, chemists recognized that phlogiston "fouled" the air. Things would stop burning when the air became saturated. Such "phlogisticated" air did not support animal respiration, either.

When methods for collecting and isolating gases emerged in the mid-1770s, chemists began dissecting the air, characterizing and labeling its various types (fixed air, inflammable air, dephlogisticated air, mephitic air, nitrous air, and so on). Eventually, Antoine Lavoisier sorted out the confusion. Air was not an element with different types. Rather, it was a *mixture* of gaseous elements—gases that could, furthermore, also combine with other elements in solid form. He thoroughly revamped the list of primitive elements—the origin of our terms hydrogen and oxygen, for example. Integral to Lavoisier's discoveries was the gas oxygen, which he considered key to things burning and was used up in the process. By measuring the mass of reactants and products—including the apparently weightless gases—Lavoisier demonstrated that oxygen combined with other elements. In combustion, oxygen combined with carbon to produce "fixed air" (carbon-oxide). It combined with hydrogen to produce water (hydrogen-oxide). In calcination, oxygen combined with the metal to yield a calx, now described as a metal-oxide. In reduction, the oxygen was removed from the ore. It was either released again as a gas (as in the case of mercury), or recombined with other elements.

So, the question became (from Lavoisier's perspective at least): is combustion explained as the *release* of phlogiston OR the *gain* of oxygen? Two apparently incompatible explanations. An interpretive ambiguity. An incongruence.

For Lavoisier and his allies, it was a manifestly stark choice. Indeed, they further underscored the dichotomy by them posturing themselves explicitly as "*anti*-phlogistonists." Yet the ultimate outcome was not strictly either-or. Both interpretations were partly valid, and both erred in some respects.

Phlogistonists properly described the relationship of the core processes, mapping the transfer of what we would today call energy (possibly using the language of high-energy electrons or reduction potential). They developed their notions before the emergence of methods to isolate and weigh gases, however. So it is not surprising that their concepts did not develop with this foremost in mind. Noting the loss of weight when things burn, some chemists conjectured that phlogiston might have negative weight, or buoyancy—why it might combine easily with air. That was an error of premature hypothesizing. But no one ever found a way to apply that concept with any quantitative rigor, and so it was never really widely accepted. The phlogistonists' focus was chiefly on the dimensions of energy and energy transfer (in today's terms), independent of weight considerations (see Figure 5.4, "Combustion, calcination, and reduction," right side).[35]

Lavoisier, by contrast, made sense of elements and elemental composition. Before his work, both water and air were regarded as elements. Those were errors, surely. An error of not being able to imagine as divisible something that could not be divided in practice: the epitome of a cryptic alternative, and another case of the failure of the heuristic of simplicity, or Occam's Razor. The principle of conservation of mass allowed him (as noted above) to characterize combustion and calcination in terms of basic compositional changes, but it did not fully explain flammability or the energy dimensions of light and heat (Figure 5.4, same box, left side).

There were also errors—on both "sides" (Figure 5.4, darkly shaded area at the bottom). For example, those who tried to conceive phlogiston in

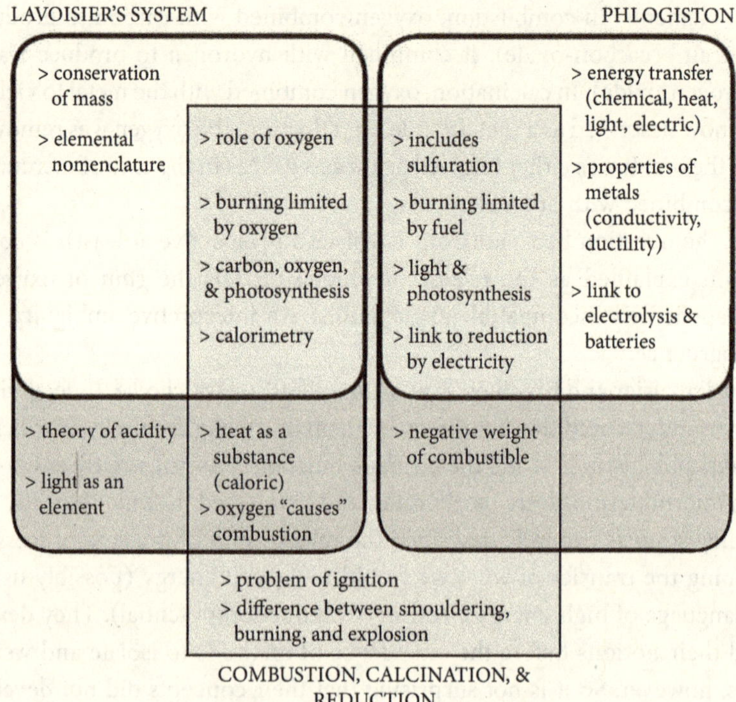

Figure 5.4 Differentiation of domains between phlogiston theory and Lavoisier's new system. Light gray box identifies phenomena related to combustion, calcination, and reduction. Dark gray highlights errors that were ultimately abandoned as viable concepts.

Lavoisier's new terms—as a material substance—ultimately could not do so. That assumption was an error of overgeneralizing the domain.

For his part, Lavoisier was also responsible for many overstatements and unjustified assumptions. His theorizing was grand. But so, too, was the scope of his errors. Lavoisier listed light and heat in his own table of elements (along with iron, silver, oxygen, and so on), although they could not be isolated or weighed either. He thereby exhibited a double standard—an error that we can classify here as a form of confirmation bias. Lavoisier also attributed the sensible heat of combustion to caloric, his newly hypothesized substance of heat. But Lavoisier also characterized caloric as responsible for changes in state. He thus confused (in today's terms) latent heat with chemical energy. He thus regarded all the heat energy of combustion as coming just from oxygen (as it changed state). Not so. That did not help, for example, in interpreting burning that involved gaseous products. Lavoisier did not have much to say about light, either, or its relation to heat. Ultimately, Lavoisier erred in his explanations of light and heat in combustion—precisely where phlogiston, as the alternative concept, was most relevant.[37]

Some phlogistic chemists found that they could accommodate Lavoisier's findings about oxygen and the conservation of mass, without sacrificing their central concept. Lavoisier, they noted, did not adequately explain light, heat, and what we call today chemical energy and reducing potential (of charcoal, for example). Later, they connected phlogiston to electricity (and to Volta's batteries). Those could not be weighed with a balance. Namely, fire is *not* properly conceived as a material substance. But that does not mean that it does not exist. But what was it? In the late 1700s, there was yet no way to concretely measure those things, only to observe their effects. So again, phlogistonists were trapped, in a sense, by the limited conceptual horizon and instrument technology. Phlogiston organized a coherent set of observations and related facts, but was unable to fit in theoretically at the time or generate tractable research questions. That part, at least, was *not* an error—and not adequately addressed by Lavoisier's new system (Figure 5.4, right side).[36]

Nor could Lavoisier really explain what *caused* combustion. Phlogistonist critics noted that some things did not combust, even in the presence of oxygen. For example, a spark was needed to ignite the combustion of hydrogen. Others substances "burn" with sulfur, not oxygen. So while oxygen may have participated in combustion reactions, it did not *cause* burning. Lavoisier's focus was tracking the material elements. So, in modern terms, he confused the compositional addition of oxygen with chemical "oxidation." In a sense, Lavoisier mistook correlation for causation. Phlogistonists occupied this shadow of Lavoisier's deficits.

In summary, the resolution of the "Chemical Revolution" was, like most sustained disagreements, mixed. Lavoisier's new system proved enormously fruitful. Much research was reoriented to pursue the many opportunities it provided (Figure 5.4, left side). With all the work to be done, Lavoisier's errors could be conveniently brushed aside for left for later investigators.

But Lavoisier had not fully eclipsed phlogiston, even if it had waned as a concept in practice. Decades later, in 1871, chemist William Odling observed how phlogistonists had clearly appreciated the nature of oxidation and reduction reactions, noting that:

> ... the truth which he [Lavoisier] established, alike with that he subverted, is now recognizable as partial truth only; and the merit of his

> generalization is now perceived to consist in its addition to—its demerit to consist in its suppression of—the not less grand generalization established by his scarcely remembered predecessors. ... Accordingly, the phlogistic theory and antiphlogistic theory are in reality complementary and not, as suggested by their names and usually maintained, antagonistic to one another.

Ultimately, combustion involves the gain of oxygen *and* the release of phlogiston. Lavoisier's error had been, in part, assuming that the two perspectives could not be reconciled. Ambiguities, as incongruences, need to be *resolved*, and their respective errors isolated, rather than subjected to either-or contests of theory appraisal.[38]

To say that the Chemical Revolution has been entirely misnamed may itself be an overstatement. Chemistry experienced a major transition. At the same time, philosophically, the conceptual dynamics of the period have been widely misconstrued. Despite Lavoisier's landmark concepts, his interpretations were not universally free from error, nor were all his criticisms of phlogiston fully justified. Accordingly, the resolution of the incongruence between phlogistonists and their critics is more aptly characterized as a differentiation of domains, than in terms of "overthrow." And in this interpretation, shaped by a philosophy of error, we may see a model for how disagreement and interpretive ambiguities are resolved in other episodes, as well. Isolation and remedy of error unfolds on two paths at once.

The puzzle of unlearning

The natural history of errors poses a fascinating puzzle. Namely, as we accrue evidence, it does not always merely fill in a pre-established pattern. It does not just enhance the resolution of a known image. Rather, when a formerly "justified" claim is rejected as ultimately unjustified and a new claim is accepted, we have wholly reconfigured the pattern established by the original evidence. In the metaphor introduced earlier, there is a gestalt switch. In *The Structure of Scientific Revolutions*, Kuhn challenged the intuitive notion of the cumulative growth of knowledge. On some occasions, he claimed, scientists do not merely "reduce" an old theory to a subset of a new, grander theory. Rather, there is a fundamental rearrangement of the constellation of facts. A study of the history of errors resonates with this view of gestalt switches. On some occasions, it seems, we must *jettison* various old "facts" as error at the same time we learn a new, deeper "truth." Sometimes, learning from the growth of evidence simultaneously involves—paradoxically perhaps—*unlearning*.[39]

The notion that knowledge is replaced, rather than always expanding, may strike many as strange, if not perverse. At the same time, it should also seem utterly familiar. It is the core feature of the modern mystery genre. Consider the canonical plot structure: a murder or crime is committed and the detective-hero sets out to collect clues. Soon suspicion falls on a particular individual: there is evidence for motive, means, and opportunity. The suspect is brought in for questioning, perhaps even charged with the crime. The whole investigative team is convinced of the person's culpability and it seems one only has to extract a confession. Then, the suspect gives testimony that reveals some detail the detectives missed. There is an ironclad alibi, perhaps. Or the suspect exposes one of the assumptions as unfounded, based on other easily documented facts. The new evidence exonerates them. So the *gestalt* changes. The viewer's/reader's expectations are unhinged. More detective work is needed. Good drama. Good narrative engagement. *Philosophically*, the lesson here is that one or a few more pieces of evidence do not just always fill in details. Rather, they may fundamentally upset the justification and, with it, the verdict.

Even more remarkably, masters of the mystery genre manage to switch the gestalt multiple times in succession in the same case. First one suspect, then another. Sometimes, a whole series. The reader (or viewer) enjoys the thrill of being surprised, or fooled in a sense, and then being "corrected." One need only recall the final scene of many novels (or dramas), where Cordelia Cupp or Hercule Poirot (say) assembles all the suspects in the same room, and artfully recapitulates the case step by step, clue by clue. As the evidence unfolds, she successively implicates different people. Eventually she throws suspicion on nearly everyone in the room before divulging how the last piece of the puzzle fits in, wherein she finally identifies the ultimate culprit. Stock dramatic formula. Good mystery thrillers epitomize the very idea of gestalt switches as the evidence accumulates and errors are revealed. What might we gain by regarding errors in science as pivot points in a good mystery story, rather than as flaws or lapses of proper practice?

6
Deeper Evidence

Evidential gaps • replicate with substitutions • increase sample size • diversify sample • expand scope, widen domain • increase resolution and differentiate • develop new technologies • fill heuristic gaps • engage alternative perspectives • happenstance • resolving error

---

So: scientists become alerted to the presence of an error when, in the course of their ordinary ongoing work, they encounter an incongruence—a discordance of data, a theoretically anomalous finding, or an ambiguity between two theoretical interpretations. There is a problem, hidden somewhere amid the many elements justifying the conclusions, including experimental, conceptual, and social-level layers. But there is often a deficit of information (at first) to resolve those emergent uncertainties. Additional epistemic work may be needed to isolate or localize the error (Chapter 6). Scientists need deeper evidence, of course. But this truism does not tell us, more importantly, how the new evidence emerges—the focus of this chapter.

The process occurs all the time in the lab, in the field, in investigations of various sorts. But the fruitful path cannot always be stipulated abstractly in advance. Thus, the resourceful philosopher (again) turns to the history of science. How does relevant new evidence originate? How does data already available come to be reconfigured? How do the patterns of justification thereby change, shifting the conclusions? Here, I provide a quick survey of some of the diverse ways deeper evidence arises (summarized in Figure 6.2 at the end of the chapter). The categories of evidence I address are all familiar. Yet it is not always clear initially which category will be relevant in any given instance. Tracking that path into the unknown is, in a sense, the fundamental work of all science.

Replicate with substitution

Surely the easiest and perhaps plainest probe once an incongruence has appeared is to redo the experiment. Researchers may already have a sense

Toward a Philosophy of Error in Science. Douglas Allchin, Oxford University Press. © Douglas Allchin (2026).
DOI: 10.1093/9780197827703.003.0006

that something was not quite right. Perhaps they were adding an aliquot of reagent when they were interrupted by a colleague or a phone call. (Did they complete the procedure, or did they perhaps move on to the next sample without adding it?) One can check by repeating the procedures with extra care and attention. One follows the prescribed protocols more consciously. One double-checks the materials to intended specifications. One may thereby isolate the source of error to basic material conditions or methods: an outdated reagent, a contaminated sample, an incorrectly measured amount, and so on. All are common—and routinely remedied without much fuss.

Of course, mere repetition does not necessarily solve the problem. Errors, such as artifacts, may simply recur. Dozens of researchers attested to the observation of N-rays.[1] Likewise for Allison's magneto-optic effect.[2] "Anomalous water" (polywater) was reproduced by many chemists.[3] Microscopists generated mesosomes (spiral membranous structures in bacteria) for 23 years before deciding that they were not authentic structures in native bacteria. Even today, mesosomes are perfectly reproducible. So, too, for polywater. Repetition itself is not necessarily informative—especially if the results remain the same.[4] While repeatability is requisite for studying stable phenomena and using reliable protocols, reproducibility of results (alone) is not informative about error. (This misperception seems to haunt many discussions in ongoing concerns about a "Reproducibility Crisis.")

On the other hand, if the results are different, which of the instances, the original or the replicated effort, is "correct"? Consider the case of Joseph Priestley's investigations on the restoration of air by plants. Originally, he reported that a sprig of mint and other plants could yield a form of air that allowed candles to burn longer and mice to breathe longer. His work was recognized by the Copley Medal. Others tried to replicate his results, yet many failed. Whose results were in error? Years later, Priestley himself, now working in a new home and laboratory, also failed to replicate his own findings. Again, which results were "correct"? After noting the absence of a window in his new lab, Priestley began to think that light alone was the relevant variable. New investigations with well water (no plants, apparently) persuaded him of that. But, *that* was an error. Priestley had noticed but not given any relevance to green scum in his vessels: algae, which filled the role of plants. Jan van Ingenhousz and others resolved that uncertainty through a series of controlled experiments. Here, the error was primarily in Priestley's *replication* efforts, where he had sadly mistaken correlation for causation.[5]

One normally conceives "replications" in terms of trying to successfully repeat unexpected but promising results. The findings seemed startling, but perhaps they were spurious? Can one, literally, reproduce them? In probing for evidence of errors, by contrast, the repetition aims to generate *different* results. It focuses on a suspect, confounding factor. The different outcome helps confirm that the discordance or anomaly was caused by an irrelevant variable: which can be confidently dismissed as a source of error.

Redoing an experiment—but with added controls on materials or methods—can thus help pinpoint artifacts. For example, when "polywater" was prepared with careful attention to protecting against human hand oils and sweat showed no anomalous properties. Weber's claims about gravity waves unraveled when an alternative computer program was used to analyze the data. Mesosomes were identified as artifacts by using different fixatives (all three cases from "Artifacts" in Chapter 2). Many prestigious labs reported having replicated the bogus phenomenon of tabletop "cold" fusion in 1989. The team at CalTech "succeeded," too, but only when they deliberately failed to stir the fluid in the fuel cell: violating a basic control customary in calorimetry. The original temperature measurements had been misleading: an artifact of placing a thermometer in the wrong place and imagining that vigorous bubbling indicated active mixing. The substitution, not the replication itself, was the key to debunking the pretensions of Pons and Fleischman's amateurish efforts.[6]

"Deeper evidence" is produced by varying the materials or methods—and obtaining telltale different results.

Increase sample size

As noted in Chapter 2, as a result of variation in nature, our sampling of it may be unrepresentative of the whole. One may be led to generalize incorrectly. As one encounters more instances, anomalies to the presumed pattern may arise. Clearly, more samples are needed to establish a consistent and reliable pattern: to either confirm the old one, or develop a new one, or to reconcile two conflicting patterns. By increasing the number of samples, one increases the chances that they will be more representative of the whole, and hence more accurate.

When is a sample large enough? This itself may be an empirical question. We saw in the case of the "oops-Leon" particle (Chapter 2) that the team at

FermiLab was beguiled by less than 1-in-50 odds that the cluster in their data could arise by chance alone. But with more collisions and more observations, the blip disappeared. And particle physicists learned to be more wary, and to use more extreme statistical cut-offs.

Yet the same illusion occurred again in 2015, when—now using data from two different instruments operated by two independent research teams—physicists "discovered" another deviation from the expected curve of energy readings. This time, the odds were closer to 1 in 3,000. The unspoken consensus was that this provided much anticipated evidence of a new particle that might reveal more about the structure of the universe. Still, with more data, the cluster disappeared. Increased sample size revealed another error. In the Lederman case, at least, the continued observations eventually led to finding the predicted particle, but at a noticeably different mass (energy level). Unfortunately, perhaps, it had escaped detection in the earlier, smaller sample: also an error? Increased sample size also solved that problem, too, on this occasion.[7] Statistical "significance" (a probabilistic measure of sampling error) may thus be considered an arbitrary threshold—a fact worth recalling in ongoing concerns about a "Reproducibility Crisis."

Increased sample size can have significant theoretical implications. Consider the case of E. O. Wilson's proposed explanation for eusociality—why honeybees and other animals have a cooperative social structure, which includes shared reproductive efforts. Originally, such societies were a puzzle for evolutionists because natural selection seems to function only "selfishly." This form of cooperation involved helping others to reproduce, while not reproducing oneself. How could this behavior persist if it was not passed on to offspring? The presumptive explanation, developed in the mid-1960s, was that the helpers share genes with the offspring they helped raise. Their reproductive "interests" were thus met indirectly through the genes of their relatives. The genetic structure of honeybees and other insects, in particular, seemed to support this notion of "kin selection." The distinctive haplodiploid form of sex determination dictated the organisms' relatedness and thus, it seemed, their evolutionary tendency to breed cooperatively. The idea, promoted in Wilson's landmark 1975 book, helped launch the field of sociobiology to prominence.

Yet in 2010, Wilson prominently renounced his earlier interpretation. His new argument rested on studies of many more cases of eusocial species, which his earlier work had, ironically, helped inspire. Three decades of research had revealed many cooperative breeding societies (such as termites) that do not exhibit the required haplodiploid genetic structure. At the same time, many species that do (including sawflies and horntails) are not social. With these additional cases now documented, it became clear that the original explanation did not align well with the relevant examples. The map had apparently been misframed: the genetics of sex determination did not lead to eusociality. Rather, the societies—from ants and honeybees to ambrosia beetles, snapping shrimp, gall-making thrips, and naked mole rats—all seemed to have nests with restricted access, guarded by just a few individuals. The social cooperation seems just an "ordinary" adaptation to certain environmental conditions. The striking genetic structure, Wilson now contended, was an evolutionary consequence—not a primary cause—of the social organization. Further research has underscored the relative importance of the ecological factors, compared to the genetic structures of relatedness. It also matters, for example, whether the non-breeding helpers are working to defend the colony or assisting in brooding the young: that indirectly determines the ratio of males and females (not the genetics). The explanation involving haplodiploid-based kin selection was mistaken, Wilson and his colleagues contended, and should be abandoned—based on the increased sample size of societies studied.[8]

Cases of larger samples undoing earlier conclusions are common, and can help reveal how bias in the interpretation of the more limited sample had once passed unnoticed. William Buckland presented his conclusions about a global (Noachian) flood based on the relics from just one cave in Kirkdale, England ("Religious bias" in Chapter 3). It was widely hailed, recognized by the Copley Medal, and inspired research elsewhere. However, when fossil assemblages in other caves across Europe were investigated, the timing of all the events did not align. Any flood had not been global. Buckland's claim had been premature, and later evidence helped indicate his religious bias.

Charles Lyell, for his part, defended a strict view of uniformitarianism in Earth's history, even as others were acknowledging that life had evolved, with some species going extinct. Lyell believed that sea serpents, modern-day descendents of icthyosaurs and plesiosaurs, still existed. Yet, as the 19th century wore on and reports of sightings were debunked, and more corners of the oceans were explored, it became painfully evident that none were to be found. Lyell had generalized too boldly from a few instances, and the bias in his anti-evolution perspective became more and more apparent.[9]

Finally, one may consider the fate of the drug rofecoxib (Vioxx). The U.S. Food and Drug Administration approved the drug, largely based on a study published in the *New England Journal of Medicine* in 2000. Ongoing monitoring, however, exposed new safety concerns, and the drug was pulled from the market. Later, it was learned that the manufacturer had data about three heart attacks that had occurred in the month after the study was formally completed. By contrast, monitoring for gastrointestinal effects continued during that same period, and helped to show benefits when compared to another pain-relieving drug used as a benchmark. That is, the study period for monitoring cardiovascular conditions had been biased by being truncated. The manufacturer had "massaged" the results in their favor by being selective in what data they chose to present. Here, the increased sample size indicated corporate malfeasance.[10]

Initial conclusions may be replaced as sample size increases. Patterns found in a subset of cases do not always match similar cases elsewhere. The generalizations can be wrong. When more instances are documented, the errors are remedied. Increased sample size can even be relevant to isolating the sources of error at the conceptual or social levels: for example, revealing how conflicts of interest or religious or communal theoretical bias have shaped the interpretation of evidence.

Diversify the sample

Overgeneralizations may easily emerge and persist when evidence is based on an insufficient sample (Chapter 2). In remedying these errors, it is not always the sheer number of samples that matters, but their breadth. Namely, a representative sample is sometimes achieved qualitatively, not quantitatively. The nature of incomplete or systematically biased sampling may be revealed through a more diverse sample.

For example, Casimir Funk named vitamins as "vital amines," based on his analysis of a handful of micronutrients (Figure 1.1). Once more vitamins were discovered, the role of nitrogen and amines proved irrelevant—although the inapt name remains.

Similarly, Lavoisier erred in his theory of acidity, the origin of our term "oxygen" (Figure 1.1). Oxygen is indeed found in many acids. But not all. Finding yet more acids with oxygen was not going to reveal that. One needed to encounter an acid without oxygen. That was the key role of Humphrey Davy's 1810 analysis of marine acid (today's hydrochloric acid). No oxygen. Suddenly, chemists had an indisputable exception to the generality Lavoisier had relied on. One clear example was enough to unravel the oxygen theory of acidity, which had little other support. That is, sometimes even one key case may be enough to significantly change whether one regards a sample as "representative."[11]

The number of samples may also be contrasted with how informative each sample is. That occurred in the history of a strange fossil creature found in the Burgess Shale. In 1911, paleontologist Charles Walcott described the species based on a single sample. Under such circumstance, one sample alone was enough to establish a new species. Walcott classified it as a segmented worm (annelid).[12]

However, when Simon Conway-Morris reexamined the Burgess Shale organisms in the 1970s, he had about 30 specimens to examine. He was able to document its seven pairs of spines, and along the other side of the body, a row of tentacles. It has a large bulb at one end, identified as the head. But it was neither worm nor arthropod. It had no identifiable phylum. Conway Morris identified it as a new genus, with a name to reflect its haunting uniqueness: *Hallucigenia*. A decade later, Stephen Jay Gould heralded the "bizarre" organism as exemplifying the early explosion of diversity in the Cambrian, yielding many basic types that no longer exist now. He accused Walcott not just of conservative taxonomic judgment, but also of shoehorning strange creatures into familiar categories. A vindication for the value of raw sample size?[13]

In 1992, a new specimen from China displayed a second leg row, which had been hidden underneath the surface. The bulbous head was missing. Lars Ramskold and Hou Xianguang reoriented *Hallucigenia* completely. It now walked, more conventionally, on clawed legs; the spines projected upwards as defenses; and its head was now at the other end. Namely, the earlier reconstruction was both upside-down and backwards! The quality of the specimen mattered.

In 2015, Martin Smith had more than a hundred new samples to work with. But one specimen now mattered more than the others. It was very well preserved, exhibiting key features with exceptional clarity. It showed an elongated head. Its eyes were plainly visible. What of the former "head"? It could now be identified as a fluid stain, from the gut emptying. Dissecting the throat, Smith found a ring of sharp teeth, which confirmed *Hallucigenia*'s affinities to modern velvet worms (onychophores). It was not an annelid (as Walcott had posited). Nor was it the "wondrous beast" that Gould once paraded as upending our view of the history of life on Earth. As the quality of individual samples increased, the interpretation of *Hallucigenia* changed dramatically and, with it, a view of large-scale patterns in evolution.[14]

The cases of "vital amines" without amines, acids without oxygen, and fossil specimens of the cryptic *Hallucigenia* all illustrate that, in addition to the number of samples, the diversity and particular composition of the samples can matter. Samples can be incomplete and unrepresentative in many ways, and early erroneous generalizations may be remedied with an appropriately diverse sample.

Expand scope, widen domain

Errors may also develop and persist when the empirical scope of consideration has been too limited (Chapter 2). Scientists mistake the partial experience as fully representing all presumably relevant phenomena. Cognitively, we may be unwitting victims (individually and collectively) of Kahneman's WYSIATI: What You See Is All There Is. More cases (in number) and even more *diverse* cases may not be as important as venturing into unexplored domain. That is, it is not always a matter of swamping statistical error with a larger sample size or of correcting narrow sampling with a greater breadth of cases. Rather, one needs to expand the whole scope of observation. "Deeper evidence" comes from new, relevant, yet unexplored contexts.[15]

Consider Boyle's law, one of the most iconic concepts of science, and a chief example of the philosophical concept of a law of nature. As a "law," one may presume it to be universal and invariant. However, almost two centuries after Boyle, Victor Regnault, Émile Amagat, and Louis Cailletet all observed the relationship of pressure and volume at high pressures. Under such conditions, the inverse mathematical relationship breaks down. In modern terms, there are limits to compression. Under very high pressures, gases behave more like a liquid than a gas. The *scope* of Boyle's law was limited. It holds only for pressures up to about 10 atmospheres (Figure 6.1). Assuming it to be universal was an error.[16]

In the 1860s, Thomas Andrews found that Boyle's reciprocal relationship also fails at very low temperatures. Again, one could not generalize Boyle's "universal" law broadly (Figure 6.1).[17]

In addition, it was well known by then that a change in temperature could affect pressure, so the mathematical relationship only held if the temperature was constant: a boundary condition (Figure 6.1).

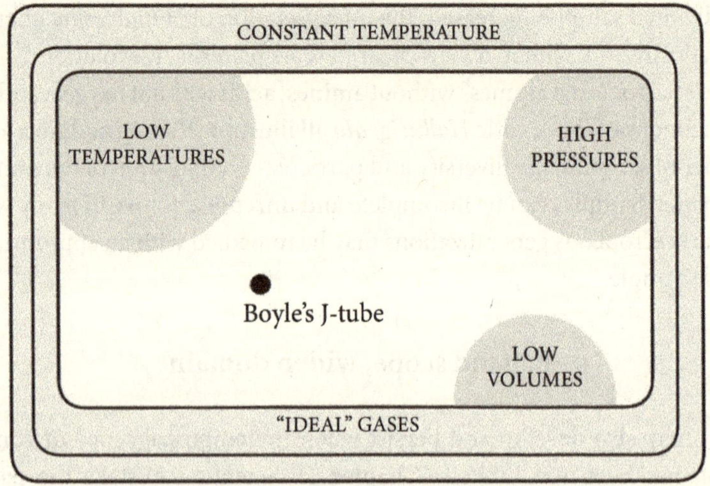

Figure 6.1 Exploration of Boyle's law in extended domains successively limited the scope of the "universal" law.

In the ensuing decades, Johannes van der Waals discovered that at very low volumes, the intermolecular London forces alter Boyle's relationship (Figure 6.1). In addition, the polarities of certain gas molecules exert intermolecular interactions of yet another kind. Boyle's law thus needed to be adjusted for each gas (using van der Waals constants *a* and *b*). What appeared invariant across all gases now varied. It depended on the type of gas. Yet another qualification to generalizing Boyle's law (Figure 6.1).[18]

Errors emerged when scientists tried to generalize the behavior of Boyle's J-tube to all temperatures, pressures, volumes, and gas compositions. Of course, all these stipulations and provisos reflected significant scientific discoveries of their own. Indeed, van der Waals' work earned him a Nobel Prize in 1910. That indicated, in a sense, just how important it was to explore Boyle's "law" with expanded scope and wider domain—to expose and resolve errors of overgeneralization.[19]

As noted in Chapter 3, similar framing errors of scope haunt the history of many familiar laws: Mendel's laws, Ohm's law, Snell's law of refraction, Newton's laws of motion (under relativistic conditions), and Galileo's "law" of the pendulum. The exceptions were all discovered—and new relationships established—through historical research that widened the scope of the original set of observations. Such laws are, ironically, universal only with boundary conditions and invariant only with provisos and exceptions. They are not really "laws" at all.[20]

Time and time again in history, scientists have discovered important general laws or regularities only to "undiscover" them some time later. Over and over, exceptions have taken them by surprise. The new cases seemed wholly unpredictable, even while upending an apparently stable bedrock of previous experience. Based on these cases, it seems misguided to imagine one could ever safely "future-proof" science. Indeed, the ubiquity of encountering errors in this way might fuel a deeply pessimistic view of science. Is no amount of evidence reliable? Still, the very unexpected nature of the newly encountered exceptional cases partly indicates that they did not matter much, at least not until they were actually encountered. They were not instrumentally important to someone's work. That is, if scientists had been active in the extended domain earlier, the anomalies would likely have appeared in the course of their day-to-day activities, and stymied their work flow. In this sense, errors of overgeneralization tend to emerge primarily at the margins. The originally elucidated regularities typically remain useful *in the limited scope of the original domain tested*. That is, the "cost" of the error seems low. Very little was lost by falsely imagining a broader domain. So long as scientists and others are aware of the circumscribed domain, Boyle's law, Ohm's law, Mendelian genetics, and so on, remain reliable guides. The challenge, of course, is to acknowledge and articulate the boundaries of that limited scope.[21]

Increase resolution and differentiate

In some cases, "deeper evidence" can be characterized as a *more precise* version of the "same" data (not just "more" of the same). Increased resolution can provide additional information from within. That allows finer distinctions or detailed discriminations. One may be able to differentiate formerly homogeneous categories: by *sorting* contrasting cases, by *filtering* independent variables, or by *teasing apart* discrete phenomena. In conventional views, the "growth" of knowledge often implies an *expanded* scope of observations, an *enlarged* body of evidence, and a *wider* field of view. "More" evidence is construed in terms of number of samples, size, magnitude, or weight. To this, we must add the role of *differentiation* (for example, as seen in differentiating the domains of clashing theories, discussed in Chapter 5). Such forms of evidence may be especially valuable in "resolving" uncertainties.

Here, we may include at least three types of resolutions. First, high-definition (or fine grain) *images* can clarify the subject with sharpness of details, articulating structures that were once vague or uncertain. For example, the increased resolution afforded by a microscope allowed Marcello Malpighi to observe capillaries in the lungs. His discovery was quite contrary to any theoretical expectation, but also helped identify as an error William Harvey's claim that no such "anastamoses" between arteries and veins existed (or could exist!). In a similar way, Walter Baade used the powerful telescope at the Mount Wilson Observatory to observe that the set of Cepheid variables consisted of two distinct sets of stars, which followed different rules relating to luminosity and distance. That led to recalibrating stellar distances, recalculating Hubble's constant, and doubling the age of the universe (Chapter 1).[22]

Second, *measurements* on a finer scale may reveal a significant difference in values that was not previously detectable. For example, the emergence of the K-Ar radioactive dating technique allowed for unprecedented precision in geological dating. Not the least of the outcomes was in dating the rocks that documented the reversal of the Earth's magnetic field and in establishing a high-resolution chronology of such reversals. That timescale was critical in interpreting the evidence for sea-floor spreading and indicating—much to the dismay of many geologists—that the wholesale rejection of Wegener's idea of continental drift had been in error. The precision of the magnetic reversals was essential to establishing the symmetry of spreading implied in the now-famous Eltanin-19 profile and, at the same time, to calibrating the rate of sea-floor deposition in core sample taken at the mid-Atlantic ridge. The ocean floor was moving laterally, while also building vertically. The crust was mobile. Continental "drift" was real. The fixist view—dogma among many for decades—was no longer tenable. One of the most significant findings of 20th-century geology had relied on the precision of those K-Ar measurements.[23]

Increased resolution may be qualitative as well as quantitative. That is (as a third category), a set of observations, once presumed to reflect a uniform phenomenon, may be recognized as two distinct phenomena. In this way, one might find and disentangle, for example, an error in overgeneralization. Or one may tease out a hidden factor accounting for discordant results. One may isolate contaminants or artifacts. Or make manifest and effectively cancel or partition off a confounding variable. One may show how "conflicting" interpretations of a body of evidence can be reconciled by articulating the scope of two independent domains.

A good example of differentiating phenomena was in deciphering the anomalous atomic weight of chlorine. That involved understanding that not all chlorine atoms were the same, even though they were still the same element. When the relative weights of elements began to be determined in the early 1800s, it soon became clear that they tended toward whole-number ratios. William Prout conjectured that all elements were multiples of a common unit, the hydrogen atom, the lightest element. It was an attractive hypothesis, and the data generally aligned with it, given a generous allowance for measurement uncertainty. Many embraced Prout's hypothesis. In reporting weights, some even (erroneously) rounded problematic experimental values to the nearest whole number. As the precision of weight determinations improved, and deviations remained, skepticism of Prout's hypothesis grew. Critical to the debate was the atomic weight of chlorine, identified as an element in 1810. Its weight was decidedly problematic. While the other elements were at least close to a whole number, Jacob Berzelius showed in 1826 that chlorine weighed in at roughly 35½. How could all the other elements exhibit a shared pattern and chlorine not? The troublesome incongruence persisted for decades.

In the early 20th century, the study of radioactive decay seemed to indicate that some atoms of the "same" element could have different weights: the origin of the concept of *isotopes*. But did such variants occur with all elements? Detecting distinct isotopes required being able to separate them and to document their different weights. But this could not be done chemically. In 1919, Francis Aston devised an effective method. He ionized the element, leaving the atoms with a positive electric charge. He could then accelerate the charged particles in an electric field into a beam. By subjecting the beam to a magnetic field, individual particles would be deflected. The angle of movement would depend on their weight. Lighter atoms would be deflected more than heavier ones. There would be a spectrum, based on mass (assuming atoms of the element did indeed have different weights). Aston calls it a mass spectrograph, and it would earn him the 1922 Nobel Prize in Chemistry.

When Aston analyzed chlorine atoms, he was able to identify two prominent isotopes, with separate atomic weights of 35 and 37. That is, Aston's mass spectrograph resolved the "single" element chlorine into two constituents: chlorine-35 and chlorine-37 (in today's terminology). And that solved the long-standing anomaly in chlorine's atomic weight. Chlorine was an element, yes. But individual atoms could differ in atomic weight, depending on the isotope. The new conclusion thereby laid to rest John Dalton's original claim (in introducing atomic theory) that all atoms of a particular element share the same weight. At the same time, Prout's view was largely resurrected from the scrap heap of error, albeit as a modified whole-number rule, not based on hydrogen as the unit. The rule applied just to the individual atoms, but not to elements (which had now been shown to be a heterogeneous category of different isotopes). Aston's analysis of chlorine was a significant discovery in its own right. Yet differentiating the isotopes was also central in resolving several outstanding anomalies and unknown errors.[24]

In general, increased resolution of evidence may lead to differentiating phenomena or concepts that help explain and resolve incongruences. One may surely see the resonance with the role of controlled experiments that help to differentiate different observational or conceptual frames. This is an important alternative to viewing "more" evidence in terms of sheer bulk alone.

Develop new instrument technologies

Another factor in generating or encountering "deeper evidence" is new technology that leads to new instruments or new methods. They often enable researchers to expand into new domains or to increase the resolution of their observations or measurements, as described in the last two sections—again, evidence of a new type, not just more samples of the same. Telescopes, microscopes, spectroscopes, optical devices, galvanometers, DNA or RNA or protein sequencing, radiometric dating, magnetometers, cryogenics, tissue culturing techniques, chromatography, electrophoresis, and so on. Nobel Prizes have often been awarded for developing such techniques, which significantly expanded the horizon of scientific exploration: for example, lasers (1964), monoclonal antibodies (1984), fluorescent proteins (2008), MRIs (2003), transistors (1956), the DNA polymerase chain reaction (1993), and the gene-editing tool CRIPSR (2020), among others. The role of technology in discovering scientific novelties is hardly controversial. Here, however, I wish to highlight its further role in identifying or clarifying errors (again, viewed as just another form of discovery).

Here, I will illustrate with a modest example, underscoring that not all technology is hardware. Today, even many nonscientists know that humans have 23 pairs of chromosomes, for a total of 46. But for decades, the number was set at 48. It was published in textbooks. The error evokes a kind of dumbstruck credulity, succinctly expressed in the title of one historical essay: "*Can't anybody count?*" In the late 1800s and for several decades thereafter, the techniques for fixing cells and observing individual chromosomes were (one might generously contend) crude.

The preparation of slides left incomplete cross sections, confusions, and genuine uncertainties. That led to highly variable chromosome counts—from 22 to 52. In the 1920s, relying on drawing, the community reached a stable consensus at 48. (Between 1930 and 1950, at least eight further studies confirmed that number.) During the 1950s, interest shifted to *individual* chromosomes. Human geneticists then borrowed a technique from plant cytology: cell squashes.

With the new technique, the chromosomes could be spread out, so that they could all be visible in one unambiguous view. After *that*, counting was easy.[25]

Instruments pose an epistemic conundrum. By themselves, instruments can be a significant source of error. Yes, the microscope could allow Malpighi to correct Harvey's error about capillaries. But microscopes could generate misleading images on their own. For decades before cell theory, microscopists thought they could see the organic matter of muscles and nerves coalesce into roughly spherical "globules." What they had observed, however, was an artifact of the optics, not something "real" in the organic tissue. One may recall Weber's gravity bar; the cell fixative that created mesosomes; the diamagnetic sample holder mistaken for the sample itself (see section on "Artifacts" in Chapter 2); and the loose time-signal cable that created an illusion of faster-than-light neutrinos (see "The Phantom of the OPERA" in Chapter 5). In other cases, noted in this section, they are the chief mechanism in remedying error. Instruments can be powerful tools, but only if they are functioning and interpreted properly: an epistemic challenge for researchers.[26]

The role of instruments in detecting or resolving errors also echoes the theme of unpredictability in science (for further discussion, see the section on "Happenstance" to follow). Beyond vague and hopeful anticipations, who can foresee new technologies? Or what unexpected surprises they may reveal? If correcting errors sometimes depends on those innovations, the notion of "future-proofing" science against error again seems quite remote.

Fill heuristic gaps

As noted earlier, investigators often use heuristics to economize on expense or time, or to tame intractable complexity. They take advantage of alternative methods that are cheaper, quicker, or simpler. They occur at all levels: observational (proxy variables, *in vitro* tissue cultures), conceptual (model organisms, computational simulations), and social (trust and credibility judgments). Of course, there is a gap between the heuristic and the fuller, surer method. That is, any heuristic substitute may fail, which, if unnoticed, may lead to error. To expose the errors in such cases, one needs to return to the longer, more costly or more complex, but ultimately more reliable, methods.[27]

For example, recall the case of high-voltage power lines and leukemia ("Proxy variables and heuristic gaps" in Chapter 2). The first study finding a correlation was published in 1979. It generated substantial public alarm. Given the significance of the topic, it prompted many "replications." Follow-up studies deliberately checked for methodological mistakes. Most, however, adopted or adapted the original methods, including the use of geographical "wire codes" as a proxy for actual EMF exposure. The apparent confirmation of these "replications" was thus misleading. They simply repeated the original error. Yet the epidemiological findings were discordant with growing laboratory research and cell biology. So, doubt focused on the proxy variable itself. That is, for many years, no one had actually measured the EMFs. The National Research Council was finally enlisted to address the problem and they did just that. Their three-year study concluded that, alas, the EMF measurements did not match the estimated values based on crude proximity. A source of error had been finally isolated—but only after nearly 20 years of policy controversy and two provocative books by a veteran journalist. Those concerns could now be put to rest, simply by having checked the original heuristic.[28]

Errors based on heuristics can often persist unnoticed for quite a while. Such was the case with enzyme chemists who mindfully developed methods to facilitate their studies, but which at the same time hid the nature of a group of intrinsically disordered proteins, or IDPs. Protein chains normally fold into distinctive "spatial geographies" critical to their function. But proteins can degrade or lose their special shape, making research problematic. So, decades ago, protein chemists adopted a custom of studying enzymes at cold temperatures (at 4°C, rather than the 37°C of human body temperature). That helped stabilize the proteins, presumably yielding more reliable results.

But that assumption, or heuristic, was challenged in the late 1990s as protein chemists began to realize through other (NMR) studies that some proteins—IDPs—may adopt multiple configurations, and that each variant in shape may be functionally important. The cold temperatures had been introducing *artificial* conditions, yielding results that did not reflect *in vivo* conditions. After discovering this "glitch," chemists had to rethink their heuristic. As a result, they developed other methods to keep the proteins from degrading—without altering the "normal" temperature (or pH or ion concentrations). For example, various inhibitors can bind to the degrading enzymes and suppress their unwanted action. Or, one may adjust the time frame of the experiment, limiting the exposure to short periods, before significant protein damage can occur. Or they may focus on temperature itself as a variable, and study the target IDP at several temperatures, to identify the range of IDP functionalities and their susceptibility to various temperatures. As noted by Guttinger and Love, cold temperature may still prove useful to investigators—sometimes. It is only "problematic if we forget that it is only a temporary crutch, and that what has been assumed has to be questioned" in other contexts. Namely, sometimes heuristic gaps have to be filled.[29]

Engage alternative perspectives

On yet other occasions, gathering deeper evidence depends on engaging alternative conceptual perspectives. Contrasting views help highlight specific deficits in the evidence, or expose conceptual blinkers. One can easily cite almost any scientific debate from history, where advocates for contrary theories exposed deficits in their adversary's claims. Alternative perspectives—from various conceptual stances, disciplines, biographical backgrounds, cultures, social classes, genders, and so on—enhance collective awareness. As addressed by Merton, the social dimension of science, with its diversity of practitioners, is integral to highlighting and thus remedying error (Chapter 4).

Alternative perspectives are particularly noteworthy in cases of communal confirmation bias—well illustrated in the case of the cause of ulcers and gastritis. From the mid-1930s to the mid-1980s, the dominant idea was that diet, stress, and/or personality induced stomach hyperacidity. Incongruent evidence that could have revealed the error—that antibiotics can provide effective treatment—was repeatedly neglected for several decades. How, then, was this intellectual inertia disrupted?[30]

The tide began to turn when a large clinical study showing the correlation of a newly identified bacterium with ulcers and gastritis (as documented with endoscopic biopsies) was published in *The Lancet* in 1984. With sufficiently persuasive (and provocative) evidence appearing in a leading medical journal, others took notice, at least. Many similar studies were initiated, which would soon yield even "deeper evidence" confirming the then-radical view. However, the critical factor was not the response to the now landmark *Lancet* publication. Rather, it was how that initial evidence ever developed in an environment permeated by conventional practices based on acid-blocking treatments and related explanations.

Here, one may note the partial contributing role of Nobel Prize winner Barry Marshall.

When Marshall first addressed the topic, he was an intern, still training as a physician. That is, he was not a specialist in gastroenterology—and not enculturated into the conventional views on the causes of ulcers. Hence, when he was approached by pathologist Robin Warren, he could respond openly to work as a clinical partner in investigating the prospective role of a new bacterium. That is, Marshall was not predisposed to dismiss the idea as ridiculous or a waste of time (as some other doctors and grantors did soon thereafter). Indeed, Marshall needed to complete a research project as part of his professional requirements, so the ready availability of a topic may have been partly welcome. Not a lot was at stake initially. Nothing to lose. Marshall took the project seriously, at least, and once the correlation of bacterium and gastritis was documented, a clinical treatment study was a natural next step. In retrospect, Marshall was ideally situated to entertain a "contrarian" hypothesis. His ready entry into an alternative perspective was instrumental in opening a trajectory that exposed the long-standing error.

Conceptual entrenchment (communal confirmation bias) also characterized the study of energy processing in the cell in the 1950s and 1960s: the ox phos debate, featured at various times in earlier chapters. The overall problem was determining how chemical energy from the food we eat is channeled into the molecule used for energy throughout the cell: ATP (adenosine triphosphate). Everyone supposed that, like other energy reactions that they had studied for the past several decades, there would be a set of chemical intermediates, conveying the high-energy bond from one molecule to the next: the chemical hypothesis. So, enzymologists approached the problem by trying to extract the protein components from living cells, separate the essential ones, identify them, and then reconstitute them in a test tube solution. But the chemists encountered unexpected headwinds. Claims to have found the central high-energy intermediates appeared one after another—16 times over 15 years—but all were found to be artifacts or the results marred by confounders.[31]

Peter Mitchell entered the discourse in 1961 with a distinctive background and a wholly different perspective. At the time, cell biochemistry was largely separated into those who studied the membranes (the structure) and those who studied enzymes (the functions). The former were mostly lipids (or fats), while the latter were chiefly proteins in solution. As the saying goes, oil and water do not mix. The chemical methods and the whole mindsets differed. Mitchell focused primarily on membranes, although he and some others were interested in trying to bridge the two styles of thinking. Mitchell became interested in ox phos because the components were embedded in the membrane of the organelle, the mitochondrion. For the protein chemists, the membrane was a bit of a nuisance, stymieing their efforts to recreate the reactions *in vitro*. Mitchell, however, viewed the membrane as essential. He envisioned the enzymatic reactions embedded in the membrane and creating an energy potential across that membrane. Namely, protons would accumulate on one side of the membrane, creating an electrical and chemical imbalance: for him, there was no high-energy chemical bond! That was dubbed the chemiosmotic hypothesis. Mitchell's alternative almost completely reversed foreground and background, generating a contentious theoretical disagreement, or ambiguity.

An enterprising junior researcher, André Jagendorf, working with chloroplasts (not the higher profile mitochondria), encountered clues that Mitchell was perhaps on to something. In 1964, he generated an artificial membrane potential in chloroplasts, and indeed it did create ATP on its own (with no input of light energy!). That result was a surprise to many. The experiment provided important evidence for Mitchell's view. The membrane potential was causally relevant. But was it central, or peripheral and indirect? More research was needed.

Mitchell, working in his own privately funded lab, developed evidence for his view over the next several decades. Other chemists eventually came to acknowledge his theory as correct, but at different times and for different reasons. Mitchell received a Nobel Prize in 1978—17 years after introducing his "chemiosmotic" theory. His unconventional approach, which sparked intense criticism at first, was integral to finding the error in the chemical hypothesis of ox phos.[32]

The cases of Barry Marshall and Peter Mitchell (and André Jagendorf) help exemplify the recurring role of "outsiders" in introducing alternative perspectives and highlighting errors in widely accepted conceptualizations. Alfred Wegener was a meteorologist and amateur explorer, not a geologist, paleontologist, or geophysicist—although his claims about evidence for continental drift ventured into those fields. Such outsiders may be variously labeled as heretics, renegades, mavericks, rebels, iconoclasts, or revolutionaries. They may be hailed as "creative geniuses," possessing exceptional insights and abilities to unravel flaws in standard conceptions. They may be lauded for exhibiting heroic courage in challenging commonly endorsed ideas. However, as the cases above may indicate, often their views are simply natural extensions of their particular backgrounds. They were not conceptual prisoners who found ways to escape from an oppressive consensus. Rather, they simply followed their own unique conceptual trajectory. If one shifts one's primary view of science from the individual to the community level, one may appreciate more fully the role of conceptual diversity, mediated through the social practices of science. Alternative perspectives arise naturally because of the heterogeneous biographical and professional roots of scientists. Some succeed. Others do not. And sometimes, an "outlier" view can facilitate the noticing of errors and the tracing of their source. Other times, they may be the very source of blind spots and errors themselves (see Chapter 7).[33]

Complementary perspectives are also important, of course, in cases of highlighting and remedying errors based on cultural bias. I have already noted how our disposition to be cognitively economical and conservative—Kahneman's WYSIATI—can foster blind spots. Essentially, there can be sampling error with respect to all the possibly informative interpretive perspectives. The social dimension of science can thus be important in widening the base of "what you see" by introducing alternatives into the collective discourse for active consideration.

Here, the history of craniology can be an informative illustration. The efforts of (male) craniologists to naturalize the inferior intellectual status of women through science are now widely known (see "Gender bias" in Chapter 4). The gendered cultural assumptions led to an error cascade of purported measurements for intelligence, each new proposal supposedly correcting the error of the former, but only leading to another error. The members of the British Anthropological Society—all males—were certainly all in agreement, blind to their own shared bias, and perhaps comfortable in their narrow "consensus."[34]

However, just before 1900, two women—Alice Lee and Marie Lewenz—entered the field. In framing her arguments, Lee shifted from comparing group averages (based on sex) to ranking specific individuals. She boldly started with 35 male anatomists who attended a professional meeting in 1898. Then she compared male instructors and female students in a London college. The rankings—with the names of the individuals published publicly for all to consider—did not seem to match the prior impressions of professional standing or the quality of the student. Using the emerging discipline of statistics (from her mentor, Karl Pearson), Lee showed that the uncertainty in measurements was 3%. That was more than enough to account for the observed differences. The flaws in analyzing cranial volume had persisted for several decades. But it had taken someone with a complementary standpoint to appreciate their gendered consequences and to pursue the plain evidence to expose their methodological deficits.

One cannot say that Lee was inherently "more objective" than others. She was simply well positioned to notice specifically the gendered errors of her male colleagues. That is, she was not wholly free from biases herself. Ironically, she advocated racial types and worked in the Eugenics Laboratory from 1895 to her retirement in 1927. Still, her case shows clearly how communally shared biases can be exposed when someone holds a specifically complementary perspective and can produce evidence to leverage the discourse.[35]

The role of alternative perspectives in regulating error has a simple but important implication for characterizing the nature of scientific justification. That is, testing scientific claims against the evidence alone is not sufficient. One also needs interpretations of the same evidence from alternative perspectives. Just as one needs a diversity of observational samples, one needs a diversity of interpretive perspectives. One needs to "control" for the various forms of conceptual bias (Chapter 3) through parallel but complementary assessments of the data. The perspective of complementary perspectives will be explored further in Chapter 7.

Happenstance

The critical evidence that helps identify and remedy errors sometimes arises through happenstance. Namely, the "deeper evidence" that resolves an incongruity does not always develop through investigation focused on the error itself. Instead, the relevant information emerges unexpectedly from a wholly separate, independent line of research. One might thereby acknowledge (sadly?) the limits to the mindful pursuit of all the "methods" elucidated in the foregoing section: replication efforts, increased sampling, added controls, more precise measurements, peer review, alternative perspectives, and so on.

The theme of contingency is certainly familiar among (and widely commented on by) historians of science. It goes by many labels. Call it accident, chance, serendipity, coincidence, fortune, circumstance, or just dumb luck. The terms *contingency* and *coincidence* help to emphasize the role of the many simultaneous "if"s that govern the unforeseen convergences at particular historical moments. The term *circumstance* helps to underscore the frequent role of a significant shift in context. Here, I enlist the term *happenstance* to focus on both the unexpectedness (or sheer unpredictability) of unfolding events (the "chance" or "accidental" encounters) and, equally, the sense that new circumstances have been introduced.[36]

To some philosophers, however, who prefer characterizing strictly rule-bound behavior or articulating methodologies in general, abstract terms, a role for happenstance (or "chance") may seem anathema to their taste. Hence, this separate section, which relies more heavily on enumerating historical examples. Not much can be said to tame contingency (as opposed to quantitative randomness). At best, one can be prepared to "notice" oddities, exceptions, or rare events, and to capitalize on the opportunities they provide. Certainly, many philosophers have developed natural selection models of science, bringing together blind variation and selective retention. So we may learn about strategies for resolving error by dissecting the detailed narratives of those cases of happenstance.[37]

Here, let us delve first into the case of Newton's formula for the speed of sound. I have already discussed how the measured speed of sound was problematic to Newton's calculated value, based on the simple mechanics of a compression wave propagating through an elastic medium, such as air or water (sub-section on "Anomalies" in Chapter 5). Over the next century, several accomplished physicists and mathematicians applied themselves to resolving the discrepancy between theory and observed value—unsuccessfully.

Meanwhile, in ostensibly unrelated events, scientists had intensified their study of the behavior of gases. In the mid-1700s, research on evaporative cooling led to the observation that air changes temperature when its pressure changes rapidly. A sudden compression leads to heating, a sudden expansion to cooling. By the late 1700s, equipped with improved thermometers and new concepts of heat, chemists were at work quantifying the effect of heat and temperature on air. For example, in 1802, Joseph Louis Gay Lussac measured the expansion of gas volume that results from an increase in its temperature (with no change in pressure) and noted that this might be relevant to sound waves. The same year, John Dalton reported his investigations on the already "well-known" phenomenon that compressing a gas increases its temperature (here, by contrast, with no change in volume).

In the early 1800s, Pierre-Simon Laplace perceived how to connect these unexpected developments with sound waves. In 1816, he articulated the basic principle that had escaped Newton and so many others. Namely, sound waves rapidly compress the air, at least across a small interval. That leads to a localized increase in temperature. (Sound waves travel so fast that the local heat from the pressure wave has no chance to expand the air—Dalton's case, rather than Gay Lussac's.) The rise in the temperature of the air, in turn, temporarily increases its elasticity, one of the variables in Newton's original equation. Sound waves change the very medium that carries them along. In so doing, the short-term change in elasticity provides an enhanced "rebound" in the air, and propels the sound wave more quickly. The opposite temperature effect occurs with the low-pressure part of the sound wave, ultimately amplifying the effect on the speed of the wave even further. Laplace recalculated the speed of sound, introducing a factor for the newly found (and perhaps counterintuitive) adiabatic effects. Newton's erroneous formula for the speed of sound was on its way to recovery.

Of course, Newton could hardly have made the correction himself. The relevant phenomenon was not yet known. Here, the remedy of the error waited until, by happenstance, physicists had expanded knowledge of physical phenomena in a neighboring field (a case of wider domain). That, in turn, was facilitated by advances in thermometry (another case of the role of new technology).[38]

Much scholarship in the history and philosophy of science focuses on major episodes of conceptual change. Many of those cases involve articulating foundational errors and dramatically altering trajectories of research. But some significant events can hinge on modest things, too. That includes the happenstance of "noticing" things that are not part of the prescribed investigation. That may be an important skill on its own, as illustrated in the next two cases.

The first case involves the unexpected and shocking discovery of cannibalism by primates. In 1971, David Bygott was a primatology graduate student. He was doing field studies at Gombe National Park in Tanzania, where Jane Goodall was leading a group researching the behavior of wild chimpanzees. One day, David was on a regular observation excursion, following a group that had formed a temporary association. For over three hours, they had been feeding and resting in the forest: a typical, uneventful day, perhaps. Then suddenly five or six adult males descended and attacked a pair of unfamiliar females. "The fighting mass of chimpanzees disappeared into dense thickets, with continuous loud screaming." David found them again two minutes later. The females had gone. The dominant male, named Humphrey, was in a tree holding a struggling 1½-year-old infant from its legs. As David watched, Humphrey intermittently beat the infant's head against a branch. A few minutes later, the chimp began to eat flesh from the infant's thigh. The infant's calls ceased. Indeed, all the chimps' vocalizations had ceased. "Then unfolded some hours of curious and ambivalent behavior, so unlike that of chimps who have killed prey of a different species." Humphrey "peered, sniffed, and poked at the carcass, groomed it, and often shook it by a leg or tapped its chest with his knuckles." Another male tore off one of the infant's feet and began to eat from it. After Humphrey finally abandoned the body, several other males in the group also ate some of the meat. After six hours, the group moved on. David had witnessed chimps cannibalizing their own.

The implications of the observation were, of course, profound. At the time, no one had reported such behavior among chimpanzees. Nor was such behavior known among any nonhuman primates. Indeed, it is exceedingly rare among mammals more generally. Goodall and others had observed chimps killing other animals (bushbucks, small monkeys) for meat, usually through concerted effort. But not their own. Later, it turned out a similar incident had been observed the same year in Uganda. The assumptions about behavior among our closest primate relatives, and the implicit errors they embodied, were duly revised. And the human observers were left to reflect on their disturbing consequences. All based on a happenstance observation.[39]

Happenstance observations are hardly rare. For example, on another occasion, they led to a striking reconceptualization of how some whales consume their prey. In the late 1980s, Alex Werth was a graduate student in cetology. That involved collecting raw information about whale diets and feeding habits through routine documentation of what they eat. Alex examined the stomach contents of lots of dead whales (a task that might be viewed as a chore, but which Alex regarded as a privilege). In some of the toothed whales—the group that includes dolphins, pilot whales, sperm whales, and others—he sometimes encountered sand dollars or stones. That's stuff off the ocean floor. "How did those get in the whales' stomachs?," he wondered. He also found many prey items—fish, squid, shrimp—still whole. Not bitten or chewed at all. That seemed in apropos for *toothed* whales, at least. Noticing those oddities set him thinking. How are these whales (the odontocetes) feeding and capturing their prey? What is the functional role of the teeth?

Pursuing that happenstance awareness opened the way to other questions. What is the nature of the teeth—ironically, not all that plentiful or sharp in a group named "toothed" whales? Maybe the teeth were not important to eating, despite being a diagnostic trait of this group? If one looks at the evolutionary history of the whales' teeth—by comparing lots of fossils (as Alex proceeded to do)—one finds reduction in number and size: more anomalies? In narwhals, functional teeth are missing entirely. Their one significant tooth has evolved into a long tusk. Fewer teeth also seemed to be correlated with changes in the shape of the head: rounder and more blunt in the front. Would that be related to the whales chasing prey and swallowing them whole? But, again, the stones and sand dollars? Alex speculated that active suction might be at work. A few conversations with workers at marine parks (who observed feeding behavior informally on a regular basis) confirmed that suction feeding was familiar to them, despite not being part of the scientific literature.

Once a speculative alternative explanation was concretely available, it could increase sensitivity (or "bias") to perceiving or "noticing" other facts that might align with it, as well as guide the search for other possibly relevant evidence. Alex reflected on the fluid dynamics involved in creating lower pressure inside the mouth. The retraction and depression of the huge tongue into the throat seemed possible. Further anatomical information was consistent with that interpretation. Throat pleats (gular grooves) now made more sense: they allowed for expansion in mouth volume. Ultimately, Alex's dissertation research focused on filming captive pilot whales and measuring the speed of ingestion and relevant distances in drawing prey in. He had video footage of whales expelling large amounts of water afterwards. Mouth shape seemed to matter, too. Round mouths were more effective. That could account for the evolutionary change in head shape: an adaptation to suction feeding. Later, when Alex presented all his evidence and conclusions—essentially identifying the errors in earlier assumptions about feeding behavior—many peers were skeptical. Once again, an error correction was partly rejected. But as still deeper evidence accumulated, it tended to confirm Alex's interpretation, and it is now accepted as uncontroversial. But gathering all that evidence stemmed from initially happenstance observations.[40]

As a final case of happenstance, we may delve further into the case of the bacterial cause of ulcers. Warren and Marshall's now landmark paper in *The Lancet*, we have noted, brought attention and empirical results to the issue and triggered research that eventually resolved the ensuing debate, about a decade later. But what led them initially to collect that critical evidence, and how did it come to be published? Happenstance had an important role on more than one occasion.

First, Robin Warren was not expecting to find any new bacterium at the outset, nor was he predisposed to think of gastritis in terms of infection. As a pathologist working in a hospital, he encountered the new bacterium in the course of his routine practice. A chance cluster of gastritis patients all showed the presence of an unidentifiable bacterium. That was anomalous, as the stomach was considered too acidic to host bacteria. Yet it was a small, haphazard sample: easily a fluke. So, Warren conducted further, more systematic sampling. He used an unconventional silver stain that helped to make this particular bacterium more prominent visually under the microscope. "Once I started looking for them, they were obvious," Warren later recalled. But the bacterium was only correlated with the gastritis: significant for treatment perhaps, but not necessarily anything causal? Although colleagues seemed indifferent, Warren was not necessarily swimming against some consensus medical view. The contingencies of ordinary professional practice led him along.

Second, Warren was fortunate in finding an appropriate collaborator. He was ready to work more closely with a clinician, but had no established partner. Barry Marshall, with his training requirement for a research project, seemed to become available at an opportune time. In addition, as noted earlier, Marshall (by chance) was not an experienced gastroenterologist and brought no strong prejudices to the investigation (that may have otherwise subtly biased the interpretation of results).

Third, in contrast to earlier decades, the technology of endoscopes had improved. The flexible tubes, with fiber-optic viewers, made it much easier to take precise tissue samples from the stomach lining and thus to conceive of sampling a significant number of patients without imposing undue hardship. Happenstance favored the timely availability of technology that enabled a clinical study at an appropriate scale.

Fourth, the hospital lab fortuitously stumbled on the unusually long incubation period needed for culturing *Helicobacter*. Collecting samples from patients was one thing. Culturing them so that they could be studied more fully was another. For a while, John Pearman, in the hospital microbiology lab, abandoned the cultures after 48 hours if they did not show growth, per standard protocol. Thirty-four consecutive efforts failed. Then the hospital was inundated with patients infected with an antibiotic-resistant *Staphylococcus*. The lab was overwhelmed and, by happenstance, one of the gastritis cultures was thus neglected over a long holiday weekend. When Pearman finally examined the culture five days later, a thriving population of *Helicobacter* had developed. He realized the misstep of disposing of the cultures prematurely. Warren and Marshall now had an important tool for investigating the bacterium further. That would be especially important when Marshall decided to experiment on himself by swallowing an active culture of *Helicobacter*.

Fifth, the final step in presenting the evidence—publication—also involved a bit of happenstance. When Warren and Marshall eventually completed their clinical study, it showed the bacterium was clearly associated with gastritis (77%) and duodenal ulcers (100%). They submitted their results to *The Lancet*, a prestigious journal where their work might get noticed more widely. At first, however, the editors could not find favorable reviewers. None seemed to believe the results. Then, Marshall contacted Martin Skirrow, a campylobacter specialist in Britain's Public Health Laboratory Service, whom—by "chance"—he had met at a conference. Skirrow, intrigued by Marshall's provocative claims, had promptly checked them at his own hospital. Skirrow now submitted a letter to the editors with his own results, confirming Warren and Marshall's, whose paper was then also published.

Many historical commentators on this episode have focused on the debate that followed. But the critical steps were all prior to the publication of the *Lancet* paper. In that context, one may note the succession of contingent events that led there: (1) the original cluster of patients alerting Warren; (2) the timing and nature of enlisting Marshall as a collaborator; (3) the history of endoscope technology; (4) a "mistake" that led to appropriate

Helicobacter culturing methods; and (5) a brief encounter that established a personal connection that enabled publication. Warren and Marshall, who eventually received the Nobel Prize, put in the requisite work surely, but (like so much science) it was buoyed by a good deal of chance along the way.[41]

These and other cases of happenstance from history are a gentle reminder of the limits of systematic efforts at error correction. One may well review experiments when observations seem contradictory, or dissect theoretical assumptions after anomalies appear, or design experiments to try to contextualize clashing explanations. But, ultimately, one cannot be assured of immediate solutions. Sometimes, science is on hold until new instruments are invented, subsidiary concepts are discovered, or someone with a unique background brings together two previously unrelated fields. History is full of contingencies.

To the degree that history writ large is uncertain, we cannot predict future discoveries. Nor can we predict the nature of every error, nor even recognize each one as they occur. That is part of the puzzling diachronic nature of error (Chapter 1). The unwelcome lesson may be that we cannot avoid error—not if we want to continue to learn and expand our scientific knowledge.

Widespread lore parrots the mantra that errors will be corrected "eventually" or "with time." The episodes of happenstance help indicate that time, by itself, is not the relevant epistemic agent. Something concrete yields deeper evidence or interpretive work. Happenstance in its various forms may "save the day" on occasions, yet it is not guaranteed.

For some, the whims of history may lead to despair. We can never know if the discovery of an unknown error lurks around the corner. So, how can we trust our current scientific knowledge? Some philosophers retreat into pessimistic cynicism, declaring that all scientific knowledge is, sooner or later, doomed to failure. Namely, nothing can really be trusted. Yet until such time as an error is encountered, the science usually seems to serve us well. Science is surely susceptible to happenstance, but when we find and resolve the errors, as much as our knowledge changes, it is also deeper knowledge (see "Negative knowledge" in Chapter 8). Articulating errors clarifies what was misconceived, while simultaneously affirming whatever prior evidence continues to justify. Some ideas are "superceded" and retired. Others persist as context-dependent models. Of course, one must be equally wary of the other extreme: imagining that some science somehow becomes "future-proof." Happenstance wreaks havoc with such blind hopes. Instead of such generalities, we should be more concerned about degrees and contexts of reliability, characterizing how conclusions have been probed for possible sources of error and thus when they are justified—and what specific uncertainties remain.[42]

The recurrence of happenstance may also lead one to believe that scientists are impotent bystanders to the chaos of history, and have no active role when contingencies arise. Not so. How an investigator responds matters. Happenstance often appears at the margin, in the guise of oddities, things out of context, or stubborn glitches. Scientists may regard them as annoyances to simply ignore or dismiss, or as puzzles to solve. A healthy curiosity about unplanned interruptions or discrepancies sometimes proves profitable. It can be as modest as transforming a "that's weird" reaction into a "that's interesting" question. Robin Warren, Alex Werth, and David Bygott, among many others, certainly took up the challenge. In this way, science can capture and capitalize on contingency, when it makes an appearance. This is certainly a strategy that has been researched and employed by innovation design firms, such as IDEO. A well-structured scientific enterprise may take advantage of "chance" events or unexpected results.[43]

Finally, episodes of happenstance tend to highlight context. Circumstances seem to situate and guide individual scientists. That seems to apply as much to the original errors as to their remedy. Namely, certain constellations of conditions or observations seem conducive to insights. But they may also foster blind spots. That fascinating entanglement is explored further in the next chapter.

Resolving error

This chapter has surveyed a variety of ways in which scientists develop deeper evidence—not just "more" evidence, but evidence that is relevant to resolving the uncertainty of incongruences at observational, conceptual, or social levels (Figure 6.2).

Again, my list is hardly complete. One could easily add such methods as: recalibrate the instruments; rebalance the samples to accommodate biases; (re)check or investigate theoretical assumptions; apply blinded observational methods, data analysis, or peer review; compare and reconfigure alternative theories; expand theory; revise theory; test correlated variables for implied causal relationships using appropriate interventions or controls; observe conditions where alternative theories differ in their predictions;

replicate with substitution
increase sample size
increase sample diversity
expand scope or widen the domain of observations
increase resolution and differentiate similar phenomena
develop new instrument technologies
fill known heuristic gaps
enlist alternative or complementary perspectives
capitalize on contingencies and happenstance

=================================

recalibrate the instruments
rebalance the samples to accommodate biases
(re)check or investigate theoretical assumptions
apply blinded observational methods, data analysis, or peer review
compare and reconfigure alternative theories expand theory
test correlated variables for implied causal relationships
observe conditions where alternative theories differ in their predictions
convene stakeholders to facilitate discussion
pre-register experimental and analytical protocols
check for undeclared conflicts of interest and others

Figure 6.2 A partial inventory of ways to search for deeper evidence toward resolving errors.

convene stakeholders to facilitate discussion; pre-register experimental and analytical protocols; check for undeclared conflicts of interest; and so on. Indeed, these are all the types of evidence that might figure in any scientific investigation. Here, the scientific problem is focused on interpreting an incongruence (Chapter 5).

My aim here has not been to be exhaustive—either of error types or of their remedies. Rather, I hope to encourage a way of thinking about error and to provide a conceptual structure to organize that thinking. Namely, rather than bemoan our epistemic lapses (or, worse, dismiss the reliability of science), we might adopt a posture that focuses concretely on how to resolve the errors that do emerge. Indeed, we might conceptualize an error in science less as a flaw or failure that needs to be "fixed," and more as a problem that needs to be solved—an opportunity for further learning.

Lists of errors that are more systematic and more complete may certainly be useful in guiding scientific practice. But they will tend to be more specific, and particular to the relevant field of study. For more, see the discussion of error repertoires and checkpoints in Chapter 8. Again, my purpose here has been to provide a structure for organizing discourse about error generally.

Conventional images of science—especially in science textbooks and public media—tend to emphasize the notion of testing theories, typically with simple confirm-or-disconfirm outcomes. Some portrayals may also include how theories or models are revised to accommodate problematic observations, and thereby gesture to how scientific knowledge evolves. However, understanding of the prevalence of dealing with error in science may help erode these entrenched stereotypes. "Troubleshooting" pervades the work of science. Namely, addressing possible sources of error, and finding the relevant evidence, constitutes a major part of scientific work. Yet there is very little philosophical work on the topic.

The significance of error (or checking for error) is reflected in daily practice. Many lab groups meet regularly—weekly, perhaps—to review and discuss ongoing work. Historians, sociologists, and philosophers of science seem not to have invested much attention on what happens at these quite commonplace gatherings. However, they typically focus on the discordances, anomalies, and ambiguities that appear in day-to-day work. These meetings might therefore help indicate more vividly just how important encountering errors is in the practical conduct of science. Accordingly, the education of future scientists (as well of nonscientists) might include more about error types and their resolution. Every student of science may learn that "to err is science."

Yet, as the last two chapters have shown, resolving error is also science. This chapter has underscored the many approaches and many possible trajectories where this happens. That might invite us to reflect on several other interesting questions. For example, what is the frequency of errors, and exactly how much time is devoted to tracking down sources of error? What is the relative frequency of different error types? We might consider how to manage the errors (or incongruences) that we do encounter: the focus of Chapter 8. By acknowledging error and how it is resolved, might we support more effective scientific practice, as well as more informed public engagement with science? We might also wonder if, with appropriate safeguards, we might ultimately be able to free science from error—the provocative topic in the next chapter.

7
The Conundrum of Bias

The cost of discovery
is the risk of error.

Insights and blind spots • bias: bane, benefit, or both? • the context dependence of bias • social checks and balances • complementary perspectives

When scientists are asked about the errors of their fellow scientists (as noted in Chapter 1), they typically attribute them to psychological and sociological factors. Indeed, as noted in Chapter 3, several forms of cognitive dispositions may indeed be sources of error, from personal theoretical preferences to the blinkers of gendered, racial, or socioeconomic standpoints, which may then be entrenched via confirmation bias. In practice (and as observed historically), individual scientists seem to function within their own particular histories and spheres of prejudice. These various perspectives—or "biases"—seem to allow conceptual errors to emerge—and then to escape detection. That is, biases seem to generate unwelcome *blind spots.*[1]

However, as I will document in this chapter, biases in perspective may also contribute to discovery. There, they are often hailed as the wellsprings of creative genius. Biases may also—counterintuitively, perhaps—lead to *insights.*

In this chapter, I explore this odd coupling. I describe a wide selection of cases where the *very same* perspective that led to a notable scientific error (on one occasion) *also* led to significant discovery (on another occasion, or in a different context): an awkward juxtaposition that we may call the *conundrum of bias.* I invite the reader to reflect (perhaps with some reluctant wonder) on the puzzling entanglement of insight and blind spot, discovery and error, in these cases.

In some visions of science, in order to reduce error, individuals should learn to transcend their biases, or personal idiosyncrasies. That is, in the

Toward a Philosophy of Error in Science. Douglas Allchin, Oxford University Press. © Douglas Allchin (2026).
DOI: 10.1093/9780197827703.003.0007

same way that we develop norms and methods to circumvent observational or interpretive errors, we should train each researcher to more closely approximate the ideal (embodied in the common stereotype of scientists) of being rigorously neutral and objective. Some may well hope that with enough effort, we could possibly eliminate bias as a source of error. However, in my view, the goal of universal "objectivity" by individual practitioners is misconceived—neither possible, nor even desirable. Ultimately, we may need to rethink the widespread intuition that bias is inevitably only harmful to science. I hope to make sense of the maxim: "the cost of discovery is the risk of error."

Accepting that biases may yield insights as well as blind spots nevertheless poses a profound challenge to the growth of reliable knowledge in science. Can one sort the helpful occasions of bias from the harmful? The conundrum is resolved (as many have already recognized) at the social level of science, as briefly sketched in the last chapter (see especially the section on "Engage alternative perspectives"). Namely, complementary perspectives—or "biases"—function as "controls" in the context of interpreting evidence. One bias becomes apparent through a countervailing perspective. Mutual criticism—peer review in its more generalized meaning—is integral to error-detection in cases where errors arise from individual factors, and helps filter the fruitful from the frivolous. The process is illustrated most vividly in cases where pairs of scientists exhibited complementary blind spots and insights—leading, fortuitously, to reciprocal corrections (each remedying the error of the other).

Bias: bane, benefit, or both?

The ambivalent nature of bias is evident in cases where insights and blind spots arise from the very same perspective, or "bias." It is hard to characterize this abstractly, in general philosophical terms. Thus, in this section, I simply present a handful of ordinary examples from history.

Michael Faraday, Sandemanian

Michael Faraday is widely known as a titan of 19th-century science. What is not so widely known is that Faraday was a member of a small Protestant sect in Britain, the Sandemanians. They espoused the harmony and unity of nature. That worldview motivated and guided Faraday's scientific investigations. He sought especially a unifying relationship between the forces of nature. Notably, that led to Faraday's celebrated discovery of electromagnetic induction. The Sandemanian belief in symmetry in nature further helped guide Faraday in establishing the reciprocal relationship. This later led to electric generators and electric motors. Those were important *insights*, partly afforded by his particular religious perspective.[2]

The very same beliefs led Faraday to investigate other relationships among natural forces, which he believed must be unified in God's world. For example, he tried to find how electricity affected polarized light or produced heat. In 1828, he projected a solar spectrum on a copper plate, expecting to show how light could induce electricity. Those investigations did not yield the same positive results as his electromagnetic research.

Even more impressive were studies done over three decades on the relationship between electricity and gravity. Faraday was convinced that a *gravielectric* effect was waiting to be demonstrated, just like the once unknown electromagnetic effect. He recorded in his diary on March 19, 1849:

> Gravity. Surely this force must be capable of an experimental relationship to Electricity, Magnetism and the other forces, so as to bind it up with them in reciprocal action and equivalent effect. Consider for a moment how to set about touching this matter by facts and trial.

By later that summer, Faraday was conducting experiments. He adapted his earlier work on electromagnetic effects, attaching a galvanometer to a wire wound around a coil, which would now experience gravitational free fall. Faraday was pleased to see the galvanometer needle register some current. However, based on his earlier work in electromagnetism, he realized that a loop in the wire was moving through a magnetic field (the Earth's magnetic field). That could account for the reading. So he twisted the wires in subsequent trials to prevent that possibility. The effect disappeared. No gravielectric effect.

Despite the negative evidence, by September he was at work on yet another test. Again, he used his earlier successful work on electromagnetic induction as a model. The apparatus followed the design of an 1831 experiment where a metal bar oscillated through a wire helix. Again, no success. And again, Faraday retained the same conviction that had promoted all his earlier work.

In presenting his findings to the Royal Society in November 1850, he indicated his "strong feeling" that the effect would yet be discovered. Indeed,

Faraday continued to reflect on the problem. Apparently undeterred, he returned to experiments yet again in 1859, anticipating that a longer free fall distance, made possible in a lead shot tower, would reveal the effect. Throughout all his gravielectric investigations, Faraday applied the same style of thinking that led to his highly celebrated electromagnetic discoveries. But his unwavering convictions of a gravielectric effect, based on a faith in the unity of nature, were in error.

Charles Darwin, Uniformitarian

From Faraday, we may turn to the equally renowned Charles Darwin. Darwin's very first scientific paper, in 1837, was a theory on the formation of coral atolls. Darwin had been greatly influenced by Charles Lyell's principle of uniformitarianism—which viewed the past as a cumulative product of gradual forces still present today. Using that lens, Darwin reasoned that coral reefs formed along the shores of mountainous islands, which then gradually eroded, leaving hollow rings. It was an act of sweeping historical imagination based on observational fragments about coral growth and location. The idea proved correct—and it helped launch Darwin's scientific career.

Darwin continued to apply Lyell's large-scale gradualist thinking in theorizing about other geological formations. In his next effort in 1839, he tackled the "parallel roads of Glen Roy," a well-known series of stony ledges lining a valley in Scotland. Here, Darwin imagined that they were the debris of successively lower shorelines, left by a receding ocean over a large span of geological time. Here, he was wrong. The ledges were glacial moraines, left by a retreating glacier, not an ocean. Darwin, to his credit, acknowledged his "great blunder" when Louis Agassiz's theory of glaciation and ice ages gained prominence. In both these cases—one insightful discovery, one error—Darwin relied on the same particular style of Lyellian reasoning.[3]

Charles Barkla, Nobelist

Next, consider Charles Glover Barkla's Nobel Prize–winning work on X-ray fluorescence in the early 1900s. Each atomic element absorbs and releases X-rays in a distinct pattern. Barkla's extensive documentation of the angles and wavelengths intersected with Henry Moseley's work identifying elements based on their atomic number (the number of protons, or positively charged particles, in the nucleus). That dimension of atomic *structure* was more exact than identifying elements based on their chemical properties, or by their atomic weight—the elements that Mendeleev had used to organize the periodic table. Moseley's work, in turn, helped identify "holes" and mistakes in the sequence of the elements, and provided a deeper explanation for the ordering of the atomic elements.

Barkla's original work focused on two series of emissions from each element, labeled the K series and the L series. But later Barkla reported finding a third series, which he called the J series. He pursued those results for two decades. After years of work, he wrote to a colleague "Yes! the J-phenomenon is very interesting and is so fundamentally new, but it may take a generation to work it out thoroughly. ... I am convinced that it is of the very greatest importance. " Alas, the J series was eventually determined by others to be "echoes" of the original series—artifacts of the Compton effect. But Barkla was biased. He admired J. J. Thomson and held to classical views about the preservation of energy in the scattering of X-rays. He refused to consider quantum principles as inherent in what he observed, as described by Compton. He thus held steadfast to his ideas and original experimental methods, which of course had served him well at first. Ultimately, he was ill-equipped to fully investigate the phenomena he labeled the J series, or to interpret it effectively. Barkla's perspective led to insights at first, but then, later, to a major blind spot and corresponding error.[4]

•

These three examples show how the very same principle can guide reasoning in different cases, fruitful on one occasion, not so in another. At the same time, one may well contend that in these three examples, the idiosyncratic pattern of reasoning that led to error was individual, and that others did not adopt the same conclusions, hence the error was limited (and perhaps inconsequential?). However, in the next three cases, the erroneous conclusions were far more influential.

Urbain Le Verrier and anomalous planetary orbits

French astronomer Urbain Le Verrier is generally celebrated for his successful prediction of Neptune, based on the anomalies in the orbit of Uranus. But the same considerations led him to predict another planet, named Vulcan, which spurred earnest search for nearly two decades. But the presumed justification for that second planet's existence was an error.

Both occasions hinged on applying Newtonian mechanics to planetary orbits. The first was based on an anomaly in the orbit of Uranus. After its discovery in 1781, subsequent observations charted Uranus' orbit around the Sun. But its pathway did not accord with the calculations based on Newton's theoretical equations. By 1821, there was no escape in hoping that more measurements might resolve the dilemma: a conspicuous incongruence. Well, perhaps Newton's gravitational "constant" was not so constant after all and was subject to some further subtle variation—say, based on distance? A denser ether, perhaps? Or, as Alexis Bourvard suggested, there might be some other unknown massive object exerting additional gravity on Uranus, perturbing its orbit. For example, did Uranus have an undiscovered moon? Or was there a nearby comet? Uncertainty. Still, no one at the time seemed to consider it prudent to regard Newton's theory itself as the source of the error.

About two decades later, Le Verrier assumed that another massive body might be lurking nearby. He completed the complex calculations, making a precise prediction of where it should be, according to Newton's equations. After failing to inspire his colleagues in Paris, Le Verrier wrote to a German astronomer who, within hours of receiving the letter, found the new planet just about where it "should" have been: the discovery of Neptune in 1846. The anomaly was resolved and Newton's theory triumphantly vindicated.

However, another orbital mismatch soon arose. Its resolution was not so accommodating for Newton's theory. At virtually the same time as the discovery of Neptune, a rare transit of Mercury across the Sun had allowed precise measurements of its movement. According to the latest calculations—now taking the gravitational pull of nearby Venus into account—Mercury was 17 seconds behind schedule. In context of the precision available, that was problematic. Later, the problem was refined. Mercury's orbit has a "wobble" (its perihelion precesses). Le Verrier now calculated the relative gravitational influence of other planets on Mercury: Venus could account for half of the wobble, Jupiter another quarter, and Earth about 15%. That left 7% of the total discrepancy unexplained. A new planetary orbit anomaly. At least a solution seemed conveniently at hand: based on experience with Uranus and Neptune, there must be another planet in the vicinity of Mercury. In 1860, Le Verrier named it Vulcan.

Other astronomers seemed to agree. And the search began. Yet many promising efforts failed. Some prospects turned out to be sunspots; others, known stars. Yet other reports could not be confirmed. So, despite persistent efforts, it could not be found. Vulcan (the solution to Mercury's anomaly) became the very anomaly itself.

After exceptional opportunities in 1878 yielded no concrete observations, confidence in ever finding Vulcan waned, although some astronomers continued to look for it over the following decades. (What alternative was there?) Even without a suitable explanation for Mercury's wobbling orbit, Vulcan gradually dwindled to the status of an error. The analogy with Neptune had been too eagerly accepted. An instance of hasty overgeneralization, perhaps? As for Mercury, the uncertainty persisted into the next century. In 1915, Einstein's general theory of relativity explained Mercury's puzzling orbit. This time, Newton's theory of gravity and mechanics did indeed prove wrong, at least in the limit. For some specifics and particular domains (including Mercury's orbit), they do not apply.

Le Verrier's method for predicting planets based on anomalous orbits, which had proved so remarkable in the case of discovering Neptune, failed in the case of Vulcan. Insight and blind spot.[5]

Spectroscopy and new elements

Next, consider the "non-discovery" of two elements. In the early 1800s, chemists learned that elements emit distinctive colors. With the development of the spectroscope by Robert Bunsen and Gustav Kirchoff in 1860, those colors could be associated with distinct wavelengths. Spectroscopy thus became a powerful perspective for chemical analysis: for identifying elements in samples of unknown mixtures based on unique spectral lines. It also quickly became the basis for discovering new elements: cesium, rubidium, and thallium (each named, notably, for their distinctive spectral color) and, later, scandium. Helium was identified in 1868 on the basis of unfamiliar spectra lines from the Sun ("helios"), and with three decades, it was found on Earth.

In 1864, William Huggins found that light from the Cat's-Eye nebula (seen in the constellation Draco) was found to exhibit yet-unknown spectra lines. Following the pattern of spectroscopic reasoning, he attributed that to a new element. It was given the name (appropriately enough) *nebulium.*

Five years later, during a solar eclipse, Charles Young and William Harkness each observed other anomalous spectra lines in light from the Sun's corona: evidence of yet another element. Its name? Yes: *coronium.*

Yet those puzzled because they do not recognize these elements (and cannot find them on the periodic table) are correct. They do not exist. In 1927, Ira Bowen showed that the particular lines of "nebulium" were emitted by highly ionized forms of oxygen and nitrogen (as yet unrecognized). In the 1930s—after decades of confirming and refining the measurements of the "coronium" spectra lines—the lines were traced to highly ionized forms of iron and nickel—at temperatures that were simply not precedented on Earth. So the elements quietly receded from view. Spectroscopy yielded the discovery of new elements on some occasions, and on others, using precisely the same reasoning, errors.[6]

Isaac Newton and chemical affinities

Finally, consider Isaac Newton's notion of chemical affinities, wherein he adapted his perspective on mechanics and gravity to atomic matter: conceived similarly as discrete particles attracted by linear forces on a minute scale. Newton's monumental credibility, in particular, gave legitimacy to the interpretive perspective, which guided work for nearly a century, before it was abandoned as unworkable and erroneously framed.

Newton's law of gravitation was enormously powerful and effective, as illustrated in the prediction of Neptune. Newton conceived matter analogously, as composed of corpuscles and imagined them in the same context. In an addendum to his 1717 *Opticks*, Newton speculated on how a gravity-like force might characterize the interactions of atomic particles. Each substance, he proposed, had a characteristic attractive force—its *affinity*—which determined how it reacted with other substances. He illustrated this with a series of replacement reactions, showing the relative power of each substance to displace others when combining with nitric acid in solution. That is, Newton envisioned simple quantitative laws governing chemical reactions—an atomic physics, of sorts.

Others followed Newton's suggestive (but perhaps somewhat vague) lead, treating his list of replacement reactions as an exemplar, or paradigm (in a narrow Kuhnian sense). In 1718, Newton's table was expanded to include other substances, each adding another column to an emerging affinity table, now with 16 columns. Then, yet more columns were added. More substances also meant adding new rows to each column. It became a useful summary table—displayed on the walls of many laboratories—for what chemical reactions to expect when combining particular substances. Over the century, the affinity tables became more complex. By 1783, Torbern Bergman's grand version encompassed 59 columns.

However, the order of replacement was not always the same from one column to another. The sequence also varied based on whether the substances were mixed in water (wet) or were heated (dry). Results varied with temperature. The concentration, or saturation, of the substances mattered in some cases. Later, the role of volatility and solubility were noted. Additional substances (mixtures of four or more) exhibited further complications. In short, over time, understanding of context riddled the table with anomalies. All these informed chemical practice, but the prospects of fulfilling Newton's vision by establishing simple laws of affinity became increasingly problematic.

Nevertheless, Newton's vision, especially of *quantifying* affinities, persisted throughout. Mid-century, Comte de Buffon helped articulate the basic view: "The laws of affinity ... are the same with that general law by which the celestial bodies act upon one another. The exertions are mutual, and proportional to their masses and distances." In 1758, the Academy of Rouen offered a prize for the best essay on affinities, expecting entrants to elucidate "the physico-mechanical system" on which they were based. The prize was divided between George Louis Le Sage and Jean Philippe de Limbourg. Le Sage drew the analogy with gravity, but to maintain consistency among his observations, he had to suggest that attractions depended on density, not absolute masses. Limbourg, for his part, appealed to three factors in attraction (not one), but even so had to acknowledge many contradictions and exceptions. Despite the difficulties, both invoked Newton's authority and used Newtonian language, without challenging his basic framework. In 1772, French chemist Guyton de Morveau echoed the basic view again, appealing to Newton's inverse square law of gravitation, while endeavoring (unsuccessfully) to quantify affinities in a consistent way. By 1783, Bergman was distinguishing between "contiguous" attraction (affinities) and "remote" attraction (gravity), but still treating them as expressions of a uniform Newtonian system. Despite the accumulated anomalies encountered in chemical practice—all reflecting the importance of context—he did not abandon the hope for a unifying quantification. In Ireland, Richard Kirwan, too, tried to measure affinities. His efforts earned him the prestigious Copley Medal from the Royal Society in 1782. But he also struggled to identify which factors were relevant to "that power by which the invisible particles of different bodies intermix and unite." Was it specific gravity? Concentration? Saturation? As exemplified in these efforts over many decades, Newton and his original idea remained a benchmark for comparison, even though his concepts were suggestive and not based on systematic experimental work. The analogy based on his more famous physical concepts proved quite powerful.

None of the efforts at quantification or explaining affinities succeeded. Eventually, after nearly a century of work, Newton's concept was abandoned. Chemists shifted to quantifying the relative proportions of each substance that would be required to react with each other: a more practical measure for the chemical laboratories of pharmacy and industry. Later, with the emergence of electrochemistry (and, with it, electrolysis), chemists began to conceive chemical combination instead in terms of the attraction between positive and negative charges. While one could still conceive affinities loosely, Newton's concept of unique, quantifiable forces, akin to gravity, was in error.[7]

The context-dependence of bias

The implicit assumption among many is that errors associated with perspectival bias (by virtue of the error itself) exhibit irrationality, or a lapse in "scientific" thinking. The corresponding notion is that anyone who perceives and corrects such an error is inherently more objective, has demonstrated the ability to exercise skepticism productively, and thus (by virtue of their action) reflected more fully the scientific norms. There is little consideration of context.

For example, in this view, one is liable to say that the advocates of N-rays (the notorious case discussed in the section on "Observer bias" in Chapter 2) were *self-deluded.* Robert Wood (who exposed the sad tale) was, by contrast, a champion of objectivity. Likewise, critics, like Irving Langmuir, who chastised the likes of Fred Allison for his eponymous magneto-optico effect, are the rare but essential champions of (true) science. Alas, a more complete account of the historical details yields a quite different conclusion.[8]

From N-rays to the Allison effect to cloud seeding

Recall the case of N-rays in the early 1900s. René Blondot claimed to have detected a new kind of radiation. It was confirmed by the observations of many French colleagues. However, the images on the screen were very faint, and not everyone attested to seeing them. Did the concurrence among many observers bolster their legitimacy, or was the phenomenon an artifact of communal expectations and subtle, but shared perceptual bias? The key factor in answering that question was someone with a complementary perspective, who did not share the French physicists' potential for bias. Enter Robert Wood, visiting from America. Wood could not see the rays, and his posture quickly turned to one of probing the circumstances. He mischievously tampered with the experimental setup afforded by the secrecy of the dark room—adopting questionable professional ethics, but certainly achieving the blinded conditions for testing the observers themselves with suitable controls. The observers did not pass the test. Many historical commentators credit Wood for exhibiting a more rational, skeptical stance. But that judgment seems largely shaped by the outcome ("N-rays were imaginary, so *of course* Blondot and others must have been prejudiced, inferior scientists—and the debunker, a hero"?). It seems more appropriate to say that Wood simply brought to the occasion an alternative viewpoint that, critically, fostered collecting "deeper evidence" than what was already available. It was the evidence from blinded observation—not some abstract measure of "rationality"—that concretely helped isolate a hidden source of error.[9]

One of the great ironies of the N-rays affair is that Blondot's "debunker," Robert Wood, himself came to endorse erroneous claims of a quite similar nature decades later. There, he was not so critical. In 1930, Fred Allison had apparently developed a powerful method of chemical analysis that could detect subtle variations of composition based on an element's atomic weight, even at low concentrations. The research had begun as a somewhat routine investigation into the Faraday effect (the ability of an ionic solution to rotate polarized light in a magnetic field). Chemists were eager to know about any time delay, which might provide clues about the atomic nature of the phenomenon. Allison's strategy involved using two complementary samples whose rotational effect could cancel each other, but would be induced at slightly staggered times. The amount of light transmitted through the apparatus would vary with the samples' respective time delays. Like the N-rays setup, Allison's apparatus also involved a spark generator. It emitted short bursts of light as the primary light signal, which simultaneously triggered the electromagnetic field around the samples. The experimenter could vary the length of the electric wires, and thus the time difference between the two samples, and search for the critical time that minimized the intensity of the light signal. It may sound complicated, but it was simple enough that others, including many graduate students, were able to copy it and study the "Allison effect" for themselves.

After setting up the basic apparatus, Allison investigated a diversity of samples, documenting the time delays for each. He soon concluded that the delay time varied with the atomic weight of the element, reflected in smooth graphs comparing a series of related compounds. In 1931, he encountered results that indicated hydrogen had an unknown second, heavier isotope (depicted in graphs slightly displaced from the previous ones). The following year Harold Urey announced spectroscopic evidence for the same—an apparent validation of Allison's method. Encouraged, Allison reasoned that his method could detect the presence of new elements, marked indirectly by unfamiliar atomic weights. He went searching for element 87, and found evidence of it in ores containing (as one might expect) the related element, cesium. He named it virginium. Soon thereafter, he announced the apparent discovery of element 85, which he named alabamine. The periodic table was duly revised in textbooks and classroom posters. Ironically, on this occasion, Robert Wood (Blondot's critic) summarized Allison's work favorably in the 1934 edition of his popular textbook, *Physical Optics*.

Many chemists replicated Allison's findings, and even announced the discovery of other new elements. However, sometimes they could later not repeat their own initial successes, or they encountered puzzling results. For example, Francis Slack observed the critical minima in arbitrary places on the scale, not just at the values determined earlier by others. Then, like Wood, he began to replace or remove various parts of the apparatus—with no apparent effect. He tried a constant source of illumination—which should have eliminated any punctuated delay—and found no change. Independently, two other labs also encountered many minima, randomly distributed. Both also introduced new methods for assessing the brightness of the light. They split the light beam from the original spark: one went through the samples, while the other served as an unmodified reference, for comparison (if the spark should flicker, say). They also used photographs or photometric technology to measure the difference in light intensity. Both labs documented variation in the intensity of the spark that could occasionally (but unpredictably) decrease the signal intensity and that, without the direct comparison, could easily have been mistaken as caused by the samples. Slack added how, during his own visit to Allison's lab, the assistant's informal chatter had included subtle indications about where he expected the critical minima. Those comments could easily coax an observer into how to interpret the very weak signals. Ultimately, the Allison magneto-optic method of analysis proved an error. Research on it was abandoned. It originated in a diabolical combination of random variability in the spark generator, faint light signals at the physiological limits of human vision, and the psychological power of theoretical expectations and of subtle cues that could unconsciously shape basic sensory perceptions. Under these conditions, the human observer was simply an unreliable instrument, and a source of observational error. In this case, Wood was *not* the "rational skeptic." Rather, he endorsed the phenomenon that later proved illusory.[10]

Years later, Nobel Prize–winning chemist Irving Langmuir would characterize these two episodes as examples of "pathological science." He portrayed them as patently "bad science"—lapses of a skeptical attitude. He warned against the dangers of psychological bias in science: "wishful thinking," "fantastic theories," "subjective" phenomena, "ad hoc" excuses, and the like. Langmuir's verdict of "self-delusion" in such cases has been widely echoed by others. They follow the pattern of disparaging the scientists who erred as inherently less-than-competent individuals. For them, error inevitably results from "unscientific" attitudes, not just from bad luck or the vagaries of individual blind spots.[11]

Ironically, Langmuir's own research would later exhibit the same sort of blind spot. Like those he criticized, he seemed to readily attribute error to others, while claiming error-correcting skills for himself. Yet he had his own Achilles heel, once featured on the cover of *Time* magazine (August 28, 1950): cloud seeding. Here, then, is the story of how Langmuir himself succumbed to the very same error he disparaged in others.

Irving Langmuir certainly earned his formidable reputation in chemistry and physics in the 1920s–40s. His achievements are numerous. He originally gained renown for his work on the chemistry of adsorption on surfaces—the occasion for the 1932 Nobel Prize in Physics. His extensive ongoing work in that field has been recognized with an eponymous unit of measure (in ultra-high vacuum physics) and the naming of the American Chemical Society's premier journal on such topics. In addition, his work led to the understanding of Langmuir waves and Langmuir probes (plasma physics), Langmuir circulation (ocean currents), Langmuir-Taylor detectors (ionizing beams), and many other eponymous phenomena and instruments (24 in total are listed by Wikipedia). One might imagine that Langmuir developed a sense of self-confidence that his insights could help master any problem he encountered, even if venturing into a new field.

After World War II, Langmuir turned his attention from icing on the surface of airplane wings to the related problem of rain formation. Working with a home freezer in a lab, his assistant stumbled upon a role for dry ice. It dramatically turned condensed water vapor—an artificial cloud—into indoor snow. Soon, Langmuir was at a nearby airfield, sending up a plane to inject dry ice pellets into a nearby cloud to make it rain. Langmuir was impressed by the results, although professional meteorologists far less so. For example, the snow all evaporated before reaching the ground. Still, it made news headlines. Thinking again in terms of surface chemistry, Langmuir and his assistant began looking for a chemical with a crystalline structure similar to ice, which could serve as a seed for crystallizing water and thus coax clouds to release their stored moisture. That led to silver iodide as an alternative method for cloud seeding. More tests, more rain.

However, Langmuir had ambitious visions for controlling the weather. The following year, he was working with the U.S. military in an effort to neutralize hurricanes. In October 1947, with the help of a sturdy Air Force B-17 bomber, 180 pounds of dry ice were dumped at the edge of the eye of a hurricane heading out to sea. He hoped the disruption would calm the fierce winds. Instead, the hurricane veered and headed back to shore, stronger than ever. In retrospect, there was no firm evidence that the cloud seeding had changed the course of nature. But that also seemed true for all the earlier "tests." The weather system was too complex and Langmuir had not considered adequate controls to support any definitive conclusions. Meteorologists who did random tests found no clear effect from cloud seeding. Moreover, later understanding of the pressure dynamics of hurricanes indicated that Langmuir's strategy, based on normal rain clouds, should not have worked, even theoretically. It also later turned out that the crystalline structure of silver iodide was not the key factor responsible for seeding ice crystals. So, Langmuir was wrong. He had made some grandiose claims based on little data and had certainly rallied research funding with a good deal of attractive promises. But his claims far outstripped the available justification, both conceptually and experimentally.

Undeterred by the hurricane failure and other inherent uncertainties, Langmuir avidly continued his cloud-seeding research. Given the prospect of controlling the weather (even with remote chances), sources of funding were readily available. A new endeavor, Project Cirrus, funded by the Army, moved to New Mexico, where the arid climate provided some hope, at least, of more controlled tests. For several years, Langmuir led various efforts to induce rain by sending silver iodide-laden smoke up into the atmosphere. In 1949, Langmuir observed one thunderstorm that he claimed he had initiated. Meteorologists later attributed it to a naturally occurring weather front that moved in from the Gulf of Mexico at the same time. Later, he claimed that his actions in New Mexico had changed rain patterns in Ohio several days later. Despite cautions from his colleagues, he published the statistics in *Science* magazine, hoping to claim credit for large-scale influence of the weather. But others did not interpret the correlation as evidence of causation—widely known as a notorious error type. Research on cloud seeding waned over the next decade (although a project on disrupting hurricanes emerged again, from 1963 to 1983, without any documented successes). Cloud seeding may, under appropriate conditions, help clouds "weep." But Langmuir's engineering-type vision virtually promised that large masses of water could be moved through the atmosphere at will, allowing it to rain wherever one wanted, whenever one wanted. However, the results of the very first test—where the seeded precipitation had merely evaporated—may have indicated the limits of the laboratory knowledge in a complex world. Still, even until his death in 1957, Langmuir never wavered in his beliefs that the concrete promise of cloud seeding, based on surface chemistry, was fully justified. Such was Langmuir's blind spot. And perhaps his epistemic hubris, too [12]

Langmuir's story includes a cautionary epilog. In 1953, Langmuir gave a presentation at Princeton University in which he proposed the notion of "pathological science." He did not neatly define the concept, but was clearly inspired by concern for fellow researchers who seemed to succumb unwittingly to motivated reasoning, rationalization, and some of the cognitive errors presented in Chapter 3. "These are cases," Langmuir commented, "where there is no dishonesty involved but where people are tricked into false results by a lack of understanding about what human beings can do to themselves in the way of being led astray by subjective effects, wishful thinking or threshold interactions." "Criticisms," he continued, "are met by ad hoc excuses thought up on the spur of the moment. They always had an answer—always." Langmuir was clearly intrigued by the cognitive lapses: "To me, the thing is extremely interesting, that men, perfectly honest, enthusiastic over their work, can so completely fool themselves."

Sadly, of course, this syndrome of short-sightedness largely described his own work on cloud seeding over the previous half decade. His enthusiasm had always been ahead of the data. He inevitably highlighted favorable results, while peripheralizing apparent anomalies and ambiguous data. When others politely challenged his conclusions, he always found reasons to discount their criticism. For instance, cloud seeding only works on certain clouds, he would say, and under certain conditions: no one should expect it to be uniformly successful. He would appeal to laboratory models and data to justify his claims, even when real-world circumstances were far more complex. The alternative interpretations from professional meteorologists were merely plausible, not necessarily proven, he contended. His own conclusions, based on statistical correlations, could not be ruled out. And so on.

In his presentation on pathological science, Langmuir expressed profound regret at the blindness of his peers: "If things were doubtful at all, why they would discard them or not discard them, depending on whether or not they fit the theory. They didn't know that, but that's the way it worked out." Langmuir's most significant blind spot, then, was perhaps that he seemed to regard himself as an exception to what he plainly recognized as the very human denial of error. He readily saw it in others but, sadly, not in himself (see the section on "Epistemic Hubris" in Chapter 3).[13]

Even so, I do not wish to imply that Langmuir's errors were pathological. Nor were the errors in the episodes Langmuir critiqued pathological either. Error in science is not pathology.[14] Indeed, bias and error are part and parcel of the process of science, viewed as a whole. Insights often come with blind spots.

Ultimately, this series of cases—cascading from Blondot (N-rays) to Wood (textbook) to Allison (magneto-optic effect) to Langmuir (cloud seeding)—helps demonstrate that error is not corrected by some scientist's superior objectivity or rigorous methodology, but by contingent, complementary perspectives. Insight on one occasion was coupled with blindness on another. The value of each perspective depended on the context of the case.

Social checks and balances

As portrayed in this chapter, scientific insights (the seed of valuable discoveries) and blind spots (the root of many individual errors) may arise from the same source. Both stem from the distinctive interpretive perspectives of individual scientists. So, eliminating those particular differences (often construed as harmful "biases") in an effort to reduce errors would not necessarily benefit science. Diversity in ways of thinking fosters discovery. Paradoxically, perhaps, "bias" can be fruitful. At the same time, individual bias also seems to introduce opportunities for error. The cost of discovery seems to be the risk of error.

To promote the growth and deepening of knowledge, therefore, managing error rooted in individual perspectives becomes paramount. Hence: the critical discourse of science, which functions as a system of checks and balances. In Merton's characterization (Chapter 4), fruitful science depends on "organized skepticism." Here, we can emphasize the *organized* element. Namely, let us set aside the conventional stereotypes of skepticism based on solopsistic Cartesian doubt or blind raw criticism. Instead, we should envision skepticism *socially*, in terms of the interaction of contrasting perspectives.

David Hull underscored an important dimension in conceptualizing the social system of science by focusing on what *motivated* scientists. He highlighted the roles of curiosity and credit. I have little further to add, other than to note how motivation is intimately linked to individual perspective. *Individuals* may well exhibit interestedness, partiality (noncommunalism) or political leveraging (nonuniversalism)—an apparent affront to Merton's norms—at the same time that the *social* structure of science regulates and channels these motivations toward more productive collective ends. Ultimately, "bias" is managed via the reciprocal criticism when appropriately motivated.[15]

Feminist critiques of science have highlighted yet another dimension: the role of an adequate sampling of relevant perspectives within the scientific community. Originally, this was closely allied with identity politics, primarily based on gender or race. However, in retrospect, we can be more generous and simply note the role of diverse perspectives and their relevance to shedding light on the topic at hand. Ultimately (as noted briefly earlier), we need to control for interpretive perspective by comparing different standpoints side by side, which may well be based on cultural dimensions, such as class, religion, or political ideology. In the remainder of this section, I consider historical cases that illustrate the significance of alternative perspectives, specifically in highlighting flawed justifications and in remedying the corresponding errors. Again, the philosophical "argument" is made indirectly, through the analysis of a pattern of repeated particulars in history, rather than through abstracted generalities.[16]

Gender and primatology

Prior to the 1960s, the field of primatology was dominated by male researchers. In the 1960s and 70s, that changed significantly. As a result of the new inclusion of women, gendered (counter)perspectives began to expose some of the field's earlier overgeneralizations and errors.

For example, in the 1930s, (male) zoologist Solly Zuckerman had studied captive baboons. His attention was drawn to their dramatic conflicts, chiefly between males. His work focused on how males vie for dominance. He documented the males' various aggressive behaviors and the resulting political hierarchies, which Zuckerman presented as the fundamental basis of the group's social organization. In the 1960s, Irven DeVore (also a male researcher) confirmed those patterns in field studies. Those studies were taken to represent all primates—including humans—in a sense, giving a "natural" context (and implicit justification) to patriarchal cultures and the violent behavior of human males. At the time, there seemed no hint of any shortcoming or gendered blind spot.

In the 1960s, many women became primatologists, and their observations differed markedly. Perhaps they were just as "biased" in noticing female behaviors, but their complementary focus could notably limit the generality of the earlier conclusions, at least. Thelma Rowell, working in Uganda, and Shirley Strum, in Kenya, found that their baboons did not follow the competitive, male-dominance-based model, now seen as an error of premature overgeneralization. Women also elected to study other, less obviously aggressive species. Phyllis Jay found her langurs in India "laid back and peaceful." Infant care was central. Moreover, most of the "policing" functions that regulated social behaviors, ascribed to males in the baboon troops, were performed primarily by female langurs. Sarah Hrdy also worked with langurs and showed that it is not always the case that males compete for reproductive access to females. Indeed, sometimes it was the other way around, with females exercising strategies to choose males. In some cases, they mated with multiple males—leaving them to guess who fathered a particular infant, thereby protecting that offspring from harm. Alison Jolly studied lemurs. Their societies, by contrast, proved to be matriarchal. Jane Goodall's observations of chimps highlighted the dynamics of maternal care and familial relationships, the social economy of exchange, and the politics of cooperation.

In 1974, Jeanne Altmann reflected on the imbalance of observations. She noted that conspicuous behaviors were more likely to be noticed in a haphazard method of observing. Unstructured observation was undisciplined and susceptible to bias. So she proposed a basic methodological shift: do not look at the "group" indiscriminately, but focus on individuals and follow each one successively for a standard period of time. That method—clearly a more representative form of sampling behavior—began to shift the emphasis from dramatic male interactions and to behaviors that had gone unnoticed (by the male researchers, at least).

Women also introduced longer studies. That allowed them to observe, for example, seasonal differences in behavior, and to track interactions of specific individuals over time—with new insights that showed the errors in more truncated studies. Longitudinal studies helped to clarify the reproductive biology, and to map the familial relationships that proved foundational to the structure of most primate societies.

With their complementary gendered perspectives, women transformed the nascent field of primatology. By asking different questions and using different observational methods, they exposed misleading assumptions, overlooked behaviors, and erroneous conclusions. Who participates in science seems to matter significantly to the content of its conclusions, and certainly to its reliability.[17]

Cancer, molecular biology, and gender

Complementary gendered perspectives have been shown to expose adverse or limiting biases in science well beyond the "nature" of females or what some may consider overt expressions of feminist politics.

For example, Penny Gilmer studied cancer and autoimmune diseases in a way that she considered distinctly different from her male colleagues. She did not focus on how immune cells recognize tumor cells (an "attack"-centered model). Rather, she considered how each participant cell had a significant role—a complementary perspective she ascribed to her enculturation as a woman. Thus, she focused on the cell-to-cell recognition mechanisms and how the tumor target cells can inhibit or block recognition by the immune cell. Considering only one side of the interaction halved the relevant information and opened the way to error. Gilmer's approach has since been the basis for certain anti-cancer drugs, which weaken the tumor cell's ability to resist the body's natural defenses.[18]

For her part, Bonnie Spanier found gender bias at work in molecular biology. She apprenticed with Nobel Prize winner David Baltimore, but in the 1990s she disagreed sharply with the conceptualizations expressed in his popular textbook. For example, she noted the gendered tendency to characterize causality in terms of "active" versus "passive" elements (which, as noted above, led to misleading conclusions about primate behavior), here applied to the parts of a cell. Genes and the nucleus were considered dominant. They supposedly determined what happened in the rest of the cell. This view was "based on the highly questionable assumptions that complex and variable behaviors can be abstracted or reified into a single" determinant. Namely, interactive cell processes were reduced to the control of genes: a gendered model of centralized power and hierarchical influence. The "political consequence may at first seem negligible," she noted. "But the power of science to define what is natural has serious repercussions." Spanier, like Gilmer, also saw the influence on narrow and distorting conceptions of cancer and its causes. For Spanier, a complementary gendered perspective "functions as a necessary experimental control to eliminate pervasive and unconscious gender bias." Namely, an alternative view opens awareness of the limits and misleading consequences of adopting just one particular view—exposing the types of errors described in Chapters 2–4. New complementary views suggest where or how to collect the relevant evidence that can fill the blind spots.[19]

Tuberculosis, pellagra, and class

As noted in Chapter 3, there are many forms of cultural bias. Those based on economic class may seem among the most obscure or improbable—and perhaps least likely to be remedied. Yet by viewing science through a complementary (here, Marxist) perspective, the biasing influences of commercial interests and class privilege can be readily appreciated. Such critiques can show just *how* the evidence has deficits and *how* interpretations involve perhaps hidden assumptions, while being able to highlight the features of "deeper evidence" that provides for a more balanced overall view.

Consider something as "simple" as the cause of tuberculosis. It may seem unproblematic to say that the cause is *Mycobacterium tuberculosis*, the bacterium famously identified by Robert Koch in 1882. But as Richard Lewontin noted (adopting a Marxist perspective), "It is certainly true that one cannot get tuberculosis without a tubercle bacillus. ... But that is not the same as saying that *the* cause of tuberculosis is *the* bacillus." Here, a Marxist standpoint considers who suffered from tuberculosis and why. Thus, Lewontin noted that "tuberculosis was a disease extremely common in the sweatshops and miserable factories of the nineteenth century, whereas tuberculosis rates were much lower among country people and in the upper classes." We may well be justified in claiming, by the same style of reasoning, that *the* cause of tuberculosis was unregulated industrial capitalism. Namely, "if we did away with that system of social organization, we would not need to worry about the tubercle bacillus."

Further support for this notion comes from the history of the incidence of tuberculosis. It declined remarkably over the course of the 19th century—well before Koch's discovery of the bacillus. Indeed, "by the time chemical therapy was introduced for tuberculosis in the earlier part of [the 20th] century, more than 90% of the decrease in the death rate from that disease had already occurred." Much of this could be attributed to nutrition, and the wages that allowed those who tended to suffer from tuberculosis to attain a better diet. Thus, we may distinguish between the bacillus as the proximal *agent* of tuberculosis, versus a more holistic view of its ultimate *cause*.

Lewontin concluded that "although one may say that the tubercle bacillus causes tuberculosis, we are much closer to the truth when we say that it was the conditions of unregulated nineteenth-century competitive capitalism, unmodulated by the demands of labor unions and the state, that was the cause of tuberculosis." The Marxist's complementary perspective helps us rethink what evidence is relevant, at least, and see how the conventional microbiological answer is an error, in the sense of being misleading and/or significantly incomplete.[20]

The history of tuberculosis is certainly not an isolated case of such an analysis. What caused pellagra in the southern United States in the early 1900s (Chapters 2 and 3)? The chief theories were either a nutrient deficiency or some unidentified microbial infection. A study by Joseph Goldberger and

Edgar Sydenstricker published in 1916 certainly implicated diet. But they went on to attribute the dietary problem to a more fundamental factor: low wages. Their data showed that "the proportion of families affected with pellagra declines with a marked degree of regularity as income increases." They blamed the whole agricultural system, the reliance on cotton as a cash crop, and the corresponding reduced availability and high cost of vegetables. For the next decade, they clearly focused on the plight of the tenant farmers and, from that perspective, the cause of pellagra was the status of the agricultural economy. Was poverty a "confounder" or a causal co-factor?

After 1927, yeast supplements proved an effective practical solution. It had the essential nutrients that were lacking in cases of pellagra. But it was *also* cheap. The economic dimension was just as important as the nutritional one, although perhaps hidden in how one tells the story.[21]

•

Reciprocal criticism, capitalizing on different standpoints or backgrounds—namely, Merton's "unending exchange of critical judgment"—is essential to discovering certain errors and achieving reliable knowledge. That is, the system of epistemic checks and balances works diffusely at the social level. One cannot always expect individuals to be disinterested, or wholly neutral (or "objective"). One cannot always expect individuals to be open and communal in spirit. One cannot always expect individuals to "universally" respect others. Indeed, one cannot always expect individuals to be rigorously skeptical—especially of their own work. Still, the norms of disinterestedness, communalism, and universalism still apply. Science relies on the organizational structure to bring those blinkered perspectives together, where dialogue can begin to effectively address the scattered deficits. It is for this reason, of course, that philosophers of science—and scientists themselves, through their institutions—emphasize (once again) the importance of consensus. Individual scientists err, yes. Scientific communities, less so (see also the section on "Critical consensus" in Chapter 4).[22]

There is a caveat, however. A very important one. The ensemble of perspectives, or voices, must be sufficiently diverse. As illustrated in the cases in this section, the specter of communal confirmation bias and/or communal cultural bias (Chapter 4) always haunts science. For this reason, it is not just "alternative" perspectives that are important, but *complementary* ones. Again, as Merton noted, a well-structured scientific community "affords both commitment and reward for finding where others have erred or have stepped before tracking down the implications of their results or have passed over in their work what there is to be seen by the fresh eye of others." That scrutiny can only be complete (and the corrective system function effectively) if the appropriate contrasting perspectives are available. As Sandra Harding noted, the role of this diversity is the difference between ordinary objectivity ("mere" consistency with the empirical evidence) and what she called *strong objectivity*. The cases in this section illustrate how that principle has functioned in history through contrasting voices or standpoints.[23]

Complementary perspectives are especially critical when the science has implications for politics or social justice. Many of the historical cases presented here involved justifying (directly or indirectly) power or privilege. The scientists who benefited from particular scientific conclusions tended to be blind to the biases that shaped those very conclusions. By contrast, the complementary perspectives that were significant in correction acknowledged those who were disempowered, disenfranchised, or otherwise marginalized. Correcting cognitive biases may thus involve, counterintuitively perhaps, attention to the political context that helps subconsciously shape cognition. That places an important condition on what counts as "diverse" or "complementary."

Complementary perspectives

In this chapter, I have highlighted how alternate perspectives are integral to remedying the errors that arise from perspectival bias. I have endeavored to establish two principles of error analytics. First, no single perspective has any inherent epistemic privilege. All perspectives are susceptible to error, and all may potentially help discern errors in others. It depends on context. Second, to be effective, the alternate perspective will be *complementary*. One needs to engage a *countervailing* standpoint. This second notion is especially prominent in Miriam Solomon's view of social empiricism, although she expresses it in terms of an equal distribution of (or overall balance between) "decision vectors."[24]

When taken together, the two benchmarks I have profiled imply that pairs of scientists may sometimes be able to exhibit effective reciprocal criticism. That is, complementary perspectives should potentially be able to expose and correct each other's errors. In principle, at least. In this section, I provide historical examples of just such reciprocal interaction, where pairs of individuals exhibited tightly corresponding insights and blind spots. Both exhibited error, and each corrected the other. This form of checks and balances is especially striking, of course, when (as seems typical in these cases) the investigators posture themselves as mutually exclusive rivals or lay stake to what are considered to be incompatible claims.

For example, consider the 1908 Nobel Prize in Physiology or Medicine, which was shared for a pair of related discoveries in immunology. Paul Ehrlich had characterized various immune reactions—agglutination, bacteriolysis (via complement), and hemolysis—all chemical in nature. His work embodied the then popular approach based on blood chemistry. At the same time, Ehrlich denigrated cell-oriented approaches as utterly misguided—his blind spot. He erroneously excluded any role for white blood cells engulfing waste, say, or for immune action mediated by what we now know as T-cells. Yet such processes had already been observed and investigated by Ilya (Elie) Metchnikov—who received the other half of the prize, along with Ehrlich. Metchnikov was a champion of the cell-oriented approach that showed the importance of various white blood cells and phagocytes, as the Nobel Committee acknowledged. But Metchnikov, for his part, dismissed the relevance of approaches based on chemical elements carried in the blood—his complementary blind spot. Both immunological mechanisms—humoral and cellular—are now recognized as important and functionally integrated. Historically, the Nobel Prize committee seemed to conspicuously recognize the respective insights and their respective erroneous overgeneralizations.[25]

Consider, next, the famous Volta-Galvani debate on "animal electricity." Luigi Galvani observed the twitching of a freshly dissected frog leg stretched between two different metals. As an anatomist, he interpreted it in terms of an "animal electricity." Alessandro Volta, a physicist, was skeptical of that vitalist notion. He showed that the electric current could be produced with the metals alone, thereby dispensing with the need to appeal to any special organic power or energy—Galvani's blind spot, perhaps. But in focusing on the physical phenomenon only, Volta simultaneously failed to explain what caused the frog muscle to contract or how nerves worked. That was his blind spot. Galvani was justified in concluding that the movement of the frog's leg was more than just electricity in action. Electricity was only a stimulus. That implied that nerves must have an electrical nature, although not necessarily by virtue of metals (as in a Voltaic pile, or battery). It would be decades before biologists could articulate what that meant at a cellular level, however. Contrary to Volta's claims, organisms are more than "mere" physical machines. Two complementary insights, two complementary blind spots.[26]

A similar story can be told for the Great Devonian Controversy in early 19th-century geology. Henry de la Beche initially reported his very basic field observations about the geological strata in Devon, England. His sequence, however, implied that fossils (from the layer known as the "Coal Measures") were older than any known previously. That was a significant insight. However, de la Beche was apparently blind to the large-scale theoretical implications. For Roderick Murchison, the presence of fossils that old was strictly not possible, based on his own observations of strata of a corresponding position and age. So (he claimed), de la Beche must have been incompetent and made a cardinal error in his elementary observations. At the same time, Murchison had proposed his sharp historical boundary on somewhat arbitrary foundations. That was his blind spot, exposed by de la Beche's mundane field work. Veteran geologist William Buckland was able to reconcile the two views by interposing a new geological period, the Devonian. The fossils were old, yes, but not quite so old as to violate Murchison's views on geological history. Again, complementary insights and blind spots, accommodated in a synthesis.[27]

Charles Darwin and Alfred Russel Wallace on human evolution

For a more nuanced case, we may turn to the cofounders of the theory of evolution by natural selection: Charles Darwin and Alfred Russel Wallace. Both endorsed the evolution of humans from primates, but whereas Darwin gave it a fully naturalistic explanation, Wallace saw a spiritual element. On the other hand, Darwin's view had racist overtones, which Wallace starkly rejected.

Darwin's view was itself a fascinating combination of insight and blind spot, all at once. Darwin reasoned about human descent from his gradualistic perspective, but also in the context of his social status. British society was stratified, and Darwin enjoyed membership in the upper class. He was also a white European at a time when Europeans (notably the British) dominated the globe. This context shaped perceptions of other races, easily construed in a hierarchy. While voyaging on the *Beagle*, for example, Darwin was appalled by the habits of the natives of Tierra del Fuego:

> It was without exception the most curious and interesting spectacle I ever beheld: I could not have believed how wide was the difference between savage and civilized man: it is greater than between a wild and domesticated animal, inasmuch as in man there is a greater power of improvement.

Improvement there was. One of the Fuegians had been taken to London, educated, and entered into elite society. When he returned, however, he seemed content to revert (as Darwin saw it) to his "primitive" habits. It was all too easy for Darwin to consider racial differences as innate and to rank them on a scale from "savage" to "civilized." That conception proved both fruitful and dramatically misleading.[28]

When Darwin began considering human ancestry, he saw immediately that the problem was not primarily anatomical. Humans had long been classified as primates. The challenge was accounting for the origin of mental faculties and moral sensibilities. Darwin's early musings turned to the Fuegian episode. He wrote to himself in the fall of 1838:

> Nearly all will exclaim, your arguments are good but look at the immense difference. between man,—forget the use of language, & judge only by what you see. compare, the Fuegian & Ourang & outang, & dare to say difference so great ... "Ay Sir there is much in analogy, we never find out."[29]

Darwin essentially cast the Fuegians as intermediates between orangutans and "fully developed" humans, such as himself and his elite British peers. That is, Darwin's view of stratified races facilitated his conception of a transition, linking apes and humans through a series of gradual changes. "Savages" became convenient transitional forms in moral and mental development. That is, Darwin arrived at his insight of human evolution in part because of his cultural view of human races.[30]

The relevance of Darwin's social class in shaping this conclusion comes into relief by comparing him with Wallace. Wallace, by contrast, was middle class and had to work to earn his living. While collecting in the Malay Archipelago, he depended heavily on the local natives. He learned to respect their knowledge and willingness to help. In 1855, he wrote to a friend: "The more I see of uncivilized people, the better I think of human nature and the essential differences between civilized and savage men seem to disappear." For Wallace, if even such "brutes" could show kindness, then all humans seemed to share the same moral sense. He echoed his sentiments in 1873:

"We find many broad statements as to the low state of morality and of intellect in all prehistoric men, which facts hardly warrant." Wallace, in contrast to Darwin, experienced moral and mental discontinuity between man and beast. Wallace certainly acknowledged that humans had primate ancestry—anatomically. Still, he maintained that the human mind was unique. The human brain, he contended, emerged by some guided process, *not* by natural selection.

Wallace erred in viewing the human mind as an exception to evolution. He never considered, as Darwin did, that morality itself might evolve. At the same time, however, Wallace's views, arising from his social status, help expose Darwin's unfortunate bias of ranking races hierarchically. Insight here, blind spot there, neatly fitting together as a pair.[31]

Peter Mitchell and Paul Boyer on cell energetics

My final example involves another pair of Nobel Prize winners whose insights accommodated each other's blind spots—here, on deciphering how cells process energy (the ox phos controversy, encountered several times in earlier chapters).

The first error was made by Paul Boyer in 1963 (see "Confounders" in Chapter 2). In the 1950s and 1960s, biochemists were looking for a set of high-energy molecules that transferred energy from the electron transport chain to the final energy molecule (ATP). After a decade of failed claims from several labs, Boyer employed a technically demanding method using a radioactive tracer. He thereby isolated an intermediate and identified it as phosphohistidine. He reported the apparent triumph in the prestigious journal *Science.* Relief rippled through the community. The high-profile discovery soon fizzled out, however. When additional controls were applied, Boyer's lab attributed the results to other energy reactions in the cell. A discovery on its own, but not the imagined discovery: a framing error. "I was wrong," Boyer frankly admitted later.

Boyer was actually wrong on two levels at once. Phosphohistidine was not the intermediate. Boyer acknowledged as much. But the very concept of the intermediates, for which everyone had been searching so earnestly, was also mistaken. Boyer soon reached that conclusion, as well. (If he hadn't found the intermediate using his methods, he boldly believed, no one else would.) Boyer hypothesized instead that the energy must be transferred through energized changes in protein shape (like a pair of interacting molecular springs). This concept, too, would eventually prove mistaken, at least at this level.

Here, the unexpected solution was introduced by Peter Mitchell. Mitchell was guided in his thinking by a novel principle that he developed while thinking about the transport of ions and substrates across cell membranes (sections on "Cryptic alternatives" in Chapter 3 and "Engage alternative perspectives" in Chapter 6). In his conception, chemistry was vectorial, not just scalar. That is, enzymatic reactions included an important spatial dimension. For example, an "ordinary" reaction might span a cell membrane, with reactants on one side and products on the other. The movement of molecular components would be an integral part of the reaction, not recorded in conventional chemical equations. Drawing on a collage of findings, Mitchell conceptualized the intermediate energy state as a proton gradient across a membrane (a chemiosmotic potential). That revolutionary idea ultimately earned Mitchell a Nobel Prize in 1978. Mitchell had solved Boyer's error, in a sense.

Mitchell's own claims, however, were hardly free from error. In the first formulation of the theory, for example, the direction of the gradient across the membrane was completely reversed! Mitchell also specified one proton as adequate, when two were needed. Such "minor" errors were soon remedied. But the unrealistic quantitative analysis had already convinced many chemists that Mitchell's credibility was questionable and his notions fundamentally flawed.

Most dramatically, Mitchell had a vision about how ATP was energized from the proton gradient. Using his foundational principle of vectorial chemistry, he insisted that protons flowed into the interior of the ATP enzyme and there participated directly in forming the high-energy phosphate of ATP. That concept was creative, but never fit comfortably with the data. Here, it was Boyer's concept, rather, that prevailed. Boyer had adapted his ideas on the role of protein shape. He reasoned how ATP formed on the surface of the enzyme, and was then released through an energy-requiring change in the enzyme's shape. The energy was provided by protons, now in accord with Mitchell's overall scheme. The ATP enzyme worked like a miniature molecular motor. Boyer's insights were recognized in a 1997 Nobel Prize. So, on this occasion, Boyer solved Mitchell's error.

Ultimately, Boyer and Mitchell had both been right (on different occasions). And both had been wrong (again, in separate cases). Their insights, arising from separate backgrounds and orientations, neatly complemented each other's blind spots.[32]

•

In summary, this handful of cases—Metchnikov and Ehrlich, Galvani and Volta, de la Beche and Murchison, Darwin and Wallace, Boyer and Mitchell—illustrates how the perspectives and individual "biases" of different scientists can complement each other. One person's blind spot may also be the occasion for insight into someone else's blind spot. *And vice versa.* Accordingly, the errors that arise from individual "biases" can be resolved—with "more" bias, not less. In these cases, one can also see the differentiation of domains, described in Chapter 5. Discourse among complementary perspectives is integral to regulating error, while also providing the opportunity for discovery. Insights and blind spots function in tandem.

8
Managing Error

The chief aim of science is not to open the door to infinite wisdom but set a limit to infinite error.

—Bertold Brecht, *Galileo*

Philosophy of error • isn't science self-correcting? • negative knowledge • methodological norms • progress through error • error repertoires • error signatures • checklists and checkpoints • error probes • prospects

To recap: Sometimes, scientists find that claims once considered justified are, in the light of deeper evidence, unwarranted (Chapter 1). Ironically, blind spots may occur for the very same reason that insights occur (Chapter 7). There are many error types (Chapters 2–4). There are also many ways to find and resolve errors (Chapter 6). But errors do not solve themselves. Resolving the incongruences that signal a latent error requires more scientific work (Chapter 5). All this might motivate us to study error in science more formally and systematically. A kind of "errorology." Call it the Philosophy of Error, which might incorporate a set of practices known as Error Analytics.[1]

A Philosophy of Error is ready to embrace numerous efforts. In this chapter, I can only begin to sketch the shape of the many opportunities and their potential fruitfulness. For example, from a philosophical perspective, how should we characterize what we learn from past errors? Is it all just water under the bridge? (See section below on "Negative knowledge.") Based on a well-developed knowledge of error types, might we be able to articulate corresponding methods designed specifically to reduce the incidence of error—for instance, the adoption of blinded analysis (Chapter 2) or controlled experiments (Chapters 2, 5, and 6)? (See section on "Methodological norms.") If history documents numerous episodes of wholescale conceptual change, what (if anything) can we say about the nature of progress in science? (See section on "Progress through error.") All these projects might give us a deeper abstract understanding of error in science.

Toward a Philosophy of Error in Science. Douglas Allchin, Oxford University Press.
DOI: 10.1093/9780197827703.003.0008

A second major thread of the Philosophy of Error pursues what can be done to improve scientific practice: Error Analytics. For example, can we list the relevant error types for particular fields or areas of study, to use as a point of reference? (See section on "Error repertoires.") Do particular errors leave telltale signs, allowing one to recognize the error type? (See section on "Error signatures.") Might we develop fruitful habits or guides for practicing scientists? (See section on "Checklists and checkpoints.") How might we reconceptualize the aims of scientific practice, integrating the conventional aim of developing evidence, with the corresponding goal of actively reducing errors? (See epigram and section on "Error probes.") Error in science may be inevitable. But that need not deter us from finding ways to *manage* error. All these considerations may lead us to map the road ahead for a philosophy of error (see "Prospects").

But before we delve into such details, perhaps we should pause. Why fuss about the nature or error or its dynamics at all? Isn't science *self-correcting*? That's what you hear, at least, from many philosophers, editors of *Science*, and practicing scientists.[2]

Isn't science self-correcting?

In a word: *no*. Yes, science can—and frequently does—find errors and remedy them. Every case reported in this volume is testimony to correction at some level. But that fact alone does not tell us how the error was corrected, or with what timeliness.[3]

The error types are diverse (Chapters 2–4), as are the various incongruences that signal latent errors (Chapter 5), as well as the forms of evidence that help resolve them (Chapter 6). That is, there is no prescription for identifying, localizing, or testing sources of error. So, just as there is no universal "scientific method," there is no uniform "error-correction method." The process is haphazard and somewhat blind, not systematic. There is no *mechanism* for self-correction, in the sense sketched by philosophers (inputs leading via organized activities to predictable outputs).[4]

As a result, some errors linger for years, decades, or even centuries. Some errors are compounded into other errors downstream before the original error is corrected. Some errors are "corrected," but soon lead to another error because the source of error was not properly identified (error cascades—Chapters 5 and 6). Some errors, when corrected, are (ironically) not acknowledged by the community. Some errors are corrected by happenstance (Chapter 6). All these instances (which I have discussed and documented more fully elsewhere) reflect a lack of systematic "self-correction."[5]

Indeed, one typically finds the promise of self-correction hedged. Errors, advocates say, are corrected "eventually." Or "ultimately." Or "in time," or "sooner or later," or "in the long run."[6] There is no guarantee. The language is more like a convenient escape clause. Yet timeliness seems essential for any bona fide label of "self-correction." An expected time horizon must be specified if we are to assess the resilience of any particular scientific claim. Namely, the prospect of "eventual" correction cannot warrant trust in *current* scientific knowledge, the context where public appeals to self-correction in science are most frequently found.

As discussed in Chapters 4 and 7 (and illustrated throughout), various social practices contribute to detecting and remedying errors through critical discourse and reciprocal review. But even those social practices are not error-free. Two social processes, in particular, are typically cited as the "mechanism" of self-correction: peer review and replication. Peer review, however (as noted in Chapter 4), is largely ineffective in this regard. Prepublication review can improve the quality of publications and can act as a coarse filter against egregious error. But it does not (and, structurally, cannot) serve the self-correcting function that many imagine. Replication, likewise, is an important tool, but does not yield unambiguous "tests" (see "Replication with substitution" in Chapter 6). As molecular biologist Walter Gilbert once cautioned, "you can reproduce artifacts very, very well."[7] The comedy of errors that resulted in replicating Priestley's work on the restoration of air by plants (Chapter 6) exemplifies how there is much more to interpreting results than merely reproducing them, or not. The two highly vaunted "mechanisms" of self-correction are, in fact, not stopgaps at all.[8]

The social practices of science do help science, in general, *accommodate* to new evidence, once it is available. Science *adapts*. But this is far from a system that dependably generates the critical new evidence or finds hidden errors as a matter of course. The system is responsive, not (yet) probative (see "Error probes" below).

Perceiving and resolving errors hinges on new evidence. And that remains unpredictable. Having reviewed all the ways of developing remedial evidence (Chapter 6), one may well observe that the strategies are not different from ways of developing *any* evidence in science. Error-correction may thus seem like merely an exercise in ordinary science. And perhaps, in a sense, it is. But this comparison merely underscores the many contingencies at work. Historically, evidence was sometimes pursued deliberately, yes. But in many other cases, detecting errors (or isolating them) waited until there was new technology, new collateral knowledge in other fields, new perspectives, or sheer happenstance (Chapter 6). Those are all contingent events. Thus, there can be no guarantee of systematic, timely "self-correction." The relevant question becomes rather, "*When* does science correct its errors?," and that is more subtle and conditional.[9]

Philosophers interested in characterizing the trustworthiness of science in public discourse, as it informs public policy and personal decision making, thus have a challenge. First, they must set aside the normative conception of science-as-it-should-be-ideally and focus instead on science as it is actually practiced and experienced. That means jettisoning the widespread rhetoric of "self-correction," which oversells science. It means admitting that, yes, scientists can err. Yes, some scientists may even express cultural bias (Chapters 3 and 7). But that does not lead to doomsday. Acknowledging error can be coupled with clarity about how scientists find and remedy their errors (Chapters 5 and 6). What are the various possible error types (Chapters 2–4), which have been probed, and thus what are the limits of our current knowledge? Acknowledging possible error or uncertainty is not inconsistent with confidence (to varying degrees) in current claims. As articulated recently by a few philosophers, disagreement and error-correction is a hallmark of strong, active science-in-the-making, not a weakness.[10] Making the whole system of error-correction transparent and portraying it as a standard scientific practice seems an essential first step. And philosophers seem well-positioned to do this.

Negative knowledge

The resolution of error poses a puzzle, of sorts. Identifying a source of error allows one to rule out a certain conclusion, even one formerly considered justified. That would seem to be a negative outcome. At the same time, one has *learned* this. That is a "positive" outcome. What was once known, then an object of uncertainty, is now known again, in its inverse. The scope of observations or evidence has been widened and the pattern of justification clarified—even if the conclusion was upended in the process. Something concrete has been added to our knowledge, albeit in a negative sense of

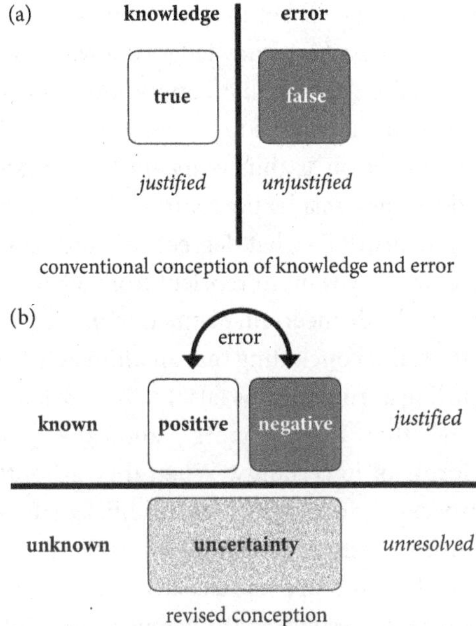

Figure 8.1 Comparison of conventional and revised view of knowledge, error, and uncertainty.

what is not the case. It occurs at the level of (deeper) justification (a view of scientific progress nicely articulated by Jabob Stegenga[11]).

Accordingly, we should regard both "positive" knowledge and "negative" knowledge as forms of knowledge—in contrast, say, to uncertainty or sheer ignorance. Characterizing "negative" knowledge as *knowledge* invites us to rethink our basic conception of knowledge itself. Traditionally, philosophers have defined knowledge as "justified true belief," as depicted in Figure 8.1(a). Conventionally, knowledge stands in contrast to error, as false (left vs. right in the diagram). But, as we have seen in Chapters 5 and 6 (and throughout all the cases in this book), *knowing* that a claim is false requires evidence. It requires a justification all on its own. Negative knowledge is thus not "error." Nor is a false claim error, by itself. Rather, error is in mistaking a false claim for a true one (Chapter 1). Or vice versa (arrow in Figure 8.1[b]).

Hence, we may couple both "true" and "false" claims together as knowledge, acknowledging that each represents a justified conclusion. Positive knowledge *and* negative knowledge (what is known) may then be contrasted to what is *unknown*: not to error, but to *uncertainty*—as depicted in Figure 8.1(b). The key distinction marking knowledge (known vs. unknown) is thus not between true and false (the familiar convention), but between what is justified and what remains unjustified (top vs. bottom in the revised diagram). The critical dimension is thus one's *epistemic posture* toward the claim (Chapter 1): does one consider the claim—whether as true *or* false—to be justified by the evidence? (To what degree?) Accordingly, in many cases in scientific practice, we may want to reorient from a true/false distinction as fundamental to a resolved/uncertain distinction instead. Error is misconstruing the *justification* and concluding that an ultimately false claim is true (or vice versa: regarding a true claim as false) (Chapter 1).

Incongruences—whether discordances, anomalies, or ambiguities (Chapter 5)—are forms of uncertainty. When they arise, they signal that the justification process is incomplete, or insufficient to sort true from false (or the real from illusion, or fact from artifact). The challenge to the investigator is to find the appropriate evidence relevant to deciding among the alternatives, to resolve the inherent uncertainties (at least at some level that satisfices our demands). When the process is complete, an earlier error is thereby, in retrospect, identified (Chapter 1). Namely, we have resolved—or sorted (or *differentiated*)—positive knowledge from negative knowledge (Chapter 5). By reframing what constitutes knowledge in this way, a study of error may dramatically transform how we think philosophically.

But is "knowing" about errors anything more than a shadow or an unfortunate residue—waste, in a sense—an unwelcome but inevitable byproduct of learning by trial and error? That is the typical attitude, recounted in Chapters 1 and 5.

Is negative knowledge epistemically important? Well, Charles Darwin thought so. He wrote to a colleague expressing appreciation for helping him identify an error: "To kill an error is as good a service as, and sometimes even better than, the establishing of a new truth or fact."[12]

What errors are to science, failures are to engineering. And engineers (Henry Petroski often noted) take heed of their failures. They investigate why they occurred, hoping to learn from them. Why did this particular dam or bridge collapse? Why did the Space Shuttle explode or the 1982 Air Florida Flight 90 crash shortly after takeoff? Investigations into failures often yield important principles to guide future design or operating protocols.

They are not merely forgotten or brushed under the rug. They become part of the knowledge base. And (Petroski has argued) they should be integrated into the education of future engineers, along with the principles they illustrate.[13]

Reflecting a similar way of thinking, Marvin Minsky (noted pioneer in artificial intelligence) has defined experts as people who, in part, never make mistakes. Experts know, of course, how to achieve certain goals. But *also* how to avoid disasters. They know what actions will invite trouble. That requires knowing what *not* to do. It entails familiarity with exceptions and past mistakes. Hence, they can *avoid* behaviors *known* to lead to unwanted outcomes. For example, we post signs to warn people about sharp turns in the road, thin ice, or animals that might bite. Namely, *negative* knowledge has a positive role. In science, that might include (Minsky observes) "knowing how to intercept and interdict unproductive lines of thought." It is not exclusively about knowing the standard methods for solving problems.[14]

Past errors are guides. Negative knowledge can be fruitful in many ways. For example, it helps us discard apparently reasonable, but mistaken assumptions. It may help us recognize when results are misleading, or "red herrings" (for example, faux fossils such as *Eozoon*). For example, the "negative" results of experimental controls are integral to eliminating the possible role of confounding variables or in rejecting a null hypothesis. Errors, once encountered, may also indicate ineffective methods or procedures not to be used again or relied on. They help us discover exceptions and the limits of certain laws or concepts (Boyle's law, Ohm's law, Mendel's law, etc.). They may circumscribe the scope of models or heuristics—to be heeded judiciously in future instances. Or they may help caution us against jumping to conclusions from small samples or incomplete evidence. Errors constitute useful negative knowledge.[15]

However, memory is essential. Errors must be effectively documented and communicated. This is how we transform trial and error into trial and *learn*. If one is trying to negotiate one's way through a maze, knowing which unsuccessful paths not to repeat is critical. This subtle lesson is expressed somewhat comically in a sketch by Peter Cook and Dudley Moore. One portrays the owner of a restaurant who admits that his whole venture has been a "catastrophic" failure. The other interviews him, asking (innocently), "Well, do you feel you have learned from your mistakes?" The failed entrepreneur replies, nodding, "Oh, certainly, certainly, I've learned from my mistakes.

And I'm sure that I could repeat them *exactly*." Others of us, of course, might choose *not* to repeat our mistakes. In either case, we at least need to remember them. To be of general value, errors must be communicated and become part of the public record: error—explicitly identified *as error*—in the scientific literature. And that may pose new challenges for how to publish and archive negative results and errors. Such initiatives are already underway. But it may be profitable to continue to articulate just how important such efforts are, and to conceive how to sufficiently fund the relevant archiving and to incentivize (or reward) researchers to publish their negative knowledge.[16]

Here, then, one may formally acknowledge the significance of assembling an inventory of error types (Chapter 2). It is more than a philosophical formality of being exact or completing a taxonomy. We need to remember the experimental errors—contaminants, lack of controls, confounders, and so on—to avoid repeating them. We need to be cognizant of our conceptual errors—fallacies in reasoning or cognitive biases—to sidestep them, or to remedy them before they become significant. We need to attend to the social-level errors—conflicts of interest or shared biases creating a collective blind spot—if scientists, as a community, want to build reliable, consensual knowledge. The inventory of error types (Figure 2.2) establishes functional sign posts, or guardrails, to guide and facilitate trustworthy scientific work.

The communal memory of scientific errors may be realized in at least two ways (besides a comprehensive list of general error types): specific error repertoires (for particular fields) and general methodological principles, each addressed in sections below.

Methodological norms

The most significant outcome of remembering past errors has been developing new ways of practicing science that can prevent or reduce such errors in the future. Most of the standard methodological norms of science did not arise from abstract thinking about reasoning or experimentation, or from *a priori* idealization of its principles. Rather, they have a history. They emerged in response to concrete encounters with errors. Such experiences led to meta-reflection, and to devising methods to safeguard against repeating the missteps. Controlled experiments, double-blind clinical trials, calibration, standardized measurement protocols, analysis of sample size

Error	Methodological Norm	*Reference*
non-comparable results	standardization (standard weights and measures; std. organisms; reference samples)	e.g., Ankeny & Leonelli (2021); Keeling (1998); Rader (2004)
instrument malfunction	calibration	Franklin (1998)
coincident variables / confounders	control	Allchin (2021); Boring (1954), Lilienfeld (1982)
observer bias; placebo effect	blinding	Haygarth (1800); Herr (2005); Kaptchuk (1998); Kaptchuk et al. (2000); Tröhler (2011)
observer effect	double blinding	Glanz (2000); Shapiro & Shapiro (1997)
analyzer effect	triple blinding	MacCoun & Perlmutter (2015)
sampling error	statistical analysis of power choice of p-value cut-off	Hacking (1990), Mayo (1996); Porter (1986)
biased sampling	randomized clinical trials weighting of samples	Gould (1981); Hall (2007); Student (1931)
appropriate inferences from results & use of methods	peer review (I. pre-publication)	Benos, Bashari, & Chaves (2007); Shapin (1996)
post-study "statistical gerrymandering"	preset cut-offs blind data reanalysis	Klein & Roodman (2005); Michaels (2008; 2020)
confirmation bias	peer review (II. critical discourse)	Hull (1988); Merton (1973)
disagreement; social-level ambiguity	expert panels consensus conferences	Jacoby & Simopoulus (1986); Perry (1988); Thagard (1999)
gender, racial, or other cultural bias	appropriately diverse scientific community	Harding (1991); Oreskes (2019); Solomon (2001)
publication bias (positive results)	new editing practices and/or repositories for archiving negative results	e.g., Nimpf & Keays (2020)
conflicts of interest	disclosure requirements	Resnik (2023); Ruff (2015)
credibility bias (fraud)	(a) [more] internal review; (b) whistleblower protections;	Johnson (2002); Office of Research Integrity (2025)
communal confirmation bias or communal cultural bias	peer review in diverse communicities—yet to be fully developed	Solomon (2001)

Figure 8.2 Methodological norms and their historical origins in errors.

and statistical power, randomized sampling, blinded peer review, and editorial requirements to disclose conflicts of interest—all are historical products, I contend, of science having gone awry at some earlier point (see Figure 8.2).

Consider, for example, Blaise Pascal's work on the weight of atmospheric air, an early instance of controlled observation. The 17th century witnessed the emergence of modern science, based foremost on giving experimentation a central role. Natural philosophers conspicuously appealed to concrete empirical demonstrations, denigrating purely logical arguments from arbitrary assumptions, or refusing to defer to the supposed wisdom of the Ancients. Observation, witnessing, and *doing* all became important hallmarks for authoritative knowledge. All transformative. But few investigators in this era focused systematically on checking for possible errors (Bacon's "idols" notwithstanding!).

For example, in one of the earliest examples of this new genre, *De Magnete*, in 1600, William Gilbert reported case after case from 18 years of observing how magnets and the compass work. But, almost exclusively, they were trials that *demonstrated* certain effects under various conditions. There was little exploration of ways that things might go "wrong" or be misinterpreted. Likewise, William Harvey's great treatise *On the Motion of the Heart and Blood* (1628), which argued for the unidirectional flow of the blood, was filled with fastidious observations. Harvey performed vivisections in order to watch hearts still beating. He observed and compared heart motions in dozens of different organisms and in embryos. He crudely quantified the amount of blood being pumped through the heart. He also argued by analogy. Harvey persuasively argued that many of the claims of the revered Galen were incorrect. But there was little by way of anticipating alternative interpretations (or confounders) in how he arrived at his own conclusions or systematically ruling them out. Even Robert Boyle's landmark 1660 *On the Spring of Air* was largely a compilation of observations. What are the various effects of a vacuum (or near-vacuum)? On sound, on tied-off bladders, on cool cloth and hot irons, on a pendulum swinging, on a burning candle, on gunpowder exploding, on smooth stones adhered to each other, on corrosive liquors, on the bubbling of water, on freezing liquids. These all illustrated the pressure of air, but did not probe alternative interpretations.

Pascal's investigation was notably different from others in the era. At issue was why a column of mercury stays in a close-ended tube when inverted and its open end submerged in a bowl of mercury. Was it because "nature abhors a vacuum"? Why did the mercury not flow out? In fact, the height of the column of mercury was limited, and an "empty" space appeared at the top of the tube, provoking much speculation about its nature. Evangelista Torricelli, who devised this experiment, claimed that the mercury was held up by the weight of the air pressing down on the pool of mercury in the bowl. The height was limited by the amount of air above it. That is what Pascal set out to test in 1648.

Pascal reasoned (based on hydrostatic principles) that if one could move the Torricelli apparatus to an elevated position, there would be less atmosphere above it, less weight bearing down on the pool of mercury, and the height of the mercury column would fall. He enlisted his brother-in-law, Florin Perier, to test this hypothesis. Perier dutifully carried a Torricelli tube up a substantial lava dome mountain in central France, about 500 toises (or approximately 1,000 meters) tall. Much to his surprise (and to those who accompanied him), the height of the mercury column did indeed decrease—by almost 12%! That *demonstration* was impressive and, in that era, might have sufficed (on its own) to convince many contemporaries. But Pascal was aware that the explanation was susceptible to possible flaws. He knew that the height of the Torricelli experiment varied with time and place. One could easily have attributed the outcome to natural variation. To guard against the possibility of that source of error, Pascal had thus specified a relevant benchmark for comparison: his control. At the outset of the day's journey, a similar tube had been set up at a monastery in the local town and monitored all day long. Its height had not changed. And when the adventurers returned at the end of the day, the height in the two tubes matched again. The critical variable had thus been the elevation change. Not a possible confounder. That is perhaps the earliest example of a controlled experiment in modern times (although not labeled as such). And it was inspired and governed by anticipating potential criticism and the specter of error.[17]

Historians have traced the origin of other important methods in science, as listed in Figure 8.2. These norms may seem like they have always been a part of "good science," and justified just by plain good thinking. But they all have roots in the scientific practices of the time they emerged. They were designed to address certain types of errors. In this way, errors have been "composted" into improving scientific methodology. And that has contributed importantly, in a sense, to the growth of science.

More recently, there has been much consternation about perceived "failures" in the social system of checks and balances in science (Chapter 4). There are various calls to tighten and/or expand the peer-review system (on the one hand) and to open access to publication (on the other); to overhaul professional rewards and incentives (too focused now on mere quantity, not quality of publications); to hold investigators accountable to pre-registered protocols; to share data and code; to increase disclosures on conflicts of interest; and so on. The response has contributed to a widespread impression that science is in "crisis." From my perspective, however, these efforts all reflect a venerable historical pattern of learning how to manage error (Figure 8.2 again). Scientists have (finally perhaps?) become more attuned to the social practices of science and their epistemic significance. They are noting the flaws in the system and (appropriately) engaged in developing new methodological norms. Science is evolving and adapting, as it always has. (Not a crisis, in my view, but an occasion to celebrate epistemic responsibility.)

Progress through error

The emergence of error, especially after scientists have achieved a consensus, provokes questions about the progress of science. The conventional image, of course, is of cumulative growth. One fact builds upon another, like bricks in a wall, gradually creating the monumental and imposing edifice we know as science. Any error thus seems to threaten the integrity of that image. Removing an established fact implies that the edifice may crumble. And if one fact fails, are there others? Trust in science would seem at risk.

The view of errors as a form of negative knowledge might ameliorate this rather severe impression. If errors (once they are resolved from earlier uncertainty) count as knowledge, too, then as long as scientists continue research, and sort fact from fiction, illusion from reality, then knowledge continues to grow. As noted above, science will incorporate both "positive" and "negative" knowledge. If negative knowledge is not construed as a "loss," then the fretting and handwringing of the pessimistic induction (Chapters 3 and 6) simply dissolves.[18]

Moreover, errors frequently emerge when scientists apply more rigorous evidential or argumentative standards. Higher levels of rigor can expose weaknesses in the original conclusions. In a sense, our earlier judgments were overstated and misplaced to begin with. Who would want to endorse an illusory claim, based on a faulty assumption, incomplete data, mistaken observations, logical fallacy, or hidden conceptual bias? Such flawed conclusions may have seemed plausible once, and even eminently reasonable under some circumstances. Yet we are better off when we realize that the claims were premature and perhaps context-dependent, and our initial confidence in their reliability was perhaps short-sighted or ill-advised. Enforcing a higher level of rigor should inherently be regarded as progress, even if it means abandoning some earlier claims in the process.

Thomas Kuhn introduced an alternative, more challenging view of conceptual change in 1962 in his now landmark *Structure of Scientific Revolutions.* Kuhn focused on scientific revolutions. He noted that new theories do not merely encapsulate older theories, like nested Russian dolls. Rather, they rearrange the facts. The conceptual gestalt switches. Sometimes, some of the formerly accepted "facts" are jettisoned. That is, the new theories may replace—or perhaps more accurately, *dis*place—earlier theories. Further, that shift may include adopting new methods and new ways of doing science, based on new exemplars. That change implies that the once-accepted theories, and not just the individual facts, may come to be viewed as "wrong"—as errors to be rejected or cast aside. For example, a Sun-centered solar system dispenses with an Earth-centered cosmology. In Kuhn's account, scientific change may sometimes include a significant *loss* of theoretical (explanatory or puzzle-solving) knowledge: what has come to be known (appropriately enough) as "Kuhn-loss." (Still, Kuhn maintained, adamantly, that there was "progress through revolutions.") The puzzle of Kuhn-loss has deeply troubled many philosophers of science, who are uncomfortable with the apparently adverse implications for scientific progress and, again, for long-term trust in *any* scientific claim.[19]

The analysis of errors here opens two ways to (re)interpret Kuhnian revolutions in ways that may accommodate a view of progress. First, as noted above, if error is a form of negative knowledge, then Kuhn-loss is inappropriately characterized as a *loss*. Namely, if the error is indeed fully resolved, whether observational or theoretical, then we can add that to the balance sheet of knowledge gained. We learn to steer clear of some reasonable but ultimately misleading concepts: for example, chemical affinities, caloric (or heat as substance), teleological evolution, geological fixism, the luminiferous ether, a universal genetic code—all plausible, but eventually found unworkable as evidence amassed. Negative knowledge, but still knowledge nonetheless.

Indeed, in many cases, the original theory remains valid in a limited domain, as a qualified model (for example, *Mendelian* genetics, the *Bohr* model of the atom, *Newtonian* mechanics, or *geocentric* celestial navigation). The new theory helps expose those limits and the error that would arise if one extended certain theories beyond their newly described limits (for example, Boyle's "law"; Figure 6.1). In other contexts, different (newer) theories apply. Distinct phenomena, once conflated, have now been differentiated. We see the world with increased resolution. Kuhn later came to see new theories (or "paradigms") as divergent alternatives, rather than outright successor replacements. Old and new could possibly coexist, even if they were "incommensurable" alternatives that could not be seamlessly knitted together.[20]

Second, during a Kuhnian revolution, anomalies (or other incongruences) may resolve into new, unprecedented ways of seeing things. We may *discover* inside an initial anomaly a previously unrecognized error. Echoing comments above, elucidating a new source of error (in a specific field or domain) may help establish a new, higher standard of justification for what counts as reliable knowledge. Again, rejecting claims that do not meet that revised standard—even if the claims were once fully accepted—should hardly be regarded as a "loss." We do not mourn the loss of mesosomes (Chapters 2 and 6), or the high-energy intermediates of ox phos (Chapters 2, 3, 5, and 7). Rather, the error corresponds to a deeper realization of how one's reasoning can be blinkered, and how one can avoid erroneous, parochial conclusions. Some ideas (and explanatory power) may be lost, but that hardly matters if overall we have a clearer, sharper view of the world. Once again, it is all about the justification, not the claims themselves. We may perhaps "lose" conceptual content, but we have *gained* a higher quality of knowledge, through deeper or more finely resolved justification.[21]

Here, analysis of a historical example of reported Kuhn-loss may help clarify the dynamics. Let us revisit the rather notorious case of phlogiston (see the section on "The Chemical 'Revolution'?" in Chapter 5), comparing "before" and "after." What knowledge was lost, and what was gained? Basically, at first, combustion was explained as the release of phlogiston; after, as the gain of oxygen. With the adoption of Lavoisier's theory of combustion (along with his system of nomenclature and the conservation of mass as an organizing principle—which he conspicuously characterized as an "antiphlogistic" doctrine), chemists largely abandoned phlogiston. So, all the features of phlogiston, in explaining the properties of metals and acids, the basis of combustibility, how burning and reduction related to electricity, and so on, were presumably lost. Kuhn presented this episode as "paradigmatic" of his concept of a revolutionary gestalt-switch. It thus becomes particularly apt for reflecting on Kuhn-loss and scientific progress.

First, as illustrated in Figure 5.4, the episode was much more complex than a simple either-or switch.[22] Much of the "loss" appears so only when one focuses on Lavoisier's perspective as the sole relevant benchmark. In a larger view, many explanatory elements of phlogiston persisted, but fragmented, under different labels, and apart from the new primary concerns about elements and measuring elemental composition. That is, the phlogistic concepts were typically redistributed, not all lost.

For example, phlogiston helped clarify a link between chemical reactions and electricity. Phlogiston, considered the source of light and heat in fire, was also regarded by some as the basis for sparks and electrical current (or electric fluid). Based on that interpretation, electricity was successfully applied in reducing calxes to their corresponding metals. Oxygen theory did not explain that. Humphrey Davy, in his early work on chemical batteries, interpreted the electrolysis of water in terms of the release of phlogiston from the battery's cathode, or negative terminal (giving hydrogen its combustibility). He continued his work on electrolysis, but later shed the language of phlogiston. George Gibbes proposed that phlogiston was "the principle of the negative side of the galvanic apparatus" (a view that fits the interpretation of phlogiston as electrons). These elements of phlogiston may seem to have been lost, at least in the short term. But the connections between chemical energy and electricity were not entirely abandoned. They would be elucidated many years later, although without phlogiston as a concept—*nor* with the concept of oxygen in its stead.

While many chemists accepted the role of oxygen as an element in combustion, they noted that Lavoisier did not adequately explain flammability: why some things burn, and others not. Yes, oxygen is a limiting factor in combustion. But so, too, is fuel. By focusing on mass, Lavoisier had dismissed phlogiston as a *substance*. But phlogiston was essentially a form of energy (in today's terms, chemical energy, or reduction potential). This dimension of combustion remains—for example, in the concept of "the fire triangle."

Thus, phlogistonists had not, as Herschel so virulently claimed, "impeded the progress of science" with a "false theory" (Chapter 1). As Alexander Crum Brown noted decades afterwards:

> There can be no doubt that this [potential energy] is what the chemists of the seventeenth century meant when they spoke of phlogiston. . . . we have only to regret that the valuable truth embodied in it [the phlogistic theory] should have been lost sight of; that the antiphlogistonistic chemists, like other reformers, destroyed so much of what was good in the old system.[23]

Those insights survived, nonetheless, piecemeal.

There was loss, of a kind. But it accompanied the (positive) epistemic work of identifying and excising errors, based on deeper evidence (Chapters 5 and 6). Many of Lavoisier's "successes" proved to be errors. Most notably, his theory of acidity based on oxygen (still evident in the vestigial term "oxygen") and his view of heat as a substance (caloric) were wrong. Some combustion can occur without oxygen (e.g., sulpheration). Lavoisier also imagined the heat of combustion originating from oxygen changing state from a gas to a solid: an error, and a reflection of his erroneous confusion of latent heat with chemical energy (in today's terms). Even Lavoisier's achievements could not be uniformly aligned with gains in knowledge.

In assessing any notion of Kuhn-loss or progress, one must consider what knowledge was lost, what was gained, both in the short term and the long term. The episode known as the Chemical Revolution exhibited plentiful error, but among multiple perspectives. One may construe those errors as "losses," but they reflected rising standards of proof. For example, it was no longer plausible to imagine phlogiston in terms of weight (or even negative weight). A theory of acidity had to account for newly discovered acids that did not contain oxygen. Discovery of a former error is not really a loss, but a clarification of a premature conclusion. Ultimately, the outcome of this famous debate was a differentiation of complementary explanations (Figure 5.4)—and, to the degree that this reflects increasing standards of evidence and justification, that may generally be interpreted as progress.

As illustrated in the case of phlogiston, and other historical cases of conceptual change, standards of what counts as reliable knowledge sometimes escalate. Along the way, some concepts are contextualized and circumscribed, while others are abandoned as not meeting the new standards. Their former justifications are recognized, albeit belatedly, as unwarranted. We *know* now not to rely on them. The concepts that persist are thus of higher quality, and worth "trading" for the former ones. Negative knowledge, along with its conventional partner, positive knowledge, fills the gap where once there was uncertainty or ignorance (again, see Figure 8.1). That's progress.

Error repertoires

Here, we may shift focus to the more pragmatic concerns of the lab and field: developing a set of practical principles we may call Error Analytics. How can our knowledge about the dynamics of error guide and possibly bolster the practice of science? Recently, there have been renewed public concerns about the reliability and trustworthiness of scientific knowledge. Many commentators have focused on reproducibility and replication failures, or fraud and questionable research practices, as key issues. But I hope this book underscores the importance of a more expansive approach. A historical perspective alerts us to the significant role of a wide spectrum of error types. And it provides us with critical information about when and how scientists have effectively resolved those errors. Our chief focus should thus be not successful replications (or not), but how to interpret incongruences and how to localize errors. How do we minimize the incidence of errors of all sorts? How do we *manage* error?

Memory is our first tool. The memory of error may be fruitful in multiple contexts. As noted earlier, it may be in the form of a vast inventory of general error types (see Figure 2.2). Or it may be in the form of methodological norms that are designed to avert patterns of reasoning or justification that have proven susceptible to error (section above, Figure 8.2). Yet another way of formalizing the memory of errors accommodates the needs of specific fields of study. Scientists focus more on the vulnerabilities of the particular circumstances and types of investigative work within that field. We may refer to these local and context-specific catalogs of error types as *error repertoires.*

The concept of an error repertoire was introduced by Deborah Mayo in 1996. "The history of mistakes made in a type of inquiry," she noted, "gives rise to a list of mistakes to either work to avoid (before-trial planning) or check if committed (after-trial checking)."[24] Namely, in the course of successive investigations by a particular research group, say, or in a particular field, incongruences inevitably arise. As they are resolved in succession, one learns what not to do (again). The list of possible errors is a reference guide.

As an error repertoire grows, it simultaneously raises the standards of what counts as an acceptable solution. That is, the criteria for reliable claims become more stringent. Consider the classic oil-drop experiment of Robert Millikan. Millikan was mindful of many sources of error, such as temperature, convection currents, the role of a fan in distributing the temperature evenly, the position of his floating drop relative to the edges of the electric plates, and so on. At first, Millikan's results were widely variable, reflecting the influence of those factors. As Millikan became more familiar with the sources of error in his apparatus and was able to regulate them, the variability decreased. From October 28, 1911 to February 13, 1912, Millikan observed some 68 drops. None of those appear in his published work. After that date, however, the experiment seemed to be more "under control" and the calculated values of *e* more uniform. That was when data collection began in earnest: a tribute, in a sense, to exploring—and heeding—his error repertoire.[25]

Successive determinations of many physical constants exhibit a similar pattern. The historical sequence of values converge over time as errors are successively encountered and resolved, as unreliable methods are discounted, and as corresponding precautionary measures become institutionalized. That includes the value for Hubble's constant (Chapter 1). As each error in the classification of celestial objects and their brightness was discovered, it set a new standard. And historically, one sees the values begin to stabilize over time.[26]

Some error repertoires are implicit. Others are quite explicit. Consider the medical problem of determining death. In the late 1960s, conceptions of death shifted toward "brain death": cessation of mental function. Easily said, perhaps. But what were the observable medical criteria? A Harvard committee of experts attempted to define an "irreversible coma." Attention focused on three forms of unresponsiveness and the role of an electroencephalograph (EEG) in measuring lack of electrical activity of the brain. But how does one ascertain if the loss of function is permanent, the dimension that essentially signifies death? There were plentiful examples of people who had revived after a coma or loss of consciousness. For the sake of trustworthy medical practice and professional ethical conduct, the criteria needed to be error-free—and transparent. So, the committee included an explicit caveat (acknowledging a possible error): repeat the tests after 24 hours. Also, ensure "that the [EEG] electrodes have been properly applied, that the apparatus is functioning normally, and that the personnel in charge is competent." Those were important requisite checks, of course, however cursory. The Committee also noted key exceptions, leading to "additional safeguards"—other possible errors to avoid. "The validity of such data," they noted, "... depends on the exclusion of two conditions: hypothermia (temperature below 90 F [32.2 C]) or central nervous system depressants, such as barbituates." Indeed, a subsequent 1977 study by the National Institutes of Health found that a flat EEG, coupled with unresponsiveness, led nearly universally to cardiac arrest. But not completely. Of 187 patients, 2 recovered. Yet those two patients were suffering drug intoxication, and they exhibited reactive pupils, conditions that the Committee had specified in their error repertoire.

Debates on the philosophical and legal meanings of "brain death" in relation to death continued in the ensuing decades. But the criteria for ascertaining the condition clinically, at least, remained relatively stable. The reliance on an EEG waned, as physicians gained more confidence in clinical means of definitively gauging unresponsiveness and loss of reflexes mediated by the central nervous system. Still, in 2010, the American Academy of Neurology offered further clarity. While they noted that there was "insufficient evidence" to answer some subtle, detailed questions, they were able to establish an explicit checklist for use by practicing physicians. Eight separate brainstem reflexes were specified—including pupils' response to light, gag response, and cough reflex: *all* needed to be checked. They also added such qualifications as: "no evidence of residual paralytics (electrical stimulation if paralytics used)"; "absence of severe acid-base, electrolyte, endocrine abnormality"; "systolic blood pressure ≥100 mm Hg"; and "no spontaneous respirations": other conditions where the diagnosis might prove mistaken. That is, the error repertoire had not only been useful, but it had evolved further: another case reflecting progress via negative knowledge.[27]

Error repertoires may seem like inconsequential tools. But they can be quite significant. In the case of sequencing the Neanderthal genome, it led to a Nobel Prize. Over the course of roughly two decades, Svante Pääbo endeavored to extract the ancient human DNA that could be studied in relation to modern humans. Originally, the process was plagued by problems of contamination and uninformative fragmentary samples. The error repertoire grew, and guided each subsequent effort. Each challenge was documented and met with a new corresponding experimental tool. Modern DNA? Use clean rooms and other hygienic techniques. Small samples of DNA? Use PCR. Artifacts from the PCR amplification? Run multiple extractions and develop a consensus sequence. Small fragments? Search for overlaps in the segment ends and reconstruct longer sequences. More modern DNA introduced inadvertently by the workers? Tag the original DNA with diagnositc "molecular bar codes" to differentiate the original from any later contaminant. More modern DNA from lab workers? Identify their own sequences and probe the Neanderthal results for (and delete) those specific modern sequences. Each error in the error repertoire was leveraged to mitigate later errors.[28]

Error repertoires can be put to work in other ways. They may also guide fruitful error analysis. For example, models of climate change often disagree in their predictions. Although the models are simply models—and thus acknowledged as incomplete and not intended to represent reality faithfully in all respects—the incongruence between different models indirectly indicates significant errors somewhere. Modelers are interested in diagnosing the errors in their models, toward identifying weaknesses or limitations that can inform the development of more realistic or powerful models. However, the complexity of the various models and the diversity of their assumptions and structure make fruitful comparison extraordinarily difficult. Ryan O'Loughlin observes how a reference list of errors—an error repertoire—may assist in error diagnostics. For example, some models have shown sensitivity to the parameters used to represent clouds, a variable in reflecting solar radiation. Others are sensitive to the choice of parameters for convection units—and how they affect precipitation trends. These known sources of errors can be used to assess other models. Background knowledge of physical relationships has also proven useful on occasions, and by assembling more examples, this strategy of error diagnosis can become more consistent and exact. This is a project yet to be fully realized, but based on the work of some climate modelers, shows promise.[29]

Error repertoires may function just to avoid mistakes again. But this can be an indicator of expertise, as noted above. By knowing how to avert disaster, one thereby ensures a more efficient and reliable research enterprise. One may also put the negative knowledge to active work. They can also be important diagnostic tools in troubleshooting incongruences, and in trying to isolate putative errors. Errors may be inevitable. But savvy scientists can catalogue the errors, and learn how to mitigate such interruptions and detours moving ahead, using error repertoires to inform future research.

Error signatures

Some errors carry characteristic *error signatures*. That is, the results are not merely "abnormal." They are irregular in a particular way. If one has encountered and deciphered the same irregularity before (again, relying on memory of error), the distinctive pattern can provide concrete evidence of what exactly has gone awry.

This is what happened when Greg was investigating superconductivity in his metal alloy ("The case of the mislabeled hafnium" in Chapter 5). When his collaborator encountered discordant results, he and his advisor reexamined their original results. The advisor—exhibiting perhaps his expertise through intimate knowledge of errors—noticed likely evidence of impurity phases in the X-ray diffraction image. That was the critical clue for Greg to track down the material error in his supposedly pure hafnium, tainted (he found) by traces of zirconium.

An error signature also helped in the case (above) of analyzing Neanderthal DNA. One sample showed that dating discrepancies were limited to longer DNA segments, but not to shorter segments. On the assumption that older samples would more likely be degraded and fragmented, compared to modern DNA, the researchers reasoned that the bias indicated that modern human sequences had indeed crept into the ancient sample. Here, the distinctive "signature" of the error helped indicate what error had occurred: namely, contamination. It even allowed estimating the likely proportion of contaminated DNA.[30]

Error signatures can obviously be valuable diagnostic tools in research. More work needs to be done, not only on making error repertoires explicit but also on articulating the error signatures that might be associated with each error type.

Checklists and checkpoints

Error repertoires resonate with the more familiar concept of checklists. As championed by Atul Gawande in *The Checklist Manifesto*, checklists are tools for "getting things right." They lead the user through an explicit compilation of diagnostic elements, each addressing a separate error type. For example: the criteria for determining brain death (noted above). Checklists have proven enormously effective at reducing the frequency of error in aviation (pre-flight checklists) and surgery (pre-op checklists), as well as in other contexts (even such specialized occasions as plating protocols for high-end restaurants or local technical requirements for touring rock bands). Checklists prove valuable in a wide spectrum of cases.[31]

There are differences, however. Checklists prove particularly effective and flourish in certain contexts. They are especially significant where complexity reigns, and where it is virtually impossible for any one person to recall every relevant detail dependably at a given moment, most notably when under stress. Checklists help ensure completeness. They are also valuable where expertise is distributed, and where knowledge from various sources has to be integrated and organized, possibly among people who are not familiar with each other. Checklists foster effective communication and coordination. And they are of utmost importance where the cost of a single error or failure is unacceptably high—as in plane crashes, or deaths by preventable surgical errors. We have seen too many skyscrapers, dams, and bridges collapse, too many wildfires sparked or fed by technical oversights, too many massive chemical or oil spills. Checklists help ensure discipline. Checklists, when followed, help stem costly "accidents" from "minor" overlooked factors.[32]

The role of error repertoires in science may not be quite so dramatic. The levels of complexity, coordination, stress, and urgency in daily scientific research are usually not extreme, and the scale of negative consequences will rarely reach a monumental scale. Still, the rigor of referring to an explicit error repertoire may improve scientific practice, by helping to protect against inadvertent mishaps. As with the case of checklists, some institutionalization of error-checking seems well advised.

While timeliness in checking for errors in science is rarely utmost, mitigating needless error is, nonetheless, important in the long run. With sources of error distributed throughout the whole process, however, it is hard to imagine one occasion where all the relevant errors would need to be addressed at the same time—say, by consulting a master checklist. Still, periodic review during the course of an investigation seems appropriate. So, the emphasis in science may shift from highly prescribed checklists to periodic *checkpoints*. Checkpoints would be occasions when explicitly revisiting the error repertoire would seem particularly fruitful.[33]

The concept of checkpoints may be illustrated biologically, in the cell cycle. As cells grow and reproduce, they pause at certain stages if the prerequisites for further development are not in place. For example, cells do not begin to reproduce until chemical signals indicate that the DNA is fully replicated (without damage) and that there are enough protein reserves. They do not proceed to division if the chromosomes are not paired and aligned to separate evenly. If the signals are awry, cancerous growth can occur. One may think analogously for science. At particular stages of the investigative process, one may envision occasions when error review seems particularly apt. In that sense, checkpoints are also like the periodic occasions for quality control in industrial manufacturing. In science, one is "producing" reliable knowledge. The checkpoints might therefore target any time an investigator is preparing for a major new investment of resources. At such moments, one is assessing whether a change in epistemic posture—namely, an increase in epistemic commitment—is warranted (see Chapter 1).

The discussion of mapping in Chapter 2 (see Figure 2.1) can provide a framework for reference. That account might easily give the misleading impression that science proceeds by simple induction. Collect the data. Develop a valid interpretation. Publish the conclusions, and then argue with peers until everyone reaches a consensus. That is a simplistic caricature, of course. But it does, nevertheless, reflect the general abstract structure *of the justification*, linking raw data to processed data, to integrated concepts, to interactive criticism and the resolution of discordances, ambiguities, and disagreements. Hence, for our purposes, one may crudely conceptualize the justification process into layers: observational, conceptual, and social level (discoursive). Any prospective "schedule" will surely vary from discipline to discipline, but let me propose one interpretation. We may generically envision major checkpoints at several key transitions:

(1) before any experiment or study actually begins:
What data need to be collected, and why? There is a subtle difference between planning (positively) to demonstrate or test a particular hypothesis, and considering (negatively) how it could all go wrong. That may involve considering the conceptual errors and theoretical alternatives. How might that prospective evidence be incomplete, or misleading, or misframed? Mindful of sampling error, one makes critical decisions about sample sizes. Aware of potential confounders, one plans appropriate controls or subsidiary studies. Graduate students are often admonished to comb the literature. Know the field's observational error repertoire. This may seem like a lot to do before even walking into a lab or the field. But it is essential to ensure that the subsequent work is not for naught.[34]

(2) as one begins the sometimes-mundane tasks of collecting data:
Here, material errors seem utmost. Checking here includes attention to any instrument or apparatus to avoid artifacts. Even rechecking the same setup: who knows what mischief may have happened overnight? In some cases, simply getting the equipment to function as intended (without error) may be the bulk of the experimental work.

(3) transitioning from (primarily) data collection to (primarily) data analysis:
Preliminary analysis of data may indicate if there have been problems at the observational stage—problems that can still be addressed through additional or supplemental observations, before the equipment or field site is decommissioned.

(4) processing and interpreting the data and preparing a formal argument for publication:
This moment of explicit reflection about error is already familiar and customary, I think. Fred Suppe's analysis of the structure of the scientific paper is valuable, here. Arguments do not only propose a series of claims based on the findings, but they also consider "associated doubts" for each. Each critical "query" offers an alternative explanation, in the implicit voice of an imagined critic. For each query, the author then provides a "rejoinder," discounting the source of error, as a way to affirm the original claim. In this way, the typical scientific paper becomes a virtual, preemptive dialogue between the author and presumptive critic. By systematically considering and dismissing the possible errors, authors underscore how reasoning about error is central to scientific justification.[35]

(5) review articles and other synoptic social-level critiques:
Vetting by peers can be just as important as the original presentation of the scientific claim. But who speaks for the community? This fraught role is often filled by leading scientists when they survey the status of research across a whole field in a review article. In that role, they frequently critique the diverse (and sometime divergent) work in the field, highlighting errors (or deficiencies) across a broad range of studies. The implicit hope is that this will spur research to address the errors or fill the gaps. Another dimension of review articles is to help mediate or adjudicate theoretical debates. The reviewer may try to adopt a bird's-eye perspective and interpret the dynamics of an active debate. They may articulate the strengths and weaknesses of alternative positions. A fruitful analysis might hint how conflict might be resolved by sorting the contexts or domains of different theories or conceptual approaches (Chapters 5 and 7). Although review articles are a regular feature of current scientific practice, they are also haphazard and *ad hoc*. An effective social system for error-checking remains a work in progress.

Deploying error repertoires at critical checkpoints is a form of discipline. But when applied prior to major transitions in the epistemic process (Figure 2.1), the practice is an investment in minimizing errors and their downstream effects.

Error probes

Physicist Robert Oppenheimer (of atomic bomb fame) once said, "science is the business of learning not to make the same mistake again."[36] As detailed in previous sections, the use of error repertoires and of error-derived methodological norms in mitigating errors seems to align with this view. However, errors—or, at least, squarely addressing the possibility of error—may also have a more constructive role: namely, in deepening the reliability of claims.

Some scientific reasoning proceeds by ruling out errors. Many philosophers—and working scientists, too—subscribe to Karl Popper's doctrine of falsification. Namely, you cannot *prove* a claim with a limited sample, no matter how hard you try. But you can (so the wisdom goes) *disprove* a claim with a confirmed counterexample. It all sounds very logical and foolproof. Falsification might be all well and good if the world was governed by the rules of mathematical logic, precise stipulative definitions, discrete and consistent natural kinds, and universal laws. But that appears not to be our world. As the cases throughout this book indicate, anomalies and exceptions abound—and they do not necessarily "falsify" anything. Falsification seems to be a convenient and beguiling model—and potent rhetorical tool. Yet discarding general theories based on individual or sporadic counterinstances proves not to be an effective knowledge-building strategy. Historically, strict falsification—of the sort propounded in the widespread rhetoric in its favor—is not a reliable model for assessing scientific theories. It is a *philosophical*, or epistemic, error type.[37]

But what if one turns this logic on its head? Suppose you test for the presence of a particular error. If you do not find the error, are you entitled to cross it off your list? Can you "falsify" the possibility of that particular error? Have you eliminated or at least reduced the viability of the alternative explanation? That is the reasoning behind Popper's notion of a *severe test*, also championed by Deborah Mayo. They both contend that surviving severe tests strengthen the claim that has been tested.[38]

The most commonplace example of a severe test is the ordinary, yet ubiquitous tool of controlled experiment. One rules out an error by testing for a single confounding variable or a specific alternative explanation, thereby persuading oneself that it is *not* relevant. Appropriately targeted controls help bolster the experimental conclusions by eliminating the possibility of specific errors. One may recall, here, any of the cases introduced earlier: John Snow's work on cholera as waterborne (Figure 5.3); Vorderman's work on rice diet as critical to beriberi in Java (Chapter 2); the Mesmer Commission's demonstration of psychological suggestion (Figure 5.2); tracking the faulty time signal cord connection in the apparent faster-than-light neutrino (Chapter 5); tracing phosphohistidine as a confounder in ox phos (Chapter 2, Figure 2.4); and so on. Ruling out error matters. Immensely.

Such logic indeed seems pervasive in contemporary scientific practice. This is illustrated most vividly in the structure of the standard scientific paper (Suppe's analysis discussed in the section above). In making their arguments, scientists habitually consider a series of prospective rebuttals (viewed as potential errors in their justification), and then proceed to discount each one in turn—aware that the whole argument could unravel if any single one of them were to be accepted as valid. On a more expansive level, eliminative induction—quaintly expressed as the "Sherlock Holmes strategy"—is a potent reasoning tool. As the fictional detective famously said, "Once you eliminate the impossible, whatever remains, no matter how improbable, must be the truth." Namely, ruling out possible alternative conclusions allows us to reason more confidently toward the remaining explanation as correct. Essentially, the process of eliminative induction applies the principle of a severe test across a variety of challenges to a proposed justification.

One may expand on this logical framework and proceed a step further. That is, we should not necessarily be satisfied with addressing and discounting just the current assortment of alternative explanations, or known pitfalls. One should actively search for other possible (yet unknown) sources of error. That is, one may go beyond exhaustively considering the errors already documented. Namely, probe more deeply. Here, the more comprehensive inventory of error types functions as an important resource, helping to structure such search and guide the imagination. We may call these more rigorous explorations *error probes* (another term introduced by Deborah Mayo).[39]

Error probes are adventures in negative knowledge. "How might the justification for this particular conclusion be in error?" (Note, again, the focus is on the justification, not the conclusion itself—see Chapters 1 and 5). For example, where a claim has been built on proxy variables, heuristics, or model simulations, one can cycle back and engage in the more lengthy and costly research that does not rely on those economical shortcuts. Or: where initial claims were based on provisional assumptions, one can redo the study, applying more rigor. Perhaps a "successful" pilot study leads to a sequel investigation with a more substantive sample size, or with more realistic conditions. Or: where there is wide variability in the data, one can seek better instrumentation and more precision. Charles Coulomb published his inverse-square "law" for electrostatic attraction in 1785 based on a mere three data points.[40] Well, that would have been an occasion ripe for review through error probes.

These strategies are found historically. For example, Millikan was aware that his calculations for the value of the charge of the electron, *e*, relied critically on the value for the viscosity of air. Indeed, he had engaged one of his students to newly determine its value, but two decades later it was shown to be inaccurate (alas, along with Millikan's original value of *e*). That physical constant might have benefitted from a more thorough error probe on that occasion. Or: after pursuing the idea of chemical affinities based on Newton's reputation, chemists might well have adopted a more critical (less deferential) stance and entertained alternative views more thoroughly (Chapter 7)? Each error probe escalates the standards of justification by engaging vulnerabilities to further errors. Every significant error that can be excluded contributes to the growth of knowledge, by strengthening the justification for the claims that pass muster. How *probative* is the evidence? I refer to the practice of probing for error as the difference between (ordinary) reliability in science and *deep reliability*.[41]

The concept of deep reliability contributes to a shift in how we think about evidence. Conventionally, scientists and philosophers tend to talk about the "weight" or "probability" of the evidence or the overall "balance" of the evidence—betraying an implicit notion that evidence is a fundamentally quantitative notion, and that *more* evidence yields a stronger justification. Attention to error and to resolving uncertainties, however, fosters a more qualitative view of evidence. Namely, evidence helps us discriminate between alternatives and thereby negate errors, or articulate the relevant negative knowledge (see again Figure 8.1). Justification improves through increased resolution or through judicious differentiation, not through sheer volume. What ultimately matters is how thoroughly sources of error have been probed and confidently ruled out.

Consider an amusing 1971 review article that compared the major biochemical theories on oxidative phosphorylation (Chapters 2, 3, 6, and 7). An unusual and striking feature of the review was a "scoreboard of experimental evidence"—framing the alternative theories (perhaps tongue-in-cheek) as rival sports teams. The author, Efraim Racker, entered his assessment for various categories of evidence, as positive, negative, or ambiguous. And he issued a verdict based ostensibly on the weight, or balance, of the evidence.

The following year, as the ostensive competition continued, he published a "revised scoreboard." Here, he announced that "the balance is shifting in favor of the chemiosmotic hypothesis." But partly what allowed his new assessment was how the evidence had become clearer and distinguished more sharply between the "competing" theories. This foreshadowed the resolution of the controversy, still several years away. Both theories had merit, but with different scope. The evidence had become divided, some decisively aligned with one theory, the rest aligned with the other. Ultimately, the resolution was not about the "weight" or volume of the evidence. Rather, it was about sorting the evidence on a finer scale and differentiating the domains of the respective theories. That essentially reflected *deeper* knowledge, not ostensibly "more" knowledge.[42]

One could equally recall other historical cases where incongruences were resolved by articulating the evidence as a finer scale of resolution: the discovery of argon from a mixed sample (Chapter 5), the discovery of different isotopes (and explaining the anomalous atomic weight of chlorine; Chapter 6), dividing Cepheid variable stars into two sub-classes, with its consequences for calculating the Hubble constant and estimating the age of the universe (Chapter 1), or determining the chemical nature of vitamins (not all "amines"; Chapter 6). Reliable solutions in all these cases depended on increasing the resolution of evidence and differentiating two claims that at first appeared similar.[43]

In summary, error probes help in deepening reliable knowledge. That may be expressed in several prospective aphorisms:

Nothing's concluded until error's excluded.
Check for flaws before concluding the cause.
Uncertainty lasts until probing error is past.
No question's solved unless error's resolved.

To err is science. But so, too, is striving to demonstrate the absence of error.[44]

Prospects

At the close of the year 2000, science journalist David Kestenbaum assembled a news report that documented some of the year's errors in science.[45] It was an entertaining and provocative variant of the routine year-end retrospectives that survey the year's top films, best books, memorable performances, or notable restaurant openings. For example, every year *Science* magazine now announces a "Breakthrough of the Year," along with nine others that together mark the year's "top ten" discoveries. Rather than focus on the most notable scientific achievements, however, Kestenbaum surveyed a handful of errors. It was a modest (but healthy) acknowledgment that missteps are integral to science and that they, too, are discoveries of a sort. They even constitute *news*. One scientist who was interviewed took exception to the whole project. "It was absurd to focus on the few things that turned out not to be so," he said, "given the overwhelming number of discoveries." I hope the reader will now appreciate quite the opposite.

Yes, focusing on "things that turned out not to be so" can indeed be informative, and analysis of them, fruitful. We need to *naturalize* and *normalize* the role of error in science. Error is an inevitable component of how science learns. It is not "pathological" (as Langmuir and others have claimed), nor is it a measure of the quality of the scientist. The cost of pursuing scientific knowledge is the risk of error.

To err is science. But so, too, is managing error effectively.

This chapter has helped to summarize the various lessons of how we learn from error on a general level. Indeed, such a focus might well change how practicing scientists practice science, how science teachers teach science, how science administrators administer science, and how consumers of science interpret the scientific consensus in informing personal and public decision making. That is the prospective horizon for a fully realized philosophy of error.

This book has only set out the groundwork for the further study of so vast a topic as error in science. Yet I hope it provides a fruitful framework for doing so and a vivid impression of its importance. A Philosophy of Error, including Error Analytics, is poised to transform our view of science. And so, I close with a list of still unanswered questions, coupled with an invitation to the reader to join in pursuing them:

- Can we construct a systematic inventory of error types that is more comprehensive and more detailed than what I have outlined here (Figure 2.2, Chapters 2–4)?
- What error types are most typical in specific fields of science? What error repertoires can be articulated for each domain of research? Can we specify error signatures for various error types (Chapter 8)?
- What is the frequency of various error types (Chapters 2–4)? Are some more common than others, hence more essential in the educational preparation of scientists?

- Can we find informative relationships between error types (Chapters 2–4)? (For example, does communal confirmation bias, at the social level, tend to correlate with an increased risk of overlooking confounders, at the observational level?)
- Can various incongruences be linked systematically to corresponding error types, facilitating the epistemic work of finding the appropriate additional evidence to resolve them (Chapters 5 and 6)?
- Can we formalize an approach to negative knowledge (Chapter 8), allowing us to structure the institutional memory of errors (or "failures") and guide how they are communicated, archived, and later searched?
- Can we articulate more fully the resolved/uncertain distinction (Figure 8.1) and the ways we rely on evidence that can differentiate, discriminate, or resolve claims, not merely adding to an overall bulk of "more" evidence (Chapters 6 and 8)?
- Can we acknowledge more fully how the difference between insight and blind spot—between discovery and error—depends on context, and is not so great as the privilege of retrospect might suggest (Chapter 7)?
- Can we curb the pessimistic view of error, nurture support for work in the Philosophy of Error and Error Analytics, and ultimately foster a view of "progress through error" (Chapter 8)?
- Can we revise the image of "the" scientific method as the guarantor of scientific reliability, and instead convey more about the importance of the social and institutional dimensions of science in addressing bias and filtering error (Chapters 4 and 7)?
- Can we clarify more about the dimensions of uncertainty relative to error (Chapters 1 and 5)?
- Can we acknowledge the pervasiveness of error in science without losing our trust in the ability of ongoing science to encounter and remedy any significant error (Chapter 1)? Can we maintain—or even strengthen—our trust in the consensus of the relevant scientific experts (Chapters 4 and 8)?
- Can we acknowledge that the cost of innovation is the risk of failure, and the cost of discovery is the risk of error (Chapters 7 and 8)?

These compelling challenges await the effort of historians, philosophers, and sociologists of science, and, of course, reflective scientists themselves.

Acknowledgments

This work spanned over two decades. I am grateful for the many (many) voices of encouragement—as well as the criticisms that exposed issues yet to resolve. This includes everyone who has endured my habitual conversational query, "So, what's your favorite blooper in the history of science, or in your own research?"

First, I am keenly aware of benefitting from the earlier work of colleagues in history and sociology of science, as well as research by journalists and teachers. Every citation is more than a customary gesture of academic courtesy or of epistemic transparency. Each is a genuine expression of gratitude and debt.

A few earn special appreciation. Lindley Darden's work on strategies for anomaly resolution and theory change has been a model and an inspiration. She has been supportive throughout, including several fruitful conference sessions together. Deborah Mayo hosted what one acquaintance dubbed a "summer camp for philosophers" (an NEH institute)—where my work on error began in earnest, and which led to my 2001 paper on error types, which forms the kernel of this volume (Chapters 2–4). Her influence will be most apparent in the last chapter. Finally, Hanne Anderson helped arrange an extended stay at Aarhus University, leading a buoyant community where I was able to transform years of haphazard notes into a detailed plan for the entire book.

In addition to these, I wish to mention those who have engaged in more extended discussion and provided poignant feedback (unjustly listed in alphabetical order): David Bygott, Hasok Chang, Carl Craver, Glenn Dolphin, Kevin Elliot, Stephen Guttinger, Peter Heering, Alexandra Karakas, Alan Love, Greg Meisner, Peter Ramberg, Alex Werth, Brian Woodcock, Gábor Zemplén, and participants in the VPI summer institute. I am also grateful for the special invitations from Roberto de Andrade Martins, Maria Élice de Brezenski Prestes, Charbel el-Hani, Cibelle Celestino Silva, and David Rudge, which became occasions to develop and present my ideas more fully.

For institutional support—not to be discounted as secondary—I am indebted to The University of Chicago and the Searle Foundation, Aarhus University, the University of Minnesota, the National Endowment for the Humanities, and my grandparents. To Libbie Henderson, my thanks for a partner's patience and support—and for posing those penetrating but insightful questions that I was sometimes too blind to notice myself.

Notes

Chapter 1

1. e.g., Rudolf Carnap, Carl Hempel.
2. e.g., John Ziman, Alvin Goldman, Robert Merton.
3. e.g., Bas van Fraassen, Wesley Salmon, Paul Thagard.
4. e.g., Imre Lakatos, Thomas Kuhn.
5. e.g., Larry Laudan.
6. e.g., Francis Bacon, Gaston Bachelard, Mario Bunge, Don Ihde, Wilfrid Sellars.
7. e.g., John Dewey, Peter Lipton, Colin Howson, and Peter Urbach.
8. e.g., Karl Popper, Massimo Pigliucci and Maarten Boudry, Lee McIntyre.
9. e.g., David Hull, Karin Knorr-Cetina, Deborah Mayo, Hans-Jörg Rheinberger.
10. See Darden (1998) and Mayo (1996). For further drifts in this direction, see McIntyre (2019) and Zimring (2019).
11. Franklin (1785). The quote is frequently attributed to Franklin. However, there was no Introduction in the original report, written in French for the king who commissioned the work. Although the aging Franklin was the nominal head of the Mesmer Commission, much of the report is now attributed to Antoine Lavoisier. There is no evidence that Franklin wrote the introduction to the English translation, which refers to him in the third person.
12. Franklin (1785). On the romanticized Enlightenment views of error, see also Bates (2002) and Boumans & Hon (2018, pp. 1–3).
13. Encyclopedia Britannica (1771); Herschel (1830, pp. 300–301).
14. Allchin (2008b 2009; Darden (1998); Holmberg (2008); Livio (2013).
15. Gilbert & Mulkay (1984, pp. 49, 65, 66, 71, 79, 81, 93, 96).
16. Gratzer (2000, pp. 1, 23, 109, 306); McIntyre (2019, pp. 155–166); Youngson (1998, p. xi, xiii).
17. Allchin (1995; 2003; 2004).
18. Exemplary corrective histories may be found in, e.g., Martinez (2011); Numbers (2009); and Waller (2002). Informed readers will know of the notorious claims of Imre Lakatos defending rational reconstructions, while leaving the real history "in the footnotes"—a practice now generally regarded as irresponsible.
19. Hull (1988).
20. Gigerenzer (2005); Gigerenzer & Selten (2001); Wimsatt (2007).
21. Buchwald & Franklin (2005).
22. e.g., Allchin & Werth (2020); Barrett (2015); Catalogue of Bias Collaboration (2020); Giere (1988); Gilovich (1991); Gilovich, Griffin & Kahneman (2002); Kahneman (2011); Piattelli-Palmarini (1994); Sackett (1979); Sutherland (1992). On cognitive regulation, see Goldman (1986).
23. Weintraub (2007).
24. See also Hon (1987, p. 382).
25. e.g., Baird (2004); Chang (2008); Craver & Tan-Cohen (2024); Franklin (1986; 2002); Galison (1987); Hacking (1984); Karakas (2022); Kohler (1994); Mari (2003); Rheinberger (1997); Rothbart & Slayden (1994); Tal (2016).
26. e.g., Harding (1991); Hull (1988); Longino (1990); Merton (1973); Solomon (2001); Ziman (1968; 1978; 2000).
27. e.g., Ankeny & Leonelli (2021); Creager, Lunbeck, & Wise (2007); Mayo (2018); Turnbull (1989); Wimsatt (2007).

28. Kragh (2008).
29. Latour & Woolgar (1979).
30. Šešelja & Straßer (2014); Whitt (1992).
31. Eriksson et al. (2008).
32. Darden (1998); Livio (2013); Pauling & Carey (1953, pp. 84–85).
33. Gilbert & Mulkay (1984); Oreskes (2019); Vickers (2022); Ziman (1968).
34. Livio (2013).
35. On conflating error and uncertainty, see, e.g., Boumans & Hon (2018).
36. Hacking (1990); Porter (1986); Stigler (1986).
37. Mayo (1996; 2018); Hacking (1990).
38. Adler (2002); Hon (1989b); Wilson (1990, Ch. 9).
39. Huckra (1992; 2008); graph: Pritychenko (2017); Trimble (1996). See also https://www.cfa.harvard.edu/~dfabricant/huchra/hubble/h1920.jpg
40. For graphical depictions of the history of various subatomic particle constants, see https://pdg.lbl.gov/2020/reviews/rpp2020-rev-history-plots.pdf
41. On Enlightenment views of error, see also Bates (2002).
42. e.g., Gratzer (2000, quote on p. 306); Rescher (2007).
43. Jastrow (1936, pp. 16–18).
44. Mayo (1996, esp. pp. 4–7).
45. e.g., Firestein (2016); Guttinger & Love (2019); Hon, Schickore, & Steinle (2009); Janis & Horowitz (1994); Karakas & Tuboly (forthcoming); Rescher (2007). For a potent critique of this perspective, see Karstens (2014).
46. Catalogue of Bias Collective; Sackett (1979).
47. Hon, Schickore, & Steinle (2009).
48. e.g., Boumans & Hon (2018 pp. 1–3).
49. Mayo (1996, pp. 2, 13–17, 146–147; 2014, p. 76).
50. This will build on my prior work in Allchin (2001).
51. e.g., Bowell & Kemp (2002); Jastrow (1936); Moore & Parker (2015).
52. e.g., Cognitive Bias Codex by Benson & Manoogian (2016); Dobelli (2013); Piattelli-Palmarini (1994).
53. e.g., Catalogue of Bias Collaboration (2020; Sackett (1979).
54. e.g., Hon (1989b; 1995).
55. Petroski (1994; 2006).
56. e.g., Alberts et al. (2015); Berg (2019); McNutt (2014); National Academies of Sciences, Engineering, and Medicine (2019).
57. Ioannidis (2005).
58. e.g., Anderson (2022).
59. Mayo (1996, pp. 5, 18, 452; 2014, pp. 68, 71).
60. Mayo (1996, pp. 437, 445; 2014, p. 68). On distinctive signs of error, see p. 5.
61. Darden (1991; 2006).
62. Jastrow (1936); Boumans, Hon, & Peterson (2018); Buchwald & Franklin (2005); Hon, Schickore, & Steinle (2009); Holton & Mack (2005); Karakas & Tuboly (forthcoming).
63. e.g., Fontani, Costa, & Orna (2015); Fox (1971); Franklin (1993); Franks (1981); Holmberg (2008); Kim (2003); Kragh (2008); Levenson (2015); Micale (1994); Ohanian (2008); Simon (2001); Weintraub (2007).
64. Boumans & Hon (2018, quotes on pp. 1, 2, 3, 10).

Chapter 2

1. Allchin (2001).
2. Bacon (1620/1902, pp. 19–22).
3. Bacon's original term, the Latin *idola*, was once commonly translated as "idols." But modern usage of this term proves highly problematic for interpreting Bacon's meaning. *Idola* denotes

"phantoms" or "false appearances," without the symbolic, religious, or devotional connotation inherent in the term "idol." Bacon characterized the *idolum* as a "corrupt and ill-ordered predisposition of mind" or "false notion"—hence "illusion," as translated here.

4. e.g., Gilovich, Griffin, & Kahneman (2002); Kahneman (2011).
5. Kahan (2013); Kunda (1990).
6. e.g., Cialdini (2007); Fennis & Stroebe (2021); Goldstein, Martin, & Cialdina (2008).
7. e.g., Lewandowsky & Cook (2020); van der Linden (2023); van Prooijen & van Vugt (2018).
8. See also review by Weingurt (2020).
9. Hon (1989b; 1995; 2009).
10. Hon (1989a; 1989b); Boumans & Hon (2018).
11. Mayo (1996, pp. 51 n.17, 150, 453–54, list on p. 18; 2018).
12. Guttinger & Love (2019).
13. Begley (2013).
14. Vazire & Holcombe (2022).
15. Sackett (1979, p. 60).
16. Darden (1991; 2006).
17. Benson & Manoogian (2016).
18. e.g., Gilovich (1991); Gilovich, Griffin, & Kahneman (2002); Gigerenzer (2005); Kahneman (2011); Piatelli-Palmarini (1994); Sutherland (1992).
19. Merton (1973); Ziman (2000).
20. Allchin (2001).
21. Ziman (1978).
22. van Fraassen (1980).
23. Judson (1980).
24. Bauer (1992).
25. Giere (1998; 2006).
26. Winther (2020).
27. Turnbull (1989).
28. Monmonier (1991).
29. Hacking (1984).
30. Hacking (1984, esp. pp. 208–209).
31. e.g., Suppe (1998).
32. Allchin (2001; 2012c); Allchin & Zemplén (2020).
33. Allchin (1999a); Latour (1987, pp. 80–83, 210–57).
34. I have omitted here the further trajectory of scientific claims relevant to the actions of citizens, customers, legislators, government agencies, judges, or corporate leaders. These claims continue their trajectory, traveling even farther from the original inscriptions in lab and field notebooks. They may appear in the news, in Congressional hearings or courtrooms, in advertising, or in the media. They may be reported further via casual conversations, blogs, webpages, or social media. Epistemic concerns here are no less important. Do the final claims ultimately reflect the consensus of professionals? Do they reflect expertise? Is the testimony, or the person reporting it, credible? Is it honest and free from the adverse effects of conflict of interest? The reliability of scientific claims in a cultural setting depends on effective science communication as much as sound research.

    Ultimately, the structure of a scientific justification involves a long trajectory—from test tubes to YouTube, from lab bench to judicial bench, from field site to website. The epistemic concern about preserving the reliability of the claims persists throughout, even beyond the scientific community. No one should understate the challenges of preserving the integrity of scientific information in the public realm, in an era when misinformation and disinformation are rampant. For the analysis in this volume, however, I focus on the development through forming a scientific consensus—that is, what happens largely within the professional scientific community. Still, one would profit from keeping the larger context in mind (Allchin, 2022b; Osborne et al., 2022).

35. Shapere (1982).
36. Ackerman (1985); Franklin (1986; 1997); Galison (1987); Hacking (1984, quote on p. 150; 1984); Kohler (1994); Pickering (1995); Rheinberger (1997; 2010; 2023).
37. Abachi et al. (1995); Campbell (1974); e.g., Hubel (1981). On naked-eye observation, see Hudson (2014, pp. xxiv, 235, 239).
38. Chang (2008); Hacking (1984); Mari (2003); Rheinberger (2010); Tal (2016).
39. Crease (1993, p. 107).
40. Alberts (2011); Cohen & Enserink (2011).
41. Meldrum (1934).
42. Borman (1990); Brackett (1990).
43. Harris (2017, pp. 93–122); Overbye (2010a; 2010b).
44. Hon (1987).
45. Harris (2017, pp. 45–47).
46. Goodman (1978).
47. Leighton & McKinley (1930); Student (1931).
48. Underwood (2015); Widom et al. (2015).
49. Lederman (1997); Yoh (1998).
50. e.g., Dutta et al. (2017); Makowski (1962).
51. Crosland (1973).
52. Carpenter (2000).
53. See, e.g., https://www.nhm.ac.uk/discover/the-origin-of-our-species.html
54. Campbell (1974); Wimsatt (2007).
55. Brodeur (1989); National Research Council (1997, quote on p. 2); SCENIHR (2015).
56. e.g., Cardot & Cavit (2012); Ghallab (2013).
57. Allchin (2013b); Carpenter (2000); Williams (1961).
58. Allchin (2016); Chase (1976); Davenport (1916); Joseph (2006, Chapter 4).
59. Allchin (2002b); Boyer (1963; 1981); Mitchell, Butler, & Boyer (1964); Attwood et al. (2006).
60. Rothbart & Slayden (1994).
61. National Weather Service (2020); Parker (2009).
62. Earman & Glymour (1980); Kennefick (2019).
63. Hanson (1958); Kuhn (1970); Zimring (2019).
64. Gratzer (2000, pp. 24, 306); Nye (1980).
65. Nye (1980); Wood (1904).
66. Farwell & Hawkes (1935); Kauffman & Adloff (2008); MacPherson (1934); Slack (1935); Woodriff (1935).
67. Bingel (1918, quote on p. 288, translated by Andrew Herxheimer); Tröhler (2011).
68. Colloff (2018); Cooper & Meterko (2019).
69. Bello et al. (2014).
70. van Wilgenburg & Elgar (2013).
71. Kozlov, Zverev, & Zvereva (2014); Zvereva & Kozlov (2019).
72. Craver & Dan-Cohen (2024); Franklin (1986; 2002); Galison (1987); Latour & Woolgar (1979).
73. Culp (1994); Hudson (1999); Rasmussen (1993); Silva (1971; 1976); Weber (2004).
74. Bascom (1972); Derjaguin & Churaev (1973, quote from p. 432); Franks (1981); Lippincott et al. (1969); Lippincott et al. (1971); Rousseau (1970).
75. Sow et al. (2020, p. 376).
76. Franklin (2002); Kaiser (2002).

77. Many sources refer to this as the Hawthorne effect. However, interpretations of the historical episode vary and many seem to oversimplify or simply misinterpret the experimental conditions and what one may justifiably conclude from them. For this reason, I refrain from this label (see Gillispie, 1991).
78. Heering (1992b); Sibum (1995).
79. Keeling et al. (1976, quote on p. 553).
80. Charles Darwin, letter to Joseph Hooker, November 3, 1873.

Chapter 3

1. On combinatorial reasoning methods, see Allchin (2020); Campbell (1960); Kantorovich (1993); Laudan (1980); Mayo (1996); Plotkin (1994).
2. Barrett (2015); Gigerenzer & Selten (2001); Gilovich (1991); Gilovich, Griffin, & Kahneman (2002); Kahneman (2011); Sutherland (1992).
3. e.g., Benson & Manoogian (2016); Piattelli-Palmarini (1994).
4. Carroll (1994); Hempel (1966, pp. 54, 58); Kosso (1992, pp. 52–60, 190); Woodward (2003, pp. 167, 236–238, 265–266); Ziman (1978, p. 32).
5. Crick (1968); Hinegardner & Engelberg (1963, quote on p. 1083).
6. Barrell et al. (1979); Elzanowski & Ostell (2019); Hanyu et al. (1986); Jungck & Allchin (2023); Krzycki (2005).
7. Leek & Peng (2015).
8. Trimble (1996).
9. England et al. (2007); Hallam (1989); New York Times (1905); Shipley (2001).
10. Mathé (1978, pp. 79–80); Vavakan (2005).
11. Allchin (1996); Carpenter (2000, pp. 24–63).
12. Bechtel & Richardson (1993); Callebaut (1993); Wimsatt (2007).
13. Ariely (2008, pp. 155–172); Giere (1988); Gilovich (1991, pp. 30–37); Kahneman (2011, pp. 79–88); Kida (2006, pp. 155–165); Mercier & Sperber (2017, pp. 212–218); Nickerson (1998, pp. 175–220); Sutherland (1992, pp. 135–142).
14. Allchin (2015b; 2016.
15. Buckland (1820; 1824; 1836); Gould (1985b)
16. Kottler (1974); Lyons (2009).
17. Stohlberg (2003).
18. Schiebinger (1989, quotes on pp. 68, 69; 1989, quotes on pp. 200, 202, 203, 214, 222); Wilkes (1810, Vol. 1, p. 560). On the naturalizing error, see Allchin & Werth (2017; 2020).
19. Fee (1979, quotes on pp. 416, 419); Schiebinger (1989, quote on p. 67).
20. Cartwright (1851, quotes on pp. 707, 709, 715); Allchin & Werth (2017).
21. Davis (1902, quotes on pp. 360–361); Fowler and Fowler (1971); Guyot (1853); Maxwell (1995); Pico (2019); Powell (1883; 1888); Shaler (1906, quote on p. 283).
22. Shaler (1906, quote on p. vi).
23. Barkan (1992); Gould (1977; 1978; 1981); Lai & Medina (2023); Saini (2020); Tucker (1994; 2007).
24. Allen (2011); Davenport (1911, quotes on pp. 4, 67, 70, 256, 257, 258; 1912a, quotes on pp. 2141, 2142; 1912b, quote on p. 1845; 1912c, quote on p. 88); Gould (1985a); Heron (1913); Kevles (19898 pp. 41–56); Morgan (1925, quote on p. 201); Witkowski & Ingris (2008).
25. Durston (1974); Garfinkel (1987, quote on p. 198); Gregor et al. (2010); Keller (1983); Reid & Latty (2016); Shaffer (1961).
26. Butler (1609, quotes in Sects. 1.4, 1.5); Johnson (2000); Jongsma et al. (2015); Sapp (1987); Seeley (2010).
27. Allchin (2008a; 2017, pp. 117–152); Allchin & Werth (2017; 2020).
28. Wimsatt (2007).
29. Ankeny & Leonelli (2021); Harris (2017, pp. 50, 71–82).

30. Hunter (2008); von Herrath & Nepom (2005); Seok et al. (2013).
31. Barry & Elith (2006).
32. Ariely (2008); Barrett (2015); Dobelli (2013); Gilovich (1993); Gilovich et al. (2002); Heick (2019); Kahneman (2011); Nisbett & Ross (1980); Piattelli-Palmarini (1994); Sutherland (1992).
33. e.g., Beveridge (1950); Kahnemann (2011); Mayo (1996); Piattelli-Palmarini (1994); Wilson (1990); Zimring (2019).
34. e.g., Salmon (1984).
35. e.g., Allchin & Werth (2020); Gilovich (1991); Kahnemann (2011); Piattelli-Palmarini (1994); Sutherland (1992).
36. Gilovich (1991, pp. 37–48); Kahneman (2011, pp. 85–88, 129–136); Nisbett & Ross (1980); Simon (1956;1969; Sutherland (1992, pp. 15–34); Wimsatt (2007).
37. e.g., Kuhn (1970); Laudan (1977).
38. Lakatos (1978).
39. Gay (1977); Sampson (2006).
40. Brower (1958); Petersen (1964); Ritland & Brower (1991); Walsh & Riley (1869).
41. National Institutes of Health (2011); Rosenson & Durrington (2020); Wein (2012; 2016).
42. e.g., Pearl (2009); Salmon (1984); Woodward (2003).
43. Hallam (1989); Montgomery (2010); Murchison (1842, quote on p. 131).
44. Allchin (1997a).
45. Allchin (1997a); Harold (1986).
46. Laudan (1981); Stanford (2006).

Chapter 4

1. Tharp, Vinson, & Yeston (2023, p. 13).
2. Merton (1973, p. 339); Oreskes (2019); Shapin (1994); Ziman (1968).
3. Hull (1988); Peterson & Jungck (1974).
4. Harding (1991); Levins (1966, quote on p. 423); Longino (1990); Solomon (2001). On demarcation, see McIntrye (2019); Zimring (2019).
5. Ziman (1968).
6. Hull (1988); Kitcher (1993); Longino (1990); Merton (1973); Solomon (2001); Ziman (1968; 1978; 2000).
7. Merton (1973, quotes on pp. 101, 339).
8. Allchin (1999a; 2012b).
9. That is, we may separate fraud into error and malpractice (Anderson, 2022).
10. Merton (1973, quotes on pp. 124, 263).
11. Harman & Dietrich (2008); Merton (1973, quotes on p. 264).
12. Soenniechsen (2008, quotes on pp. xi, 133, 179–180, 192–193, 197, 231).
13. Wimsatt (2007).
14. Harman & Dietrich (2008; 2013); Kitcher (1993); Kuhn (1977).
15. University of Cambridge (2014).
16. Davenport (1912a).
17. Allchin (2008); Allchin & Werth (2017); Fee (1979, quote on p. 416); Gould (1981).
18. Allchin (2017); Motulsky (2002).
19. Blakeslee (1970); Franks (1981); Rousseau (1970; 1992); Time (1970).
20. Judson (1980); MacFarlane (1984).
21. Tharp (2022).
22. Wadman (2021); Wise (2021).
23. Brainard & You (2018); Retraction Watch (2023).
24. Baldwin (2015); Frankel (1982); Glen (1982, "most significant paper" quote on p. 302); Morley (2003, "martini" quote on p. 84).

25. e.g., National Academies of Sciences, Engineering, and Medicine (2019); Peterson & Panofsky (2021); Science and Technology Committee (2011).
26. Bassler et al. (2016); Franco, Malhotra, & Simonovits (2014); Nissan et al. (2016); Turner et al. (2008); see also Allchin (2025b).
27. Dwan et al. (2013); Easterbrook et al. (1991); Kicinski (2015).
28. Shapin (1994).
29. Allchin (1999a); Bourdieu (1975); Hull (1988).
30. Latour & Woolgar (1979, esp. pp. 131–136).
31. Service (2023). The case continued to unfold. Allegations surfaced that the lead author plagiarized his 2013 PhD dissertation. By March 2024, the wide extent of his deception was documented, and by the end of the year, he had been dismissed from the university (Garisto, 2024).
32. Brainard (2022); Merton (1973).
33. Darden (1998); Hurd (2007).
34. e.g., Felt (2012); Harman & Dietrich (2008); Hook (2002); Oreskes (1999).
35. Allchin (2007a).
36. e.g., Bell (1992); Broad & Wade (1982); Chevassus-au-Louis (2019); Goodstein (2010); Grant (2007); Judson (2004); Kohn (1988).
37. Reich (2009).
38. Natelson (2018, quote on p. 13).
39. e.g., Huff (1993); McGarity & Wagner (2008); Markowitz & Rosner (2002); Michaels (2008; 2020); Otto (2016); Shrader-Frechette (2014); Wagner & Steinzor (2006).
40. Fainaru-Wada & Fainaru (2013, quotes on pp. 140, 180, 227); Schwarz et al. (2016).
41. Chang (2012); Erickson et al. (2017); Nestle (2018); O'Connor (2016); Schillinger & Kearns (2017); Steele et al. (2019).
42. Allchin (2021); Flaherty (2020); Held (2020); Johnston et al. (2019); Kolata (2019); Nutrition Source (2019); O'Connor (2020).
43. Oreskes (2019); Vickers (2022); Ziman (1968).
44. Englehardt & Caplan (1987); Rudwick (1985, pp. 414–428).
45. Gilbert & Mulkay (1984, pp. 112–120); Vickers (2022).
46. Thagard (1999, pp. 185–198).
47. Harding (1991); Longino (1990); Oreskes (2019); Solomon (2001); Vickers (2022).
48. Hien (1969).

Chapter 5

1. Allchin (2025).
2. Allchin (2015b).
3. Losee (2005); Grant (2006).
4. Cartlidge (2012); Reich (2011).
5. Cho (2012); Sirri (2012); Strassler (2012); Zichichi, Sirri, & Sioli (2012).
6. Suppe (1998).
7. Giunta (2001); Hirsh (1981, Nasini quote on p. 124); Kuhn (1970); Nature (1895); Rayleigh (1895, quotes on pp. 160, 163); Spanos (2010).
8. Chapman (2019); Johnson (2002).
9. e.g., Hon, Schickore, & Steinle (2009).
10. See also Latour & Woolgar (1979, pp. 75–86) on modalities of statements with varying levels of confidence, or "fact-likeness."
11. Allchin (2015a); Kuhn (1970, pp. 6, 52, 57, 65); Lightman & Gingerich (1992).
12. The analysis of incongruences here contrasts with the earlier enumeration of the varieties of "going amiss" by Hon et al. (2009). Their nomenclature reads more like a haphazard list of synonyms for "mistake," than a systematic organization based on error types.

13. Franklin (2002); Stegenga (2009).
14. Darden (1991).
15. Allen (1979); Darden (1991).
16. Darden (1995); Marcum (2007); Temin (1975).
17. Keyes (1999a, 1999b); Kimberlin (1982, p. 393); Kuhn (1970); Liu (2011).
18. Finn (1964); Kuhn (1958, quote on p. 137); Newton (1687).
19. Judson (2004); Lewin (1988); Spencer (1990)
20. Campagnari & Mulders (2022); Cho (2022). On September 24, 2024, Fermilab (2024) announced that subsequent measurements by the CMS Collaborative, following more sophisticated protocols and with exceptional precision, found the mass of the W boson to be consistent with theoretical predictions. No theory upheaval on this occasion.
21. e.g., Donovan et al. (1988); Englehardt & Caplan (1987); Kuhn (1970); Lakatos (1978); Laudan (1977); Machamer, Pera, & Baltas (2000). For a critique, see Allchin (1994b).
22. Let this be a more exacting view than the vague notion of pluralism? On the history of the cited debates, see Allchin (1991); Appel (1987); Kipnis (1987); Piccolino (1998); Racine (2013); Rudwick (1985).
23. Rheinberger (1997, pp. 4, 59, 147–164).
24. On long-standing errors, see Allchin (2015b).
25. e.g., Maxham (2025); Star & Gerson (1986).
26. Bechtel & Richardson (1993); Craver & Darden (2013); Darden (1991; 2006); Elliott (2004); Schweitzer (2015).
27. Cummins (1975).
28. Barz et al. (1980); Greg Meisner (person. commun.); Stewart, Meisner, & Ku (1982).
29. On error cascades, see Allchin (2015b; 2025b).
30. Allchin (1997a; 2007b).
31. Allchin (2020); Mill (1874, p. 280).
32. Donaldson (2014, quotes by Lavoisier on pp. 26, 27, 28); Franklin (1785).
33. Eyler (2001); Johnson (2006); Snow (1855, quotes on pp. 43, 44).
34. Best (2016); Cohen (1985, p. 201); Conant (1957); Kuhn (1970); Losee (2005).
35. The "error" of negative weight is widely overstated in many accounts, shaped in part by Lavoisier's legacy and his abrasive rhetoric. Partington and McKie's (1938–1939) exhaustive survey indicates the concept was seriously entertained by only about one-quarter of phlogistonists—hardly a consensus.
36. e.g., Allchin (1992; 1994a; 2013, pp. 184–201); Brown (1866); Chang (2009); Holmes (1985; 2000); Kim (2008); Odling (1871); Scott (1958); Siegfried (1964); Sudduth (1978).
37. Brock (1992); Levere (2001); Morris (1972).
38. Odling (1871).
39. Stegenga (2024).

Chapter 6

1. Nye (1980).
2. Slack (1935).
3. Franks (1981).
4. Allchin (2022a; 2025); Christian & Berka (1973).
5. Allchin (2012a); Schofield (2004).
6. Barnes (1989); Simon (2001); Smith (1989).
7. Hom et al. (1976, quote on p. 7); Lederman (1978); Siegel (2016).
8. Allchin (2011); Nowak, Tarnita, & Wilson (2010); Ross et al. (2014); Wilson (2005).
9. Lyons (2009, pp. 17–50).
10. Curfman, Morrissey, & Drazen (2006); Krumholz et al. (2007).
11. Crosland (1973); LeGrand (1972; 1974).

12. Walcott (1911, p. 119).
13. Gould (1989).
14. Royal Ontario Museum (2020); Smith & Caron (2015).
15. Kahneman (2011).
16. Allchin (2007c); Amagat (1883); Boyle (1662); Cailletet (1870); Regnault (1847/1852); Webster (1966).
17. Andrews (1869).
18. Gay Lussac (1802); Waals (1910).
19. Allchin (2013).
20. On Mendel, see Allchin (2002a); Bateson (1902, p. 129); Darden (1991, pp. 68, 167); Endersby (2007); Mendel (1866, quote in §8); Rodgers (1991, p. 3); Wright (1968, p. 65). On laws in general, see Allchin (2007c); Carroll (1994); Cartwright (1999); Hempel (1966, pp. 54, 58); Kosso (1992, pp. 52–60, 190); Woodward (2003, pp. 167, 236–238, 265–266); Ziman (1978, p. 32).
21. e.g., debate between Stanford (2006) and Vickers (2022).
22. Adelmann (1966); Huckra (1992; 2008); Malpighi (1661/1929).
23. Glen (1982).
24. Aston (1922); Chang (2007); Nendick, Scrancher, & Usher (2007).
25. Kottler (1974); Martin (2004).
26. Schickore (2009).
27. e.g., Gigerenzer (2005); WImsatt (2007).
28. National Research Council (1997); SCENIHR (2015).
29. Guttinger & Love (2019).
30. Kidd & Modlin (1998); Marshall (2001; 2002); Rigas et al. (1999); Thagard (1999).
31. Allchin (1997a).
32. Allchin (1991; 1996; 1997a); Weber (1991).
33. Harman & Dietrich (2008; 2013).
34. On error cascades, see Allchin (2015b; 2025b).
35. Fee (1979).
36. Allchin (2005); Kohn (1988); Rheinberger (1997); Roberts (1989); Simonton (2004); Taton (1957).
37. Callebaut (1993); Campbell (1960; 1974); Darden & Cain (1989); Kantorovich (1993); Plotkin (1994).
38. Finn (1964); Fox (1971); Kuhn (1958).
39. Many more cases have now been observed. Bygott (1972, quotes on p. 410; and person. commun.).
40. Werth (2000, and person. commun.).
41. Altman (2005, source of Warren quote); Gaudet (2014); Marshall (2005).
42. On the pessimistic induction, see, e.g., Stanford (2006); on "future-proofing," see Vickers (2022).
43. e.g., Kelley (2001).

Chapter 7

1. The term 'bias' permeates philosophical discourse, yet with no commonly accepted definition (Aronson, 2017). It is commonly construed as any source of error (e.g., Boutron, et al., 2024), or "any process at any stage of inference which tends to produce results or conclusions that differ systematically from the truth" (Sackett, 1979, p. 60). The most popular definitions exhibit several common themes: distortion, systematic error, and deviation from the truth (Aronson, 2017). Attributing error to bias (as a cause) thus often reduces to a platitude, akin to Molière's "imaginary invalid" identifying the cause of sleep as the *virtus dormativa*, or dormative principle. In my usage here, bias is any factor responsible for *unevenly distributed observation*,

usually through *selective focus*. Bias arises from the unavoidable use of a *perspective* (Giere, 2006), which embodies a point of view, which establishes a corresponding focal point, orientation, frame, and foreground and background. Accordingly, some things are highlighted and clearly delineated, while others are excluded and/or obscured (features vividly captured in the map metaphor; see Monomier, 1991: Turnbull, 1989; and Chapter 2). Thus, a bias need not express values (as stipulated by many feminist philosophers), only a particular perspective or frame of reference (see Chapter 2 on framing). In my use here, the selectivity of bias (or perspective) offers both opportunities and inherent limitations (e.g., Anderson, 2020; Giere, 2006).

2. Cantor (1991); see also Allchin (2025).
3. Adapted from Allchin (2009); see also Browne (1995, pp. 376–378, 431–433); Rudwick (1974).
4. Allen (1947, quote on p. 356); Trimble (2017); Wynne (1976).
5. Levenson (2015).
6. Bowen (1927); Fontani, Costa, & Orna (2015); Orna, Fontani, & Costa (2017).
7. Brock (1992); Holmes (1962); Hornix (1988); Kim (2003, Kirwan quote on p. 274); Lehmnan (2010); Thackray (1970, Buffon quote on p. 159).
8. e.g., Dolby (1996); Gaby (2002); Gratzer (2000); Langmuir (1953; 1989); Rousseau (1992); Turro (1999).
9. Nye (1980).
10. Allison (1933, quote on p. 73); Farwell & Hawkes (1935); MacPherson (1934); Slack (1935); Woodriff (1935).
11. see note 8.
12. Chew (1987); Kean (2017); Langmuir (1950; 1951); Time (August 25, 1950).
13. Langmuir (1953; 1989).
14. Bauer (2002).
15. Hull (1988).
16. e.g., Harding (1991); Longino (1990); Oreskes (2019); Solomon (2001). See also Haraway (1989) on ironic diptychs.
17. Altmann (1974); Haraway (1989); Morrell (1993); Tang-Martínez (2020).
18. Gilmer (1990).
19. Spanier (1995, quotes on pp. 20, 24, 110).
20. Lewontin (1991, quotes on pp. 42, 45); see also Taylor (2005).
21. Marks (2003).
22. e.g., Ziman (1968); Oreskes (2019); Solomon (2015).
23. Harding (1991); Solomon (2001).
24. Solomon (2001).
25. Adapted from Allchin (2013, pp. 72–78); see also Bibel (1988); Magner (2002, pp. 278–285); Silverstein (1989).
26. Kipnis (1987); Piccolino (1998).
27. Rudwick (1985).
28. Darwin (1845, quote on p. 218).
29. *M Notebook* (p. 153).
30. Browne (1995, pp. 234–253, 382–383); Herbert (1974/1977).
31. Adapted from Allchin (2013, pp. 79–86); Eiseley (1961, pp. 303–314, quotes on p. 303).
32. Adapted from Allchin (2013, pp. 72–78); see also Allchin (2002); Prebble & Weber (2003).

Chapter 8

1. On errorology, see Jastrow (1936).
2. e.g., Alberts et al. (2015); Bruner & Holman (2019); McIntyre (2019); McNutt (2020); National Academy of Sciences (2019); Schimmack (2019); Romero (2016); Tharp (2023); Zimring (2019).

3. Allchin (2015b; 2016; 2025).
4. e.g., Machamer, Darden, & Craver (2000).
5. Allchin (2015b; 2025).
6. e.g., McIntyre (2019, pp. 134, 138, 171, 175); Romero (2016, p. 56); Zimring (2019, pp. 79, 124, 229, 233, 234).
7. Judson (1980, p. 170).
8. Allchin (2022a).
9. Allchin (forthcoming).
10. e.g., McIntyre (2019); Oreskes (2019); Zimring (2019).
11. Stegenga (2024).
12. Letter to Stephen Wilson, March 5, 1879; Darwin & Seward (1903, 2: 422).
13. Petroski (1994; 2006; 2012).
14. Minsky (1994, quote on p. 18); see also Gartmeier et al. (2008).
15. Allchin (1999b); Mayo (1997; 2014).
16. On archiving efforts, see, e.g., Arnaout et al. (2021); Moore & Cook (1973); see, e.g., RetractionWatch.com or Nimpf & Keays (2020).
17. Pascal (1663, pp. 53–54).
18. e.g., Barwich (2019).
19. Kuhn (1970, views of progress on pp. 160–173, 206); the pessimistic view articulated by, e.g., Laudan (1977; 1981).
20. Kuhn (1990).
21. This perspective aligns with Stegenga's (2024) proposal that progress is measured by justification, rather than the amount of resultant knowledge. But while Stegenga accepts any change in (communally accepted) justification as progress, I contend that the justification itself must show "progress." As indicated in all the historical cases of resolving error, science achieves depth by encompassing a larger body of relevant evidence, greater rigor or precision, and/or more theoretical considerations. Namely, the very standards of justification escalate, and the overall quality (rather than the quantity) of knowledge increases.
22. Allchin (1992; 1994; 2013, pp. 184–201); Chang (2009); Kim (2008); Scott (1958); Schufle & Thomas (1971); Sudduth (1978).
23. Brown (1866).
24. Mayo (1996, p. 5).
25. Franklin (1981).
26. See https://pdg.lbl.gov/2020/reviews/rpp2020-rev-history-plots.pdf; Huckra (2008); Pritychenko (2015).
27. Ad Hoc Committee (1968, quote on p. 86); De Georgia (2014); Widjicks et al. (2010).
28. Green et al. (2009); Noonan (2010); Pääbo et al. (2004); Skoglund et al. (2014); Wall & Kim (2007).
29. O'Loughlin (2023).
30. Wall & Kim (2007).
31. Gawande (2009).
32. Gawande (2009); Tugend (2011).
33. Allchin (2001).
34. This analysis would complement various philosophical considerations of criteria for theory pursuit or "theory promise"; see Šešelja & Straßer (2014); Whitt (1992).
35. Suppe (1998).
36. Oppenheimer (1984, p. 177), reprint of a 1965 essay.
37. Cartwright (1999); Lakatos (1978); Losee (2005).
38. Mayo (1996; 2018).
39. Allchin (2001); Mayo (1996, pp. 4–7, 64, 184–85, 315, 445; 2018).
40. Heering (1992a).

41. Allchin (2001).
42. Allchin (1994b).
43. On Avogadro and diatomic elements, see Levere (2001, pp. 110–116); Rocke (1984, pp. 101–107, 294–299).
44. On Millikan, see Allchin (2015). On deep reliability, see Allchin (2001).
45. Kestenbaum (2000).

References

Abachi, S. and the D. Collaboration. 1995. "Observation of the Top Quark." *Physical Review Letters* 74: pp. 2632–37.

Ackerman, D. 1985. *Data, Instruments and Theory*. Princeton University Press.

Adelmann, H.B. 1966. *Marcello Malpighi and the Evolution of Embryology*. Ithaca, NY: Cornell University Press.

Ad Hoc Committee of the Harvard Medical School to Examine Definition of Death. 1968. "A Definition of Irreversible Coma." *JAMA* 205(6): pp. 85–88.

Adler, K. 2002. *The Measure of All Things*. New York: Free Press.

Alberts, B. 2011. "Retraction." *Science* 334: p. 1636.

Alberts, B., Cicerone, R.J., Fienberg, S.E., Kamb, A., McNutt, M, Nerem, R.M., Schekman, R., Shiffrin, R., Stodden, V., Suresh, S., Zuber, M.T., Pope, B.K., and Jamieson, K.H. 2015. "Self-correction in Science at Work." *Science* 348: pp. 1420–22.

Allchin, D. 1991. *Resolving Disagreement in Science: The Ox Controversy, 1961–1978*. Ph.D. diss. University of Chicago.

Allchin, D. 1992. "Phlogiston After Oxygen." *Ambix* 39: pp. 110–6.

Allchin, D. 1994a. "James Hutton and Phlogiston." *Annals of Science* 51: pp. 615–35.

Allchin, D. 1994b. "The Super-Bowl and the Ox-Phos Controversy: Winner-Take-All Competition in Philosophy of Science." In *PSA 1994*, edited by D. Hull, M. Forbes, and R.M. Burian, Vol. 1, pp. 22–33. East Lansing, MI: Philosophy of Science Association.

Allchin, D. 1995. "How *Not* to Teach History in Science." In *Proceedings, Third International History, Philosophy and Science Teaching Conference*, edited by F. Finley, D. Allchin, D. Rhees, and S. Fifield, Vol. 1, pp. 13–22. Minneapolis, MN. Reprinted in *Journal of College Science Teaching* 30(2000): 33–37. Reprinted in *The Pantaneto Forum* 7 (July, 2002). URL: www.pantaneto.co.uk/issue7/allchin.htm (accessed May 11, 2011).

Allchin, D. 1996. "Cellular and Theoretical Chimeras: Piecing Together How Cells Process Energy." *Studies in the History and Philosophy of Science* 27: pp. 31–41.

Allchin, D. 1996. "Points East and West: Acupuncture and Comparative Philosophy of Science." *Philosophy of Science* 63: pp. S107–S115.

Allchin, D. 1997a. "A 20th-century Phlogiston: Constructing Error and Differentiating Domains." *Perspectives on Science* 5: pp. 81–127.

Allchin, D. 1997b. "James Hutton and Coal." *Cadernos IG/UNICAMP* 7: pp. 167–83.

Allchin, D. 1999a. "Do We See Through a Social Microscope?: Credibility as a Vicarious Selector." *Philosophy of Science* 60(Proceedings): pp. S287–S28.

Allchin, D. 1999b. "Negative Results as Positive Knowledge, and Zeroing in on Significant Problems." *Marine Ecology Progress Series* 191: pp. 303–5.

Allchin, D. 2001. "Error Types." *Perspectives on Science* 9: pp. 38–59.

Allchin, D. 2002. "To Err and Win a Nobel Prize: Paul Boyer, ATP Synthase and the Emergence of Bioenergetics." *Journal of the History of Biology* 35: pp. 149–72.

Allchin, D. 2003. "Scientific Myth-conceptions." *Science Education* 87: pp. 329–51.
Allchin, D. 2004. "Pseudohistory and Pseudoscience." *Science & Education* 13: pp. 179–95.
Allchin, D. 2005. "Discovering the Self." In *2005–2006 Professional Development for AP Biology*, edited by C. Schofield. New York: College Entrance Examination Board.
Allchin, D. 2007a. "Albert Szent-Györgyi." *New Dictionary of Scientific Biography* 24: pp. 567–73. Detroit: Charles Scribner's Sons.
Allchin, D. 2007b. "Efraim Racker." *New Dictionary of Scientific Biography* 24: pp. 197–203. Detroit: Charles Scribner's Sons.
Allchin, D. 2007c. "Teaching Science Lawlessly." In *Constructing Scientific Understanding through Contextual Teaching*, edited by P. Heering and D. Osewold, pp. 13–31. Berlin: Frank & Timme.
Allchin, D. 2008a. "Naturalizing as an Error-type in Biology." *Filosofia e História da Biologia* 3: pp. 95–117.
Allchin, D. 2008b. "Nobel Ideals and Noble Errors." *American Biology Teacher* 70: pp. 389–92.
Allchin, D. 2009. "Celebrating Darwin's Errors." *American Biology Teacher* 71: pp. 116–19.
Allchin, D. 2011. "The Domesticated Gene." *American Biology Teacher* 74: pp. 120–23.
Allchin, D. 2012a. "A Comedy of Scientific Errors." *American Biology Teacher* 74: pp. 530–31.
Allchin, D. 2012b. "Skepticism and the Architecture of Trust." *American Biology Teacher* 74" pp. 358–62.
Allchin, D. 2012c. "Teaching the Nature of Science Through Scientific Error." *Science Education* 96: pp. 904–26.
Allchin, D. 2013. "Deciphering How Cells Make Energy: An Acid Test." *American Biology Teacher* 75: pp. 598–601.
Allchin, D. 2013. *Teaching the Nature of Science: Perspectives & Resources*. St. Paul, MN: SHiPS Press.
Allchin, D. 2015a. "Context-dependent Anomalies and Strategies for Resolving Disagreement: A Case in Empirical Philosophy of Science." In *Empirical Philosophy of Science: Introducing Qualitative Methods into Philosophy of Science*, edited by S. Wagenknecht, N.J. Nersessian, and H. Andersen, pp. 161–71. Dordrecht: Springer.
Allchin, D. 2015b. "Correcting the Self-correcting Mythos of Science." *Filosofia e História da Biologia* 10: pp. 19–35.
Allchin, D. 2016. "Is Science Self-correcting?" *American Biology Teacher* 78: pp. 695–98.
Allchin, D. 2017. "Does the Evidence Speak for Itself?: Archibald Garrod and Inborn Errors of Metabolism." *American Biology Teacher* 79: pp. 335–37.
Allchin, D. 2020. "The Counter-roll in Science." *American Biology Teacher* 82(3), pp. 188–91.
Allchin, D. 2021. "The Ecology of Meat." *American Biology Teacher* 83: 416–20.
Allchin, D. 2022a. "The Very Reproducible (But Illusory) Mesosome." *American Biology Teacher* 84: pp. 321–23.
Allchin, D. 2022b. "Who Speaks for Science?" *Science & Education* 31: pp. 1475–92.
Allchin, D. 2025. "When is Science "Self"-correcting?" Paper presented at the International Society for the History, Philosophy and Social Studies of Biology (Porto, Portugal).

Allchin, D. forthcoming. "Error Repertoires." In *Malfunction, Error, Failure: How to Learn from Scientific Mistakes*, edited by A. Karakas and A. Tuboly. Dordrecht: Springer.

Allchin, D. and Werth, A. 2017. "The Naturalizing Error." *Journal for the General Philosophy of Science* 48: pp. 3–18.

Allchin, D. and Werth, A. 2020. "How We Think about Human Nature: The Naturalizing Error." *Philosophy of Science* 87: pp. 499–517.

Allchin, D. and Zemplén, G. 2020. "Finding the Place of Argumentation in Science Education: Epistemics and Whole Science." *Science Education* 104: pp. 907–33.

Allen, G. 1979. *Thomas Hunt Morgan: The Man and His Science.* Princeton University Press.

Allen, G. 2011. "Eugenics and Modern Biology: Critiques of Eugenics, 1910–1945." *Annals of Human Genetics* 75: pp. 314–25.

Allen, H.S. 1947. "Charles Glover Barkla. 1877–1944." *Obituary Notices of Fellows of the Royal Society* 5(15): pp. 341–66.

Allison, F. 1933. "The Magneto-optic Method of Analysis." *Journal of Chemical Education* 10: pp. 71–78.

Altman, L.K. 2005. "Nobel Came after Years of Battling the System." *New York Times* (Oct. 11). https://www.nytimes.com/2005/10/11/health/nobel-came-after-years-of-battling-the-system.html

Altmann, J. 1974. "Observational Study of Behavior: Sampling Methods." *Behaviour* 49: pp. 227–67.

Amagat, E.-H. 1883. "Recherches sur la compressibilité des gaz" et seq. *Annales de Chimie et de Physique* 28: pp. 456–507.

Andersen, H. 2022. "Philosophy of Scientific Malpractice." *SATS* 22: pp. 135–48.

Anderson, E. 2020. "Feminist Epistemology and Philosophy of Science." *The Stanford Encyclopedia of Philosophy*, E. N. Zalta (Ed.). https://plato.stanford.edu/archives/spr2020/entries/feminism-epistemology

Andrews, T. 1869. "On the Continuity of the Gaseous and Liquid States of Matter." *Proceedings of the Royal Society* 18: pp. 42–45.

Ankeny, R.A. and Leonelli, S. 2021. *Model Organisms.* Cambridge University Press.

Appel, T.A. 1987. *The Cuvier-Geoffroy Debate: French Biology in the Decades before Darwin.* New York/Oxford: Oxford University Press.

Ariely, D. 2008. *Predictably Irrational: the Hidden Forces That Shape Our Decisions.* New York: Harper Collins.

Arnaout, H., Razniewski, S., Weikum, G., and Pan, J.Z. 2021. "Negative Statements Considered Useful." *Journal of Web Semantics* 71: 100661.

Aronson, J. 2017. "A word about evidence: 5. Bias—previous definitions." Catalogue of Bias. https://catalogofbias.org/2018/04/20/a-word-about-evidence-5-bias-previous-definitions/

Aston, F.W. 1922. "Mass Spectra and Isotopes." *Nobel Lectures, Chemistry, 1922–1941.* Amsterdam: Elsevier. https://www.nobelprize.org/prizes/chemistry/1922/aston/lecture

Attwood, P.V., Piggott, M.J., Zu, X.L., and Besant, P.G. 2006. "Focus on Phosphohistidine." *Amino Acids* 32: pp. 145–56.

Bacon, F. 1620/1902. *Novum Organum.* P.F. Collier & Son.

Baird, D. 2004. *Thing Knowledge: A Philosophy of Scientific Instruments.* University of California Press.

Baldwin, M. 2015. "Credibility, Peer Review, and *Nature*, 1945–1990." *Notes and Records* 69: pp. 337–52.
Barrell, B.G., Bankier, A.T., and Drouin, J. 1979. "A Different Genetic Code in Human Mitochondria." *Nature* 282: pp. 189–94.
Barkan, E. 1992. *The Retreat of Scientific Racism*. Cambridge, UK: Cambridge University Press.
Barnes, C.A. 1989. Interview by Heidi Aspaturian, Pasadena, California, June 13 and 26, 1989. Oral History Project, California Institute of Technology Archives. https://oralhistories.library.caltech.edu/212/
Barrett, H.C. 2015. *The Shape of Thought: How Mental Adaptations Evolve*. New York: Oxford University Press.
Barry, S. and Elith, J. 2006. "Error and Uncertainty in Habitat Models." *Journal of Applied Ecology* 43: pp. 413–23.
Barwich, A.-S. 2019. "The Value of Failure in Science: The Story of Grandmother Cells in Neuroscience." *Frontiers in Neuroscience* 13: 1121.
Barz, H., Ku, H.C., Meisner, G.P., Fisk, Z., and Matthias, B.T. 1980. "Ternary Transition Metal Phosphides: High-temperature Superconductors." *Proceedings of the National Academy of Science, USA* 77: pp. 3132–34.
Bascom, W.D. 1972. Polywater. *Journal of Physical Chemistry* 76: pp. 456–57.
Bassler, D., Mueller, K.F., Briel, M., Kleijnen, J., Marusic, A., Wager, A., Antes, G., von Elm, E., Altman, D.G., and Meerpohl, J.J., on behalf of the OPEN Consortium. 2016. "Bias in Dissemination of Clinical Research Findings: Structured OPEN Framework of What, Who and Why, Based on Literature Review and Expert Consensus." *BMJ Open* 6: e010024. doi:10.1136/bmjopen-2015-010024.
Bates, D.W. 2002. *Enlightenment Aberrations: Error and Revolution in France*. Ithaca, NY: Cornell University Press.
Bateson, W. 1902. "The Facts of Heredity in the Light of Mendel's Discovery." *Reports to the Evolution Committee of the Royal Society, London* 1: pp. 125–60.
Bauer, H.H. 1992. "Scientific Literacy and the Myth of the Scientific Method." Urbana: University of Illinois Press.
Bauer, H.H. 2002. "'Pathological Science' Is Not Scientific Misconduct (Nor Is it Pathological)." HYLE 8: pp. 5–20.
Bechtel, W. and Richardson, R.C. 1993. *Discovering Complexity: Decomposition and Localization as Strategies in Scientific Research*. Princeton, NJ: Princeton University Press.
Begley, C.G. 2013. "Six Red Flags for Suspect Work." *Nature* 497: pp. 433–34.
Bell, R. 1992. *Impure Science: Fraud, Compromise, and Political Influence in Scientific Research*. New York: John Wiley.
Bello, S., Krogsbøll, L.T., Gruber, J., Zhao, Z.J., Fischer, D., and Hröbjartsson, A. 2014. "Lack of Blinding of Outcome Assessors in Animal Model Experiments Implies Risk of Observer Bias." *Journal of Clinical Epidemiology* 67: 973–83.
Benson, S. and Manoogian, J., III. 2016. "Cognitive Bias Codex."
Berg, J. 2019. "Replication Challenges." *Science* 365: p. 957.
Best, N.W. 2016. "What Was Revolutionary about the Chemical Revolution?" In *Essays in the Philosophy of Chemistry*, edited by E. Scerri and G. Fisher: pp. 37–59. Oxford University Press.
Beveridge, W.I.B. 1950. *The Art of Scientific investigation*. New York: Random House.

Bibel, D.J. 1988. *Milestones in Immunology: A Historical Exploration*. Madison, WI: Science Tech Publishers.

Bingel, A. 1918. "Über Behandlung der Diphtherie Mit Gewöhnlichem Pferdeserum." *Deutsches Archiv für Klinische Medizin* 125: pp. 284–332.

Blakeslee, S. (1970, Sept. 27). "Scientist Says Mystery of Polywater Has Been Solved: Russian's Test Samples Contained Sweat." *New York Times*, p. 56.

Boring, E.G. 1954. "The Nature and History of Experimental Control." *American Journal of Psychology* 67: pp. 573–89.

Borman, S. 1990, February 19. "Benzene in Perrier Found in North Carolina Lab." *Chemical and Engineering News* 68(8): pp. 5–6.

Boumans, M. and Hon, G. 2018. "Introduction." In Boumans et al., pp. 1–12.

Boumans, M., Hon, G., and Petersen, A. (Eds.). 2018. *Error and Uncertainty in Scientific Practice*. London: Pickering & Chatto.

Bourdieu, P. 1975. "The Specificity of the Scientific Field and the Social Conditions of the Progress of Reason." Translated by Richard Nice. *Social Sciences Information* 14: pp. 19–47.

Boutron, I., Page, M.J., Higgins, J.P.T., Altman, D.G., Lundh, A., and Hróbjartsson, A. 2024. Considering Bias and Conflicts of Interest Among the Included Studies. In *Cochrane Handbook for Systematic Reviews of Interventions* version 6.5, edited by J.P.T. Higgins, J. Chandler, M. Cumpston. T. Li, M.J. Page and V.A. Welch. Cochrane. https://www.cochrane.org/authors/handbooks-and-manuals/handbook/current/chapter-07

Bowell, T. and Kemp, G. 2002. *Critical Thinking: A Concise Guide*. London: Routledge.

Bowen, I.S. 1927. "The Origin of the Nebulium Spectrum." *Nature* 120: p. 473.

Boyer, P.D. 1963. "Phosphohistidine." *Science* 141: pp. 1147–53.

Boyer, P.D. 1981. "An Autobiographical Sketch Related To my Efforts to Understand Oxidative Phosphorylation." In *Of Oxygen, Fuels and Living Matter*, edited by G. Semenza, pp. 229–44. London: John Wiley & Sons Ltd.

Boyle, R. 1662. *New Experiments Physico-mechanical, Touching the Air [...]*. London: H. Hall.

Brackett, P. 1990, February 2. "Perrier Woes Began with a Blip on a Carolina Screen." *New York Times*: p. A18.

Brainard, J. and You, J. 2018. "Rethinking Retractions." *Science* 362: pp. 390–95.

Brainard, J. 2022. "Reviewers Award Higher Marks When a Paper's Author Is Famous." *Science* 377: p. 1251.

Broad, W. and Wade, N. 1982. *Betrayers of the Truth*. New York: Simon and Schuster.

Brock, W.H. 1992. *History of Chemistry*. New York: W.W. Norton.

Brodeur, P. 1989. *Currents of Death*. New York: Simon and Schuster.

Brower, J.V.Z. 1958. "Experimental Studies of Mimicry in Some North American Butterflies. Part III. *Danaus gilippus berenice* and *Limenitis archippus floridensis*." *Evolution* 12: pp. 273–85.

Brown, A.C. 1866. "Note on Phlogistic Theory." *Proceedings of the Royal Society, Edinburgh* 5(V): pp. 328–30.

Browne, J. 1995. *Charles Darwin: Voyaging*. Princeton, NJ: Princeton University Press.

Bruner, J.P. & Holman, B. 2019. "Self-correction in Science: Meta-analysis, Bias and Social Structure." *Studies in History and Philosophy of Science Part A* 78: pp. 93–99.

Buchwald, J.Z. and Franklin, A. 2005. *Wrong for the Right Reasons*. Dordrecht: Springer.

Buckland, W. 1820. *Vindiciae Geologicae; or the Connexion of Geology with Religion Explained*. Oxford: Oxford University Press.

Buckland, W. 1824. *Reliquiæ Diluvianæ*. London: John Murray.
Buckland, W. 1836. *Geology and Mineralogy Considered with Reference to Natural Theology*. London: William Pickering.
Butler, C. 1609. *The Feminine Monarchie*. Oxford: Joseph Barnes.
Bygott, J.D. 1972. "Cannibalism among Wild Chimpanzees." *Nature* 238: pp. 410–11.
Cailletet, L.P. 1870. "Compressibilité des gaz à hautes pressions." *Comptes Rendus* 70: pp. 1131B1134.
Callebaut, W. 1993. *Taking the Naturalistic Turn: How the Real Philosophy of Science is Done*. Chicago: University of Chicago Press.
Campagnari, C. and Mulders, M. 2022. "An Upset to the Standard Model." *Science* 376: p. 136.
Campbell, D.T. 1960. "Blind Variation and Selective Retention in Creative Thought." *Psychological Review* 67: pp. 380–400.
Campbell, D.T. 1974. "Evolutionary Epistemology." In *The Philosophy of Karl Popper*, edited by P.A. Schilpp, pp. 413–63. La Salle, IL: Open Court.
Cantor, G. 1991. *Michael Faraday: Scientist and Sandemanian*. New York: St. Martin's Press.
Cardot, J.-M. and Cavit, B.M. 2012. "*In vitro–in vivo* Correlations: Tricks and Traps." *The AAPS Journal* 14: pp. 491–99.
Carpenter, K.J. 2000. *Beriberi, White Rice and Vitamin B: A Disease, a Cause and a Cure*. Berkeley: University of California Press.
Carroll, J.W. 1994. *Laws of Nature*. Cambridge University Press.
Cartlidge, E. 2012. "Loose Cable May Unravel Faster-than-light Result." *Science* 335: p. 1027.
Cartwright, N. 1999. *The Dappled World*. Cambridge University Press.
Cartwright, S.A. 1851. "Report on the Diseases and Physical Peculiarities of the Negro Race." *The New Orleans Medical and Surgical Journal* 7: pp. 691–715.
Catalogue of Bias Collaborative. 2020. A Taxonomy of Biases: Progress Report. http://catalogofbias.org.
Chang, H. (Ed.). 2007. *Chlorine: Element of Controversy*. http://www.bshs.org.uk/wp-content/uploads/file/bshs_monographs/library_monographs/bshsm_013_chang-and-jackson.pdf
Chang, H. 2008. *Inventing Temperature*. Cambridge University Press.
Chang, H. 2009. "We Have Never Been Whiggish (About Phlogiston)." *Centaurus* 51: pp. 239–64.
Chang, K. October 16, 2012. "Parsing of Data Led to Mixed Messages on Organic Food's Value." *New York Times*. p. D3.
Chapman, K. 2019. "The Element That Never Was." *Chemistry World* (June 16). https://www.chemistryworld.com/features/victor-ninov-and-the-element-that-never-was/3010596.article
Chase, A. 1976. *The Legacy of Malthus. The Social Costs of the New Scientific Racism*. New York: Knopf.
Chevassus-au-Louis, N. 2019. *Fraud in the Lab: The High Stakes of Scientific Research*. Cambridge, MA: Harvard University Press.
Chew, J. 1987. *Storms Above the Desert: Atmospheric Research in New Mexico, 1935–1985*. Albuquerque: University of New Mexico Press.
Cho, A. 2012. "Once Again, Physicists Debunk Faster-than-light Neutrinos." *Science*, doi: 10.1126/article.27262

Cho, A. 2022. "Particle's Mass May Tip the Scales to New Physics." *Science* 376: p. 125.
Christian, P.A. and Berka, L.H. 1973. "How to Grow Your Own Polywater." *Popular Science* 202(#6, May): pp. 105–07.
Cialdini, R.B. 2007. *Influence: The Psychology of Persuasion.* New York: Collins Business.
Cohen, I.B. 1985. *Revolutions in Science.* Cambridge, MA: Harvard University Press).
Colloff, P. 2018. "Blood Will Tell" (Parts I and II). New York: ProPublica. https://features.propublica.org/blood-spatter/blood-spatter-forensic-evidence-investigation/
Conant, J.B. 1957. "The Overthrow of the Phlogiston Theory: The Chemical Revolution of 1775–1789." edited by Conant, J.B. *Harvard Case Histories in Experimental Science,* Vol. 1, pp. 65–115. Cambridge, MA: Harvard University Press.
Cooper, G.S. and Meterko, V. 2019. "Cognitive Bias Research in Forensic Science: A Systematic Review." *Forensic Science International* 297(April): pp. 35–46.
Craver, C. and Dan-Cohen, T. 2024. "Experimental Artifacts." *British Journal for the Philosophy of Science* 35: pp. 381–90.
Craver, C. and Darden, L. 2013. *In Search of Mechanisms: Discovering Across the Life Sciences.* University of Chicago Press.
Creager, A.N.H., Lunbeck, E., and Wise, M.N. (Eds.). 2007. *Science Without Laws: Model Systems, Cases, Exemplary Narratives.* Durham, NC: Duke University Press.
Crease, R.P. 1993. *The Play of Nature.* Bloomington: Indiana University Press.
Crick, F.H. 1968. "The Origin of the Genetic Code." *Journal of Molecular Biology* 38: pp. 367–79.
Crosland, M. 1973. "Lavoisier's Theory of Acidity." *Isis* 64: pp. 306–25.
Culp, S. 1994. "Defending Robustness: The Bacterial Mesosome as a Test Case." In *PSA 1994,* edited by D. Hull, M. Forbes, and R.M. Burian, Vol. 1, pp. 46–57. East Lansing, MI: Philosophy of Science Association.
Cummins, R. 1975. "Functional Analysis." *The Journal of Philosophy* 72: pp. 741–65.
Curfman, G.D., Morrissay, S., and Drazen, J.M. 2006. "Expression of Concern Reaffirmed." *New England Journal of Medicine* 354: p. 1193.
Darden, L. 1991. *Theory Change in Science: Strategies from Mendelian Genetics.* Cambridge, UK: Cambridge University Press.
Darden, L. 1995. "Exemplars, Abstractions, and Anomalies: Representations and Theory Change in Mendelian and Molecular Genetics." In *Concepts, Theories, and Rationality in the Biological Sciences,* edited by J.G. Lennox and G. Wolters, pp. 137–58. Pittsburgh, PA: University of Pittsburgh Press.
Darden, L. 1998. "The Nature of Scientific Inquiry." URL: www.philosophy.umd.edu/Faculty/LDarden/sciinq/index.html.
Darden, L. 2006. *Reasoning in Biological Discoveries.* Cambridge University Press.
Darden, L. and Cain, J.A. 1989. "Selection Type Theories." *Philosophy of Science* 56: pp. 106–29.
Darwin, C.R. 1845. *Journal of Researches.* 2d edition. London: John Murray.
Darwin, F. and Seward, A.C. (Eds.). 1903. *More Letters of Charles Darwin: A Record of His Work in a Series of Hitherto Unpublished Letters.* London: John Murray.
Davenport, C.B. 1911. *Heredity in Relation to Eugenics,* Cold Spring Harbor, NY: Cold Spring Harbor Laboratory.
Davenport, C.B. 1912a. "Heredity in Nervous Disease and its Social Bearings." *JAMA* 59: pp. 2141–42.
Davenport, C.B. 1912b. "The Nams: The Feeble-minded as Country Dwellers." Project Report. Cold Spring Harbor, NY: Carnegie Institution of Washington.

Davenport, C.B. 1912c. "The Origin and Control of Mental Defectiveness." *Popular Science Monthly* 80(January): pp. 87–90.
Davenport, C.B. 1916. "The Hereditary Factor in Pellagra." *Archives of Internal Medicine* 18: pp. 4–31.
Davis, W.M. 1902. *Elementary Physical Geography*. Boston: Ginn & Co.
De Georgia, M.A. 2014. "History of Brain Death as Death: 1968 to the Present." *Journal of Critical Care* 29: pp. 673–78.
Derjaguin, B.V. and Churaev, N.V. 1973. "Nature of 'Anomalous Water.'" *Nature* 244: pp. 430–31.
Dobelli, R. 2013. *The Art of Thinking Clearly*. London: Sceptre.
Dolby, R.G.A. 1996. "A Theory of Pathological Science." In *Uncertain Knowledge*, pp. 227–44. Cambridge: Cambridge University Press.
Donaldson, I.M.L. 2014. *The Reports of the Royal Commission of 1784 on Mesmer's System of Animal Magnetism and Other Contemporary Documents*. James Lind Library and Royal College of Physicians, Edinburgh.
Donovan, A., Laudan, L., and Laudan, R. 1988. *Scrutinizing Science: Empirical Studies of Scientific Change*. Baltimore, MD: Johns Hopkins University Press.
Durston, A.J., 1974. "Pacemaker Activity During Aggregation in *Dictyostelium discoideum*." *Developmental Biology* 37: pp. 225–35.
Dutta, R., Bardhan, S., Paul, S., and Mondal, S. 2017. "The Taxonomic Status of the Genus *Hubertoceras* Spath: A New Light on Sexual Dimorphism from the Callovian Ammonites of Kutch, India." *Palaeontologia Electronica*. https://palaeo-electronica.org/content/pdfs/738.pdf
Dwan, K., Gamble, C., Williamson, P.R., Kirkham, J.J., and Reporting Bias Group. 2013. "Systematic Review of the Empirical Evidence of Study Publication Bias and Outcome Reporting Bias – an Updated Review." *PLoS One*,8: e66844.
Earman, J. and Glymour, C. 1980. "Relativity and Eclipses: The British Eclipse Expeditions of 1919 and Their Predecessors." *Historical Studies in the Physical Sciences* 11: pp. 49–85.
Easterbook, P.A, Berlin, J.A., Gopalan, R., and Matthews, D.R. 1991. "Publication Bias in Clinical Research." *Lancet* 337: pp. 867–72.
Eisley, L. 1961. *Darwin's Century*. Garden City, NY: Anchor Books.
Elliott, K. 2004. "Error as a Means to Discovery." *Philosophy of Science* 71: pp. 174–97.
Elzanowski, A. and Ostell, J. 2019. "The Genetic Codes." Bethesda, MD: National Center for Biotechnology Information. https://www.ncbi.nlm.nih.gov/Taxonomy/Utils/wprintgc.cgi
Encyclopedia Britannica. 1771. "Of the Phlogiston." 2(33): pp. 68–69.
Endersby, J. 2007. *A Guinea Pig's History of Biology*. Cambridge, MA: Harvard University Press.
England, P.C., Molnar, P., and Richter, F.M. 2007. "Kelvin, Perry and the Age of the Earth." *American Scientist* 95: pp. 342–49.
Englehardt, H.T. and Caplan, A.L. 1987. *Scientific Controversies: Case Studies in the Resolution and Closure of Disputes in Science and Technology*. New York: Cambridge University Press.
Erickson, J., Sadeghirad, B., Lytvyn, L., Slavin, J.& Johnston, B.C. 2017. "The Scientific Basis of Guideline Recommendations on Sugar Intake: A Systematic Review." *Annals of Internal Medicine* 166: pp. 257–67.

Eriksson, J., Larson, G., Gunnarsson, U., Bed'hom, B., Tixier-Boichard, M., et al. 2008. "Identification of the Yellow Skin Gene Reveals a Hybrid Origin of the Domestic Chicken." *PLoS Genetics* 4(2): e1000010. doi: 10.1371/journal.pgen.1000010.

Eyler, J.M. 2001. "The Changing Assessments of John Snow's and William Farr's Cholera Studies." *Sozial und Praventivmedizin* 46: pp. 225–32.

Fainaru-Wada, M. and Fainaru, S. 2013. *League of Denial: The Nfl, Concussions and the Battle for Truth.* Crown Business.

Farwell, H.W. and Hawkes, J.B. 1935. "Time Lags in Magneto-optics." *Physical Review* 47(1): pp. 78–84.

Fee, E. 1979. "Nineteenth-century Craniology: The Study of the Female Skull." *Bulletin of the History of Medicine* 53: pp. 415–33.

Felt, H. 2012. *Soundings: The Story of the Remarkable Woman Who Mapped the Ocean Floor.* New York: Henry Holt.

Fennis, B.M. and Stroebe, W. 2021. *The Psychology of Advertising,* 3rd ed. New York: Routledge.

Fermilab. (Sept. 17, 2024). "New Results from the CMS Experiment Put W Boson Mass Mystery to Rest" [press release]. https://news.fnal.gov/2024/09/new-results-from-the-cms-experiment-put-w-boson-mass-mystery-to-rest/

Finn, B.S. 1964. "Laplace and the Speed of Sound." *Isis* 55: pp. 7–19.

Firestein, S. 2016. *Failure: Why Science is so Successful.* Oxford: Oxford University Press.

Flaherty, C. Jan. 22, 2020. "Beef with Harvard." *Inside Higher Education.* https://www.insidehighered.com/news/2020/01/23/texas-ams-beef-harvard

Fontani, M., Costa, M., and Orna, M.V. 2015. *The Lost Elements.* Oxford University Press.

Fowler, D.D. and Fowler, C.S. (Eds.). 1971. *Anthropology of the Numa: John Wesley Powell's Manuscripts on the Numic Peoples of Western North America, 1868–1880.* Smithsonian Contributions to Anthropology, Vol. 14. Washington, DC: Smithsonian Institution Press.

Fox, R. 1971. *The Caloric Theory of Gases: From Lavoisier to Regnault.* Oxford, UK: Clarendon.

Franco, A., Malhotra, N., and Simonovits, G. 2014. "Publication Bias in the Social Sciences: Unlocking the File Drawer." *Science* 345: pp. 1502–05.

Frankel, H. 1982. "The Development, Reception, and Acceptance of the Vine–Matthews–Morley Hypothesis." *Historical Studies in the Physical Sciences* 13: pp. 1–39.

Franklin, A. 1981. "Millikan's Published and Unpublished Data on Oil Drops." *Historical Studies in the Physical Sciences* 11: pp. 185–201.

Franklin, A. 1986. *The Neglect of Experiment.* Cambridge University Press.

Franklin, A. 1993. *The Rise and Fall of the Fifth Force.* American Institute of Physics.

Franklin, A. 1997. "Calibration." *Perspectives on Science* 5: pp. 31–80.

Franklin, A. 2002. *Selectivity and Discord.* Pittsburgh, PA: Pittsburgh University Press.

Franklin, B. 1785. *Report of Dr. Benjamin Franklin, and other commissioners, charged by the King of France, with the Examination of the Animal Magnetism, as Now Practised in Paris, Translated from the French with a Historical Introduction.* London: J. Johnson.

Franks, F. 1981. *Polywater.* Cambridge, MA: MIT Press.

Gaby, J.C. 2002. "Discerning Pathological Science" [Chancellor's Honors Program project]. Knoxville: University of Tennessee. https://trace.tennessee.edu/utk_chanhonoproj/540

Galison, P. 1987. *How Experiments End.* University of Chicago Press.

Garfinkel, A. 1987. "The Slime Mold *Dictyostelium* as a Model of Self-organization in Social Systems." In *Self-Organizing Systems*, edited by F.E. Yates et al., pp. 181–212. New York: Plenum Press.

Garisto, D. 2024. "Superconductivity Scandal: The Inside Story of Deception in a Rising Star's Physics Lab." *Nature*. doi: 10.1038/d41586-024-00716-2

Gartmeier, M., Bauer, J., Gruber, H., and Heid, H. 2008. "Negative Knowledge: Understanding Professional Learning and Expertise." *Vocations and Learning* 1: pp. 87–103.

Gaudet, J. 2014. "How Pre-publication Journal Peer Review (Re)produces Ignorance at Scientific and Medical Journals: A Case Study." uO Research. http://hdl.handle.net/10393/31198

Gawande, A. 2009. *The Checklist Manifesto: How to Get Things Right*. New York: Metropolitan Books.

Gay, H. 1977. "Noble Gas Compounds: a Case Study of Scientific Conservatism and Opportunism." *Studies in History and Philosophy of Science* 8: pp. 61–70.

Gay-Lussac, J.-L. 1802. "The Expansion of Gases by Heat." *Annales de Chimie* 43: pp. 137–75.

Ghallab, A. 2013. "*In Vitro* Test Systems and Their Limitations." *EXCLI Journal* 12: pp. 1024–26.

Giere, R. 1988. *Explaining Science*. University of Chicago Press.

Giere, R. 1998. *Understanding Scientific Reasoning*, 4th ed. New York: Harcourt Brace.

Giere, R. 2006. *Scientific Perspectivalism*. University of Chicago Press.

Gigenrenzer, G. 2005. "I Think, Therefore I Err." *Social Research* 72: pp. 195–218.

Gigerenzer, G. and Selten, R. (Eds.). 2001. *Bounded Rationality: The Adaptive Toolbox*/ MIT Press.

Gilbert, G.N. and Mulkay, M. 1984. *Opening Pandora's Box: A Sociological Analysis of Scientists' Discourse*. Cambridge, UK: Cambridge University Press.

Gillispie, R. 1991. *Manufacturing Knowledge: A History of the Hawthorne Experiments*. Cambridge University Press.

Gilmer, P. 1990. "What It's Really like to Be a Woman in Science." In *More History and Philosophy of Science in Science Teaching*, edited by D.E. Herget, pp. 133–35. Tallahassee: Florida State University Department of Education.

Gilovich, T. 1991. *How We Know What Isn't So*. New York, NY: Free Press.

Gilovich, T., Griffin, D., and Kahneman, D. 2002. *Heuristics and Biases: The Psychology of Intuitive Judgment*. New York: Cambridge University Press.

Giunta, C.J. 2001. "Argon and the Periodic System: the Piece That Would Not Fit." *Foundations of Chemistry* 3: pp. 105–28.

Glanz, J. 2000. "New Tactic in Physics: Hiding the Answer." *New York Times* (August 8): D1, D4.

Glen, W. 1982. *The Road to Jaramillo*. Stanford University Press.

Goldman, A.I. 1986. *Epistemology and Cognition*. Harvard University Press.

Goldstein, N.J., Martin, S.J., and Cialdini, R.B. 2008. *Yes!: Fifty Scientifically Proven Ways to be Persuasive*. New York: Free Press.

Goodman, N. 1978. *Ways of Worldmaking*. New York: Hackett.

Goodman, N. 1965. *Fact, Fiction, and Forecast*, 2nd ed. Indianapolis: Bobbs-Merrill.

Goodstein, D. 2010. *On Fact and Fraud: Cautionary Tales from the Front Lines of Science*. Princeton, NJ: Princeton University Press.

Gould, S.J. 1978. "Morton's Ranking of Races by Cranial Capacity." *Science* 200: pp. 503–09.

Gould, S.J. 1981. *The Mismeasure of Man*. New York, NY: W.W. Norton.

Gould, S.J. 1985a. "Carrie Buck's Daughter." In *The Flamingo's Smile*. New York: W.W. Norton.

Gould, S.J. 1985b. "The Freezing of Noah." In *The Flamingo's Smile*, pp. 114–25. New York: W.W. Norton.

Gould, S.J. 1989. *Wonderful Life*. New York: W.W. Norton.

Grant, J. 2006. *Discarded Science*. Wisley, UK: Facts, Figures & Fun/Artists' and Photographers' Press.

Grant, J. 2007. *Corrupted Science: Fraud, Ideology, and Politics in Science*. Wisley, UK: Facts, Figures & Fun/Artists' and Photographers' Press.

Gratzer, W. 2000. *The Undergrowth of Knowledge: Delusion, Self-Deception and Human Frailty*. Oxford, UK: Oxford University Press.

Green, R.E., Briggs, A.W., Krause, J., Prüfer, K., Burbano, H.A., Siebauer, M., Lachmann, M., and Pääbola, S. 2009. "The Neandertal Genome and Ancient DNA Authenticity." *EMBO Journal* 28: pp. 2494–502.

Gregor, T., Fujimoto, K., Masaki, N., and Sawai, S. 2010. "The Onset of Collective Behavior in Social Amoebae." *Science* 328: pp. 1021–25.

Guttinger, S. and Love, A.C. 2019. "Characterizing Scientific Failure." *EMBO Reports* 20: e48765.

Guyot, A. 1853. *The Earth and Man*. Boston: Gould & Lincoln.

Hacking, I. 1984. *Representing and Intervening*. Cambridge, UK: Cambridge University Press.

Hacking, I. 1990. *The Taming of Chance*. Cambridge, UK: Cambridge University Press.

Hallam, A. 1989. *Great Geological Controversies*, 2nd ed. Oxford, UK: Oxford University Press.

Hanson, N. R. 1958. *Patterns of Discovery*. Cambridge: Cambridge University Press.

Hanyu, N., Kuchino, Y., Nishimura, S., and Beier, H. 1986. "Dramatic Events in Ciliate Evolution: Alteration of UAA and UAG Termination Codons to Glutamine Codons Due to Anticodon Mutations in Two Tetrahymena tRNAsGln." *EMBO Journal* 5: pp. 1307–11.

Haraway, D. 1989. *Primate Visions: Gender, Race, and Nature in the World of Modern Science*. New York: Routledge.

Harding, S. 1991. *Whose Science? Whose Knowledge?: Thinking from Women's Lives*. Cornell University Press.

Harman, O. and Dietrich, M.R. 2008. *Rebels, Mavericks, and Heretics in Biology*. New Haven: Yale University Press.

Harman, O. and Dietrich, M.R. 2013. *Outsider Scientists*. University of Chicago Press.

Harold, F.M. 1986. *The Vital Force: The Study of Bioenergetics*. New York: W.H. Freeman.

Harris, R. 2017. *Rigor Mortis*. Basic Books.

Haygarth, J. 1800. *Of the Imagination, as a Cause and as a Cure of Disorders of the Body: Exemplified by Fictitious Tractors, and Epidemical Convulsions*. Bath: J. Cruttwell.

Heering, P. 1992a. "On Coulomb's Inverse Square Law." *American Journal of Physics* 60: pp. 988–96.

Heering, P. 1992b. "On J. P. Joule's Determination of the Mechanical Equivalent of Heat." In *The History and Philosophy of Science in Science Education*, edited by S. Hills, Vol. 1, pp. 495–505. Kingston: Mathematics, Science, Technology and Teacher Education Group and Faculty of Education, Queen's University.

Heick, T. 2019. "The Cognitive Bias Codex: A Visual of 180+ Cognitive Biases." *Teachthought.com* website. https://www.teachthought.com/critical-thinking/cognitive-biases

Hein, P. 1969. *Grooks*. New York: Doubleday.

Held, J. Jan. 22, 2020. "Texas A&M Slaps Back at Harvard Critics of its Beef-industry-backed Research." *Texas Monthly*. https://www.texasmonthly.com/news-politics/texas-am-slaps-back-at-harvard-critics-of-its-beef-industry-backed-research/

Hempel, C.G. 1966. *Philosophy of Natural Science*. Englewood Cliffs, NJ: Prentice Hall.

Herbert, S. 1974/1977. "The Place of Man in the Development of Darwin's Theory of Transmutation. Parts 1 and 2." *Journal of the History of Biology* 7: pp. 217–58 and 10: pp. 155–227.

Heron, D. 1913. *Mendelism and the Problem of Mental Defect I. A Criticism of Recent American Work*. Biometric Laboratory Publications, Series: Questions of the Day and the Fray, No. 7. London: Dulau.

Herr, H.W. 2005. "Franklin, Lavoisier, and Mesmer: Origin of the Controlled Clinical Trial." *Urologic Oncology* 23: pp. 346–51.

Herschel, J.F.W. 1830/1987. *A Preliminary Discourse on the Study of Natural Philosophy*. Reprint. Chicago, IL: University of Chicago Press.

Hinegardner, R.T. and Engelberg, J. 1963. "Rationale for a Universal Genetic Code." *Science* 142: pp. 1083–55.

Hirsh, R.F. 1981. "A Conflict of Principles: The Discovery of Argon and the Debate over its Existence." *Ambix* 28: pp. 121–30.

Holmberg, S.D. 2008. *Scientific Errors and Controversies in the U.S. HIV/AIDS Epidemic*. Westport, CT: Praeger.

Holmes, F.L. 1962. "From Elective Affinities to Chemical Equilibria: Berthollet's Law of Mass Action." *Chymia* 8: pp. 105–45.

Holmes, F.L. 1985. *Lavoisier and the Chemistry of Life*. Madison: University of Wisconsin Press.

Holmes, F.L. 2000. "The 'Revolution in Chemistry and Physics': Overthrow of a Reigning Paradigm or Competition Between Contemporary Research Programs?" *Isis* 91: pp. 735–53.

Holton, G. and Mack, A. (Eds.). 2005. *Errors: Consequences of Big Mistakes in the Natural and Social Sciences. Social Research* 71(1).

Hom, D.C., Lederman, L.M., Paar, H.P., Snyder, H.D., et al. 1976. "Observation of High Mass Dilepton Pairs in Hadron Collisions at 400 GeV." *Physical Review Letters* 36: pp. 1236–39.

Hon, G. 1987. "H. Hertz: 'The Electrostatic and Electromagnetic Properties of the Cathode Rays Are Either Nil or Very Feeble.' (1883): A Case-study of an Experimental Error." *Studies in History and Philosophy of Science, Part A* 18(3): pp. 367–82.

Hon, G. 1989a. "Franck and Hertz versus Townsend: A Study of Two Types of Experimental Error." *Historical Studies in the Physical and Biological Sciences* 20(1): pp. 79–106.

Hon, G. 1989b. "Towards a Typology of Experimental Errors: An Epistemological View." *Studies in History and Philosophy of Science* 20: pp. 469–504.

Hon, G. 1995. "Going Wrong: to Make a Mistake, to Fall into an Error." *The Review of Metaphysics* 49(1): pp. 3–20.

Hon, G. 2009. "Error: The Long Neglect, the One-sided View, and a Typology." In Hon et al., pp. 11–26.

Hon, G., Schickore, J, and Steinle, F. (Eds.) 2009. *Going Amiss in Experimental Research. Boston Studies in the Philosophy of Science, No. 267.* Dordrecht: Springer.

Hook, E.B. (Ed.). 2002. *Prematurity in Scientific Discovery: On Resistance and Neglect.* Berkeley: University of California Press.

Hornix, W. 1988. "Chemical Affinity in the 19th Century and Newtonianism." In *Newton's Scientific and Philosophical Legacy*, edited by P.B. Scheurer and G. Dehrock, pp. 201–15. Dordrecht: Kluwer.

Hubel, D.H. 1981. "Evolution of Ideas on the Primary Visual Cortex, 1955–1978: a Biased Historical Account." Nobel lecture. https://www.nobelprize.org/uploads/2018/06/hubel-lecture.pdf

Huckra, J.P. 1992. "The Hubble Constant." *Science* 256: pp. 321–25.

Huckra, J.P. 2008. "The Hubble Constant." https://www.cfa.harvard.edu/~dfabricant/huchra/hubble

Hudson, R.G. 1999. "Mesosomes: A Study in the Nature of Experimental Reasoning." *Philosophy of Science* 66: pp. 289–309.

Hudson, R. 2014. *Seeing Things.* Oxford University Press.

Huff, D. 1993. *How to Lie with Statistics.* New York: W.W. Norton.

Hull, D. 1988. *Science as a Process.* Chicago: University of Chicago Press.

Hunter, P. 2008. "The Paradox of Model Organisms." *EMBO Reports* 9: pp. 717–20.

Hurd, R. 2007. "Vitamin C and Colds." *National Institutes of Health Medical Encyclopedia.* URL: www.nlm.nih.gov/medlineplus/ency/article/002145.htm.

Ioannidis, J. 2005. "Why Most Published Research Findings Are False." *PLOS Medicine* 2(8): e124.

Jacoby, I. and Simopoulos, A.P. 1986. "Nih Consensus Conferences: Guidelines and Goals." *Journal of Nutrition* 118: pp. 312–16.

Janis, A.I. and Horowitz, T. (Eds.). 1994. *Scientific Failure.* Rowman and Littlefields.

Jastrow, J. (Ed.). 1936. *The Story of Human Error.* New York: D. Appleton-Century.

Johnson, G. 2002. "At Lawrence Berkeley, Physicists Say a Colleague Took Them for a Ride." *New York Times*, (October 15), F1.

Johnson, S. 2000. *Emergence.* New York: Scribner.

Johnson, S. 2006. *The Ghost Map.* New York: Riverhead Books.

Johnston, B.C., Zeraatkar, D., Han, M.A., Vernooij, R.W.M., Valli, C., El Dib, R., et al. 2019. "Unprocessed Red Meat and Processed Meat Consumption: Dietary Guideline Recommendations from the Nutrients Consortium." *Annals of Internal Medicine* 171: pp. 756–64. doi: 10.7326/M19-1621

Jongsma, M.L.M., Berlin, I., and Neefjes, J. 2015. "On the Move: Organelle Dynamics During Mitosis." *Trends in Cell Biology* 25(3): pp. 112–24.

Joseph, J. 2006. *The Missing Gene: Psychiatry, Heredity, and the Fruitless Search for Genes.* Sanford, NC: Algora.

Judson, H.F. 1980. *The Search for Solutions.* New York: Holt, Rinehart, Winston.

Judson, H.F. 2004. *The Great Betrayal: Fraud in Science.* Houghton Mifflin Harcourt.

Jungck, J. and Allchin, D. 2023. "Was Genetic Coding a Frozen Accident?" *American Biology Teacher* 85: pp. 550–3.

Kahan, D.M. 2013. "Ideology, Motivated Reasoning, and Cognitive Reflection." *Judgment and Decision Making* 8: pp. 407–24.

Kahneman, D. 2011. *Thinking, Fast and Slow.* New York: Farrar, Straus and Giroux.

Kaiser, J. 2002. "Software Glitch Threw off Mortality Estimates.' *Science* 292: pp. 1945–46.

Kantorovich, A. 1993. *Scientific Discovery: Logic and Tinkering.* Albany: State University of New York Press.

Kaptchuk, T.J. 1998. "Intentional Ignorance: A History of Blind Assessment and Placebo Controls in Medicine." *Bulletin of the History of Medicine.* 2(3): pp. 389–433

Kaptchuk, T.J. 2000. *The Web that Has No Weaver.* Chicago: Contemporary Books.

Kaptchuk, T.J., Kerr, C.E., and Zanger, A. 2009. "The Art of Medicine: Placebo Controls, Exorcisms, and the Devil." *The Lancet* 374: pp. 1234–35.

Karakas, A. 2022. *Technical Artefacts and the Problem of Malfunction.* Ph.D. diss. Budapest: Eötvös Loránd University.

Karakas, A. and Tuboly, A. (forthcoming). *Scientific Mistakes and Mistaken Science: Malfunction, Error, Failure.* Dordrecht: Springer.

Karstens, B. 2014. "The Lack of a Satisfactory Conceptualization of the Notion of Error in the Historiography of Science." In *Error and Uncertainty in Scientific Practice*, edited by M. Boumans, G. Hon, and A. Petersen, pp. 13–38. London: Pickering & Chatto.

Kauffman, G.B., and Adloff, J.-P. 2008. "Fred Allison's Magneto-optic Search for Elements 85 and 87." *The Chemical Educator* 13: pp. 358–64.

Kean, S. 2017. "The Chemist Who Thought He Could Harness Hurricanes." *The Atlantic* (Sept. 5). https://www.theatlantic.com/science/archive/2017/09/weather-wars-cloud-seeding/538392/

Keeling, C. 1998. "Rewards and Penalties of Monitoring the Earth." *Annual Review of Energy and the Environment* 23: pp. 25–82.

Keeling, C.D., Adams, J.A., Jr., Ekdahl, C.A., Jr., and Guenther, P.R. 1976. "Atmospheric Carbon Dioxide Variations at the South Pole." *Tellus* 28: pp. 552–64. doi: 10.3402/tellusa.v28i6.11.

Keller, E.F. 1983. "The Force of the Pacemaker Concept in Theories of Aggregation in Cellular Slime Mold." *Perspectives in Biology and Medicine* 26: pp. 515–21.

Kelley, T. 2001. *The Art of Innovation.* New York: Broadway Business.

Kennefick, D. 2019. *No Shadow of a Doubt.* Princeton University Press.

Kestenbaum, D. 2000. "Science Bloopers." *All Things Considered* (28 December). Washington, DC: National Public Radio. http://www.npr.org/templates/story/story.php?storyId=1116224.

Kevles, D. 1998. *In the Name of Eugenics.* Cambridge: Harvard University Press.

Keyes, M. 1999a. "The Prion Challenge to the "Central Dogma" of Molecular Biology, 1965–1991: Part I: Prelude to Prions." *Studies in History and Philosophy of Science Part C: Studies in History and Philosophy of Biological and Biomedical Sciences* 30: pp. 1–19.

Keyes, M. 1999b. "The Prion Challenge to the "Central Dogma" of Molecular Biology, 1965–1991: Part II: The Problem with Prions." *Studies in History and Philosophy of Science Part C: Studies in History and Philosophy of Biological and Biomedical Sciences* 30: pp. 181–218.

Kicinski, M., Springate, D.A., and Kontopantelis, E. 2015. "Publication Bias in Meta-analyses from the Cochrane Database of Systematic Reviews." *Statistics in Medicine* 34: pp. 2781–95.

Kida, T. 2006. *Don't Believe Everything You Think: The 6 Basic Mistakes We Make in Thinking.* New York: Prometheus.

Kidd, M. and Modlin, I.M. 1998. "A Century of *Helicobacter pylori*: Paradigms Lost – Paradigms Regained." *Digestion* 59: pp. 1–15.

Kim, M.G. 2008. "The 'Instrumental' Reality of Phlogiston." *HYLE* 14: pp. 27–51.

Kim, M.G. 2003. *Affinity, That Elusive Dream.* MIT Press.

Kimberlin, R.H. 1982. “Reflections on the Nature of the Scrapie Agent.” *Trends in Biochemical Studies* 7: pp. 392–94.

Kipnis, N. 1987. “Luigi Galvani and the Debate on Animal Electricity, 1791–1800.” *Annals of Science* 44: pp. 107–42.

Kitcher, P. 1993. *The Advancement of Science.* Oxford University Press.

Klein, J.R. and Roodman, A. 2005. “Blind Analysis in Nuclear and Particle Physics.” *Annual Review of Nuclear Particle Science* 55: pp. 141–63.

Kohler, R. 1994. *Lords of the Fly*: Drosophila *Genetics and the Experimental Life.* University of Chicago Press.

Kohn, A. 1988. *False Prophets: Fraud and Error in Science and Medicine.* 2nd ed. Blackwell.

Kolata, G. 2019. “Eat Less Red Meat, Scientists Said. Now Some Believe That Was Bad Advice.” *New York Times* (Sept. 30): p. A1.

Kosso, P. 1992. *Reading the Book of Nature.* Cambridge: University of Cambridge Press.

Kottler, M.J. 1974. “From 48 to 46: Cytological Technique, Preconceptions, and the Counting of Human Chromosomes.” *Bulletin of the History of Medicine* 48: pp. 465–502.

Kozlov, M.V., Zverev, V., and Zvereva, E.L. 2014. “Confirmation Bias Leads to Overestimation of Losses of Woody Plant Foliage to Insect Herbivores in Tropical Regions.” *PeerJ* 2: e709.

Kragh, H. 2008. *The Moon that Wasn't: The Saga of Venus' Spurious Satellite.* Birkhauser.

Krumholz, H., Ross, J.S., Presler, A.H., and Egilman, D.S. 2007. “What Have We Learned from Vioxx?” *British Medical Journal* 334: pp. 120–23.

Krzycki, J.A. 2005. “The Direct Genetic Encoding of Pyrrolysine.” *Current Opinion in Microbiology* 8: pp. 706–12.

Kuhn, T.S. 1958. “The Caloric Theory of Adiabatic Compression.” *Isis* 49: pp. 132–40.

Kuhn, T.S. 1970. *The Structure of Scientific Revolutions*, 2nd ed. Chicago: University of Chicago Press.

Kuhn, T.S. 1977. *The Essential Tension.* University of Chicago Press.

Kuhn, T.S. 1990. “The Road Since Structure.” In *PSA: Proceedings of the Biennial Meeting of the Philosophy of Science Association*, Volume 1990, Issue 2: Volume Two: Symposia and Invited Papers 1990, pp. 1–13.

Kunda, Z. 1990. “The Case for Motivated Reasoning.” *Psychological Bulletin* 108: pp. 480–98.

Lai, K.K.R. and Medina, J. 2023. “An American Puzzle: Fitting Race in a Box.” *New York Times* (October 16). https://www.nytimes.com/interactive/2023/10/16/us/census-race-ethnicity.html

Lakatos, I. 1978. *The Methodology of Scientific Research Programmes: Philosophical Papers*, Vol. 1. Cambridge, UK: Cambridge University Press.

Langmuir, I. 1950. “Control of Precipitation from Cumulus Clouds by Various Seeding Techniques.” *Science* 112: pp. 35–41.

Langmuir, I. 1951. “Cloud Seeding by Means of Dry Ice, Silver Iodide, and Sodium Chloride.” *Transactions of the New York Academy of Sciences* 14: pp. 40–44.

Langmuir, I. 1953. “Pathological Science” [transcript]. Colloquium at The Knolls Research Laboratory, December 18, 1953. Transcribed and edited by R.N. Hall. https://www.cs.princeton.edu/~ken/Langmuir/langmuir.htm

Langmuir, I. 1989. “Pathological Science.” *Physics Today* 42(10): pp. 36–48.

Latour, B. 1987. *Science in Action.* Cambridge, MA: Harvard University Press.

Latour, B. and Woolgar, S. 1979. *Laboratory Life*. Princeton University Press.
Laudan, L. 1977. *Progress and Its Problems*. Berkeley: University of California Press.
Laudan, L. 1981. "A Confutation of Convergent Realism." *Philosophy of Science* 48: pp. 19–49.
Laudan, R. 1980. "The Method of Multiple Working Hypotheses and the Development of Plate Tectonic Theory." In *Scientific Discovery: Case Studies*, edited by T. Nickles, pp. 331–43. Dordrecht: D. Reidel.
Lederman, L.M. 1978. "The Upsilon Particle." *Scientific American* 239(4): pp. 72–80.
Lederman, L. 1997. "The Discovery of the Upsilon, Bottom Quark, and B Mesons." In *The Rise of the Standard Model*, edited by L. Hoddeson, L. Brown, M. Riordan, and M. Dresden, pp. 101–13. Cambridge: Cambridge University Press.
Leek, J.T. and Peng, R.D. 2015. "What is the Question?" *Science* 347: pp. 1314–15.
Le Grand, H.E. 1972. "Lavoisier's Oxygen Theory of Acidity." *Annals of Science* 29: pp. 1–18.
Le Grand, H.E. 1974. "Ideas on the Composition of Muriatic Acid and Their Relevance to the Oxygen Theory of Acidity." *Annals of Science* 31: pp. 213–25.
Lehman, C. 2010. "Innovation in Chemistry Courses in France in the Mid-eighteenth Century: Experiments and Affinities." *Ambix* 57: pp. 3–26.
Leighton, G. and McKinley, P.L. 1930. *Milk Consumption and the Growth of School Children*. Edinburgh: His Majesty's Stationary Office.
Levenson, T. 2015. *The Hunt for Vulcan*. New York: Random House.
Levere, T.H. 2001. *Transforming Matter*. Baltimore: Johns Hopkins University Press.
Lewandowsky, S. and Cook, J. 2020. *The Conspiracy Theory Handbook*. http://sks.to/conspiracy.
Levins, R. 1966. "The Strategy of Model Building." *American Scientist* 54(4): pp. 421–31.
Lewin, R. 1988. *Bones of Contention*. New York: Touchstone.
Lewontin, R. 1991. *Biology as Ideology: The Doctrine of DNA*. New York, NY: Harper Collins.
Lightman, A. and Gingerich, O. 1992. "When Do Anomalies Begin?" *Science* 55: pp. 690–95.
Lilienfeld, A.M. 1982. "*Ceteris paribus*: The Evolution of the Clinical Trial." *Bulletin of the History of Medicine* 56: pp. 1–18.
Lippincott, E.R., Cessac, G.L., Stromberg, R.R., and Grant, W.H. 1971. "Polywater — A Search for Alternative Explanations." *Journal of Colloid and Interface Science* 36: pp. 443–60.
Lippincott, E.R., Stromberg, R.R., Grant, W.H., and Cessac, G.L. 1969. "Polywater." *Science* 164: pp. 1482–87.
Liu, P.A. 2011. *Creating Controversy: Science Writers, Corporate Funders, and Non-expert Scientists in the Debate over Prions (1982–1997)*. Ph.D. diss. Toronto: University of Toronto.
Livio, M. 2013. *Brilliant Blunders: From Darwin to Einstein - Colossal Mistakes by Great Scientists That Changed Our Understanding of Life and the Universe*. New York: Simon & Schuster.
Longino, H. 1990. *Science as Social Knowledge: Values and Objectivity in Scientific Inquiry*. Princeton, NJ: Princeton University Press.
Losee, J. 2005. *Theories on the Scrap Heap: Scientists and Philosophers on the Falsification, Rejection and Replacement of Theories*. Pittsburgh, PA: University of Pittsburgh Press.

Lyons, S.L. 2009. *Species, Serpents, Spirits, and Skulls: Science at the Margins in the Victorian Age.* Albany: SUNY Press.

Macfarlane, G. 1984. *Alexander Fleming: The Man and the Myth.* Oxford University Press.

Machamer, P., Darden, L., and Craver, C. 2000. "Thinking about Mechanisms." *Philosophy of Science* 67: pp. 1–25.

Machamer, P., Pera, M., and Baltas, A. 2000. *Scientific Controversies.* Oxford University Press.

MacPherson, H.G. 1934. "An Investigation of the Magneto-optic Method of Chemical Analysis." *Physical Review* 47: pp. 310–15.

Magner, L.N. 2002. *A History of the Life Sciences*, 3rd ed. New York, NY: Marcel Dekker.

Makowski, H. 1962. *Problem of Sexual Dimorphism in Ammonites. Paleontologia Polonia*, No. 12. Warsaw: Polish Academy of Sciences.

Malpighi, M. [1661] 1929. "On the Lungs" [trans. by J. Young]. *Proceedings of the Royal Society of Medicine* 23: pp. 1–11.

Marcum, J. 2007. "Experimental Series and the Justification of Temin's DNA Provirus Hypothesis." *Synthese* 154: pp. 259–92.

Mari, L. 2003. "Epistemology of Measurement." *Measurement* 34: pp. 17–30.

Markowitz, G. and Rosner, D. 2002. *Deceit and Denial: The Deadly Politics of Industrial Pollution.* University of California Press.

Marks, H.M. 2003. "Epidemiologists Explain Pellagra: Gender, Race, and Political Economy in the Work of Edgar Sydenstricker." *Journal of the History of Medicine* 58: pp. 34–55.

Marshall, B.J. 2001. "One Hundred Years of Discovery and Rediscovery of *Helicobacter pylori* and its Association with Peptic Ulcer Disease." In *Helicobacter pylori: Physiology and Genetics*, edited by H.L.T. Mobley, G.L. Mendz, and S.L. Hazell, Chapter 3. Washington, DC: ASM Press. https://pubmed.ncbi.nlm.nih.gov/21290711/

Marshall, B. (Ed.). 2002. *Helicobacter Pioneers: Firsthand Accounts from the Scientists Who Discovered Helicobacters.* Victoria, Australia: Blackwell Science Asia.

Marshall, B. 2005. "Helicobacter Connections." [Nobel lecture]. https://www.nobelprize.org/uploads/2018/06/marshall-lecture.pdf

Martin, A. 2004. "Can't Anybody Count?: Counting as an Epistemic Theme in the History of Human Chromosomes." *Social Studies of Science* 34: pp. 923–48.

Martinez, A.A. 2011. *Science Secrets.* University of Pittsburgh Press.

Mathé, J. 1978. *Leonardo da Vinci: Anatomical Drawings.* Miller Graphics.

Maxham, J. 2025. *The Art of Troubleshooting.* https://artoftroubleshooting.com (accessed May 3, 2025).

Maxwell, J. 1995. "John Wesley Powell and Democracy: A Textual Politics for the Arid Region." *Journal of the Southwest* 37(3): pp. 482–94.

Mayo, D. 1996. *Error and the Growth of Experimental Knowledge.* University of Chicago Press.

Mayo, D. 1997. "Error Statistics and Learning from Error: Making a Virtue of Necessity." *Philosophy of Science* 64: pp. S195–S212.

Mayo, D. 2014. "Learning from Error: How Experiment Gets a Life (of its Own)." In Boumans, Hon and Peterson (2014), pp. 57–77.

Mayo, D. 2018. *Statistical Inference as Severe Testing.* Cambridge University Press.

McGarity, T.O. and Wagner, W.E. 2008. *Bending Science: How Special Interests Corrupt Public Health Research.* Cambridge, MA: Harvard University Press.

McIntyre, L. 2019. *The Skeptical Attitude*. Cambridge, MA: MIT Press.

McNutt, M. 2014. "Reproducibility." *Science* 343: p. 229.

McNutt, M. (2020, Dec. 16). "Self-correcting by Design." *Harvard Data Science Review* (#2.4).

Meldrum, A.N. 1934. "Lavoisier's Early Work in Science, 1763–1771." *Isis* 20(2): pp. 396–425.

Mendel, G. [1866] 1966. "Versuche über Pflanzenhybriden [Experiments on plant hybrids]." Reprinted in *Fundamanta Genetica*, edited by J. Kíížzenecký, pp. 57–92. Oosterhout: Anthropological Publications; Brno: Moravian Museum; and Prague: Czech Academy of Sciences. Translation reprinted in *The Origin of Genetics: A Mendel Source Book*, edited by C. Stern and E. Sherwood, pp. 1–48. San Francisco: W.H. Freeman. Translation by Druery and Bateson also available at MendelWeb: http://MendelWeb.org

Mercier, H. and Sperber, D. 2017. *The Enigma of Reason*. Cambridge: Harvard University Press.

Merton, R.K. 1973. *The Sociology of Science*. Chicago, IL: University of Chicago Press.

Micale, M.S. 1994. *Approaching Hysteria: Disease and Its Interpretations*. Princeton: Princeton University Press.

Michaels, D. 2008. *Doubt Is Their Product: How Industry's Assault on Science Threatens Your Health*. Oxford University Press.

Michaels, D. 2020. *The Triumph of Doubt*. Oxford University Press.

Mill, J.S. 1874. *A System of Logic*. New York, NY: Harper & Brothers.

Minsky, M. 1994. "Negative Expertise." *International Journal of Expert Systems* 7: pp. 13–19.

Mitchell, R.A., Butler, L.G., and Boyer, P.D. 1964. "The Association of Readily-soluble Bound Phosphohistidine from Mitochondria with Succinate Thiokinase." *Biochemical and Biophysical Research Communications* 16: p. 545.

Moore, B.N. and Parker, R. 2015. *Critical Thinking*, 11th ed. New York: McGraw Hill.

Moore, D. and Cook, P. 1973. *Good Evening*. New York: Samuel French.

Monmonier, M. 1991. *How to Lie with Maps*. Chicago, IL: University of Chicago Press.

Montgomery, K. 2010. Debating glacial theory, 1800–1870 [website]. St. Paul: SHiPS Resource Center. http://glacialtheory.net.

Morgan, T.H. 1925. *Evolution and Genetics*. Princeton: Princeton University Press.

Morley, L.W. 2003. "The Zebra Pattern." In *Plate Tectonics: An Insider's History of the Modern Theory of the Earth*, edited by N. Oreskes, pp. 67–85. Cambridge: Westview Press.

Morrell, V. 1993. "Seeing Nature Through the Lens of Gender." *Science* 260: pp. 428–29.

Morris, R.J. 1972. "Lavoisier and the Caloric Theory." *British Journal for the History of Science* 6: pp. 1–38.

Motulsky, A.G. 2002. "The Works of Joseph Adams and Archibald Garrod: Possible Examples of Prematurity in Human Genetics." In *Prematurity in Scientific Discovery: On Resistance and Neglect*, edited by E.B. Hook, pp. 200–12. Berkeley: University of California Press.

Murchison, R.I. 1942. "On the Glacial Theory." *Edinburgh New Philosophical Journal* 33: pp. 124–39.

Natelson, D. 2018. "Looking Back on the Schön affair." Rice University. https://doi.org/10.25611/8P39-3K49

National Academies of Sciences, Engineering, and Medicine. 2019. *Reproducibility and Replicability in Science.* Washington, DC: The National Academies Press. https://doi.org/10.17226/25303.

National Institutes of Health. 2011, May 26. "NIH Stops Clinical Trial on Combination Cholesterol Treatment" [news release]. Bethesda, MD: Author.

National Research Council. 1997. *Possible Health Effects of Exposure to Residential Electric and Magnetic Fields.* Washington, DC: The National Academies Press. https://doi.org/10.17226/5155.

National Weather Service. 2020. "8 Inch Rain Gauge." https://www.weather.gov/iwx/coop_8inch

Nature. 1895. "Argon." 51: pp. 337–38.

Nendick, J. Scrancher, D., and Usher, O. 2007. "Chlorine and Prout's Hypothesis." In *An Element of Controversy*, edited by H. Chang and C. Jackson, pp. 73–104. British Society of the History of Science.

Nestle, M. 2018. *Unsavory Truth.* New York: Basic Books.

Newton, I. 1687. *Principia Mathematica.* London.

New York Times. January 29, 1905. "Doomsday postponed." p. 6.

Nickerson, R.S. 1998. "Confirmation Bias: A Ubiquitous Phenomenon in Many Guises." *Review of General Psychology* 2: pp. 175–220.

Nimpf, S. and Keays, D.A. 2020. "Why (And How) We Should Publish Negative Data." *EMBO Reports* 21: e49775.

Nisbett, R. and Ross, L. 1980. *Human Inference: Strategies and Shortcomings of Human Judgement.* Englewood Cliffs, NJ: Prentice-Hall.

Nissen, S.E., Magidson, T., Gross, K., and Bergstrom, C.T. 2016. "Research: Publication Bias and the Canonization of False Facts." *eLife.* https://doi.org/10.7554/eLife.21451

Noonan, J.P. 2010. "Neanderthal Genomics and the Evolution of Modern Humans." *Genome Research* 20: pp. 547–53.

Nowak, M.A., Tarnita, C.E., and Wilson, E.O. 2010. "The Evolution of Eusociality." *Nature* 466: pp. 1057–62.

Numbers, R. (Ed.). 2009. *Galileo Goes to Jail.* Harvard University Press.

The Nutrition Source. 2019. "New "Guidelines" Say Continue Red Meat Consumption Patterns, but Recommendations Contradict Evidence." Harvard School of Public Health. https://www.hsph.harvard.edu/nutritionsource/2019/09/30/flawed-guidelines-red-processed-meat

Nye, M.J. 1980. "N-rays: An Episode in the History and Psychology of Science." *Historical Studies in the Physical Sciences* 11: pp. 125–56.

O'Connor, A. Jun. 16, 2020. "Scientists Examine Vitamin D in Fight vs. Covid-19." *New York Times*, D6.

O'Connor, A. Dec. 20, 2016. "Study Tied to Food Industry Tries to Discredit Sugar Guidelines." *New York Times*, B1.

O'Connor, A. Feb. 3, 2020. "Meat Increases Heart Risks, Latest Study Concludes." *New York Times.*

Odling, W. 1871. "On the Revived Theory of Phlogiston." *Proceedings, Royal Society of Great Britain* 6: pp. 315–25.

Ohanian, H.C. 2008. *Einstein's Mistakes.* W.W. Norton.

Oppenheimer, J.R. 1984. *Uncommon Sense.* Edited by N. Metropolis, G.-C. Rota and D. Sharp. Boston: Birkhäuser.

O'Loughlin, R. 2023. "Diagnosing Errors in Climate Model Intercomparisons." *European Journal for Philosophy of Science* 13(2): pp. 1–29.

Oreskes, N. 1999. *The Rejection of Continental Drift*. New York, NY: Oxford University Press.

Oreskes, N. 2019. *Why Trust Science?* Princeton University Press.

Orna, M.V., Fontani, M., and Costa, M. 2017. "Lost Elements: The All-American Errors." In *Elements Old and New: Discoveries, Developments, Challenges, and Environmental Implications*, edited by M.A. Benvenuto and T. Williamson, pp. 1–40. Washington, DC: American Chemical Society.

Osborne, J., Pimentel, D., Alberts, B., Allchin, D., Barzilai, S., Bergstrom, C., Coffey, J., Donovan, B., Kivinen, K., Kozyreva, A., and Wineburg, S. 2022. *Science Education in an Age of Misinformation*. Stanford, CA: Stanford University. https://sciedandmisinfo.stanford.edu

Otto, S. 2016. *The War on Science*. Minneapolis, MN: Milkweed Editions.

Overbye, D. 2010a. "Microbe Finds Arsenic Tasty; Redefines Life." *New York Times*, December 3, A1.

Overbye, D. 2010b. "Poisoned Debate Encircles a Microbe Study's Result." *New York Times*, December 14, D4.

Pääbo, S., Poinar, H., Serre, D., Jaenicke-Despres, V., Hebler, J., Rohland, N., Kuch, M., Krause J, Vigilant, L., and Hofreiter, M. 2004. "Genetic Analyses from Ancient DNA." *Annual Review of Genetics* 38: pp. 645–79.

Parker, W. 2009. "Distinguishing Real Results from Instrumental Artifacts: The Case of the Missing Rain." In *Going Amiss in Experimental Research*, edited by G. Hon, J. Schickore, and F. Steinle, pp. 161–77. Dordrecht: Springer.

Partington, J.D. and McKie, D. 1938–39. "Historical Studies on the Phlogiston Theory, III–IV." *Annals of Science* 3: pp. 336–71 and 4: pp. 112–49.

Pascal, B. 1663. *Traitez de L'Equilibre des Liquers et de la Pesanteur de la Masse de l'Air*. Paris: Guillame Desprez.

Pauling, L. and Corey, R.B. 1953. "A Proposed Structure for the Nucleic Acids." *Proceedings of the National Academy of Sciences (USA)* 39: pp. 84–97.

Pearl, J. 2009. *Causality: Models, Reasoning and Inference*, 2nd ed. Cambridge: Cambridge University Press.

Perry, S. 1988. "Consensus Development: A Historical Note." *International Journal of Technology Assessment in Health Care* 4: pp. 481–84.

Petersen, B. 1964. "Monarch Butterflies Are Eaten by Birds." *Journal of the Lepidopterists' Society* 18: pp. 165–69.

Peterson, N.S. and Jungck, J.R. 1974. "Problem-Posing, Problem-Solving and Persuasion in Biology Education." *Academic Computing* 6(2): pp. 14–7, 48–50.

Peterson, D. and Panofsky, A. 2021. "Self-correction in Science: The Diagnostic and Integrative Motives for Replication." *Social Studies of Science* 51(4): pp. 583–605.

Petroski, H. 1994. *Design Paradigms: Case Histories of Error and Judgment in Engineering*. Cambridge: Cambridge University Press.

Petroski, H. 2006. *Success through Failure: The Paradox of Design*. Princeton, NJ: Princeton University Press.

Petroski, H. 2012. *To Forgive Design: Understanding Failure*. Cambridge, MA: Belknap Press.

Piattelli-Palmarini, M. 1994. *Inevitable Illusions: How Mistakes of Reason Rule Our Minds*. New York: John Wiley and Sons.

Piccolino, M. 1998. "Animal Electricity and the Birth of Electrophysiology: The Legacy of Luigi Galvani." *Brain Research Bulletin* 46: pp. 381–407.Pickering, A. 1995. *The Mangle of Practice*. Chicago: University of Chicago Press.

Pico, T. 2019. "Landscapes and Scientific Racism. GeoContext. https://geo-context.github.io

Platt, J.R. 1964. "Strong Inference." *Science* 146: pp. 347–53.

Plotkin, H. 1994. *Darwin Machines and the Nature of Knowledge.* Cambridge, MA: Harvard University Press.

Porter, T.M. 1986. *The Rise of Statistical Thinking, 1820–1900.* Princeton University Press.

Powell, J.W. 1883. "From Savagery to Barbarism." *Transactions of the Anthropoligical Society of Washington* 3: pp. 173–96.

Powell, J.W. 1888. "From Barbarism to Civilization." *The American Anthropologist* 1(2): pp. 97–124.

Prebble, J. and Weber, B. 2003. *Wandering in the Gardens of the Mind: Peter Mitchell and the Making of Glynn.* Cambridge, UK: Cambridge University Press.

Pritychenko, B. 2017. "A Nuclear Data Approach for the Hubble Constant Measurements." In *The European Physical Journal Conferences* 146, 01006.

Racine, V. 2013. "The Cuvier-Geoffroy Debate." Embryo Project Encyclopedia. Arizona State University School of Life Sciences. http://embryo.asu.edu/handle/10776/6276.

Rader, K. 2004. *Making Mice.* Princeton University Press.

Rasmussen, N. 1993. "Facts, Artifacts and Mesosomes: Practicing Epistemology with the Electron Microscope." *Studies in the History and Philosophy of Science* 24: pp. 227–65.

Rayleigh, L. 1895. "Argon." *Nature* 1337: pp. 159–64.

Regnault, H.V. 1847. "Relations des expériences entreprises par ordre de Monsieur le Ministre des Travaux Publics, et sur la proposition de la Commission Centrale des Machines a Vapeur, pour déterminer les principales lois et les données numériques qui entrent dans le calcul des machines a vapeur." *Mémoires de l'Académie Royale des Sciences de l'Institut de France, Vol. 21.*

Reich, E.S. 2009. *Plastic Fantastic: How the Biggest Fraud in Physics Shook the Scientific World.* New York: Palgrave Macmillan.

Reich, E.S. 2011. "Speedy Neutrinos Challenge Physicists." *Nature.* https://doi.org/10.1038/477520a

Reid, C.R., and Latty, T. 2016. "Collective Behaviour and Swarm Intelligence in Slime Moulds." *FEMS Microbiology Reviews* 40(6): pp. 798–806.

Rescher, N. 2007. *Error.* Pittsburgh: University of Pittsburgh Press.

Resnik, D. 2023. "Disclosing and Managing Non-financial Conflicts of Interest in Scientific Publications." *Research Ethics* 192: pp. 121–38.

Retraction Watch. March 15, 2023. "Top 10 Most Highly Cited Retracted Papers." https://retractionwatch.com/the-retraction-watch-leaderboard/top-10-most-highly-cited-retracted-papers

Rheinberger, H.-J. 1997. *Towards a History of Epistemic Things.* Stanford University Press.

Rheinberger, H.-J. 2010. *An Epistemology of the Concrete.* Durham: Duke University Press.

Rheinberger, H.-J. 2023. *Split and Splice: A Phenomenology of Experimentation.* Chicago: University of Chicago Press.

Rigas, B., Feretis, C., and Papavassiliou, E.D. 1999. "John Lykoudis: An Unappreciated Discoverer of the Cause and Treatment of Peptic Ulcer Disease." *The Lancet* 354: pp. 1634–35.

Ritland, D. and Brower, L.P. 1991. "The Viceroy Butterfly Is Not a Batesian Mimic." *Nature* 350: pp. 497–98.

Roberts, R.M. 1989. *Serendipity: Accidental Discoveries in Science.* New York: John Wiley.
Rocke, A. 1984. *Chemical Atomism in the Nineteenth Century: From Dalton to Cannizzaro.* Columbus: Ohio State University Press.
Rodgers, J. 1991. "Mechanisms Mendel never knew." *Mosaic* 22(3): pp. 2–11.
Romero, F. 2016. "Can the Behavioral Sciences Self-correct? A Social Epistemic Study." *Studies in History and Philosophy of Science Part A* 60: pp. 55–69.
Rosenson, R.S. and Durrington, P. 2020. "HDL Cholesterol: Clinical Aspects of Abnormal Values. v.45.0." In *UpToDate*, edited by T.W. Post. Waltham, MA: UpToDate Inc. https://www.uptodate.com
Ross, L., Gardner, A., Hardy, N. and West, S.A. 2014. "Ecology, Not the Genetics of Sex Determination, Determines Who Helps in Eusocial Populations." *Current Biology* 23: pp. 2383–7.
Rothbart, D. and Slayden, S.W. 1994. "The Epistemology of a Spectrometer." *Philosophy of Science* 61: pp. 25–38.
Rousseau, D.L. 1970. "Polywater: Polymer or Artifact?" *Science* 167: pp. 1715–19.
Rousseau, D.L. 1992. "Case Studies in Pathological Science." *American Scientist* 80: pp. 54–63.
Royal Ontario Museum. 2020. "The Burgess Shale: *Hallucigenia sparsa.*" https://burgess-shale.rom.on.ca/en/fossil-gallery/view-species.php?id=60&m=3&
Rudwick, M. 1974. "Darwin and Glen Roy: A "Great Failure" in Scientific Method?" *Studies in the History and Philosophy of Science* 5: pp. 97–185.
Rudwick, M.J.S. 1985. *The Great Devonian Controversy.* Chicago: University of Chicago Press.
Ruff, K. 2015. "Scientific Journals and Conflict of Interest Disclosure: What Progress Has Been Made?" *Environmental Health* 14: p. 45.
Sackett, D.L. 1979. "Bias in Analytical Research." *Chronic Diseases* 32: pp. 51–63.
Saini, A. 2020. *Superior: The Return of Race Science.* HarperCollins India.
Sampson, M.T. 2006. "Neil Bartlett and Reactive Noble Gases." American Chemical Society. https://www.acs.org/content/dam/acsorg/education/whatischemistry/landmarks/bartlettnoblegases/neil-bartlett-reactive-noble-gases-commemorative-booklet.pdf
Sapp, J. 1987. *Beyond the Gene: Cytoplasmic Inheritance and the Struggle for Authority in Genetics.* New York: Oxford University Press.
SCENIHR (Scientific Committee on Emerging Newly Identified Health Risks). 2015. *Potential Health Effects of Exposure to Electromagnetic Fields* (EMF). https://ec.europa.eu/health/scientific_committees/emerging/docs/scenihr_o_041.pdf
Schickore, J. 2009. "Error as Historiographic Challenge: The Infamous Globule Hypothesis." In Hon et al. (2009), pp. 27–45.
Schiebinger, L. 1989. *The Mind Has No Sex?: Women in the Origins of Modern Science.* Cambridge: Harvard University Press.
Schiebinger, L. 1990. "The Anatomy of Difference: Race and Sex in Eighteenth-century Science." *Eighteenth-Century Studies* 23: pp. 387–405.
Schiebinger, L. 1993. *Nature's Body: Gender in the Making of Modern Science.* Boston: Beacon Press.
Schillinger, D. and Kearns, C. 2017. "Guidelines to Limit Added Sugar Intake: Junk Science or Junk Food?" *Annals of Internal Medicine* 166: pp. 305–06.
Schimmack, U. 2019. "Psychological Science Is Self-correcting" [blog]. Schimmack, (2019). Psychological science is self-correcting (accessed December 13, 2024).

Schofield, R. 2004. *The Enlightened Joseph Priestley*. University Park, PA: Pennsylvania State University Press.

Schufle, J.A. and Thomas, G. 1971. "Equivalent Weights from Bergman's Data on Phlogiston Content of Metals." *Isis* 62: pp. 499–506.

Schwarz, A. Bogdanich, W., and Williams, J. 2016. "N.F.L.'s Flawed Concussion Research and Ties to Tobacco Industry." *New York Times* (March 24): A1.

Schweitzer, B. 2015. "From Malfunction to Mechanism." *Philosophia Scientiæ* 19(1): pp. 21–34.

Science and Technology Committee. 2011. *Peer Review in Scientific Publications. Eighth Report of Session 2010–12*, Vol. I. London, UK: House of Commons.

Scott, J.H. 1958. "Qualitative Adequacy of Phlogiston." *Journal of Chemical Education* 29: pp. 360–63.

Seeley, T. 2010. *Honeybee Democracy*. Princeton University Press.

Seok, J., Warren, H.S., Cuenca, A.G., Mindrinos, M.N., et al. 2013. "Genomic Responses in Mouse Models Poorly Mimic Human Inflammatory Diseases." *Proceedings of the National Academy of Sciences, U.S.* 110: pp. 3507–12.

Service, R.A. 2023. "Superconducting Crystal May Be 'Revolutionary.'" *Science* 379: pp. 966–67.

Šešelja, D. and Straßer, C. 2014. "Epistemic Justification in the Context of Pursuit: A Coherentist Approach." *Synthese* 191: pp. 3111–41.

Shaffer, B.M. 1961. "The Cells Founding Aggregation Centres in the Slime Mould *Polysphondylium violaceum*." *Journal of Experimental Biology* 38: pp. 833–49.

Shaler, N.S. 1906. *Man and Nature in America*. New York: Charles Scribners.

Shapere, D., 1982. "The Concept of Observation in Science and Philosophy." *Philosophy of Science* 49: pp. 482–525.

Shapin, S. 1994. *A Social History of the Truth*. Chicago: University of Chicago Press.

Shipley, B. 2001. "'Had Lord Kelvin a Right?': John Perry, Natural Selection and the Age of the Earth, 1894–1895." In *The Age of the Earth: from 4004 BC to AD 2002*, edited by C.L.E. Lewis and S.J. Knell, pp. 91–106. London: The Geological Society.

Shrader-Frechette, K. 2014. *Tainted: How Philosophy of Science Can Expose Bad Science*. Oxford University Press.

Sibum, H.O. 1995. "Reworking the Mechanical Value of Heat: Instruments of Precision and Gestures of Accuracy in Early Victorian England." *Studies in History and Philosophy of Science* 26: pp. 73–106.

Siegel, E. 2016. "How Did We Fool Ourselves into Believing in a New Particle That Wasn't There? *Forbes*. https://www.forbes.com/sites/startswithabang/2016/08/09/how-did-we-fool-ourselves-into-believing-in-a-new-particle-that-wasnt-there/.

Siegfried, R. 1964. "The Phlogistic Conjectures of Humphry Davy." *Chymia* 9: pp. 117–24.

Silva, M.T. 1971. "Changes Induced in the Ultrastructure of Cytoplasmic and Intracytoplasmic Membranes of Several Gram-positive Bacteria by Variations in OsO_4 Fixation." *Journal of Microscopy* 93: pp. 227–32.

Silva, M.T., Sousa, J.C.F., Polónia, J.J, Macedo, M.A.E., and Parente, A.M. 1976. "Bacterial Mesosomes: Real Structures or Artifacts?" *Biochimica et Biophysica Acta* 443: pp. 92–105.

Silverstein, A.M. 1989. *A History of Immunology*. San Diego, CA: Academic Press.

Simon, B. 2001. *Undead Science: Science Studies and the Afterlife of Cold Fusion*. Rutgers University Press.

Simon, H. 1956. "Rational Choice and the Structure of the Environment." *Psychological Review* 63(2): pp. 129–38.

Simon, H. 1963. *The Sciences of the Artificial.* Cambridge: MIT Press.

Simonton, D.K. 2004. *Creativity in Science: Chance, Logic, Genius and Zeitgeist.* Cambridge University Press.

Sirri, G. March 12, 2012. "Measurements and Cross Checks on OPERA Timing Equipments." LGNS. https://slidetodoc.com/measurements-and-cross-checks-on-opera-timing-equipments

Skoglund, P., Northoff, B.H., Shunkovc, M.V., Dereviankoc, A.P., Pääbo, S., Krause, J., and Jakobsson, M. 2014. "Separating Endogenous Ancient DNA from Modern Day Contamination in a Siberian Neandertal." *Proceedings of the National Academy of Science, U.S.* 111: pp. 2229–34.

Slack, F.G. 1935. "The Magneto~optic Method of Chemical Analysis." *Journal of the Franklin Institute* 218(4): pp. 445–62.

Smith, D.L. 1989. Quest for fusion. *Engineering and Science* 52(4): pp. 2–15. https://calteches.library.caltech.edu/3616/1/Fusion.pdf (accessed November 14, 2024).

Smith, M.R. and Caron, J.-B. 2015. "Hallucigenia's Head and the Pharyngeal Armature of Early Ecdysozoans." *Nature* doi:10.1038/nature14573.

Snow, J. 1855. *On the Mode of Communication of Cholera*, 2nd ed. London: John Churchill. Facsimile edition available online at https://archive.org/details/b28985266.

Soek, J., Warren, H.S., Cuenca, A.G., et al. 2013. "Genomic Responses in Mouse Models Poorly Mimic Human Inflammatory Diseases." *Proceedings of the National Academy of Sciences USA* 110: pp. 3507–12.

Soennichsen, J. 2008. *Bretz's Flood: The Remarkable Story of a Rebel Geologist and the World's Greatest Flood.* Seattle: Sasquatch Books.

Solomon, M. 2001. *Social Empiricism.* Cambridge, MA: MIT Press.

Solomon, M. 2015. *Making Medical Knowledge.* Oxford University Press.

Sow, C., Yonezawa, S., Kitamura, S., Oka, T., Kuroki, K., Nakamura, F., and Maeno, Y. 2020. "Retraction." *Science* 368: p. 376.

Spanier, B.B., 1995. *Im/Partial Science: Gender Ideology in Molecular Biology.* Bloomington: Indiana University Press.

Spanos, A. 2010. "The Discovery of Argon: A Case for Learning from Data?" *Philosophy of Science* 77: pp. 359–80.

Spencer, F. 1990. *Piltdown: A Scientific Forgery.* Oxford University Press.

Stanford, P.K. 2006. *Exceeding Our Grasp: Science, History, and the Problem of Unconceived Alternatives.* Oxford: Oxford University Press.

Star, S.L. and Gerson, E.M. 1986. "The Management and Dynamics of Anomalies in Scientific Work." *Sociological Quarterly* 28: pp. 147–69.

Steele, S., Ruskin, G., Sarcevic, L., McKee, M., and Stuckler, D. 2019. "Are Industry-funded Charities Promoting 'Advocacy-led Studies' or 'Evidence-based Science'?: A Case Study of the International Life Sciences Institute." *Globalization and Health* 15: p. 36. doi: 10.1186/s12992-019-0478-6.

Stegenga, J. 2009. "Robustness, Discordance, and Relevance." *Philosophy of Science* 76: pp. 650–61.

Stegenga, J. 2024. "Justifying Scientific Progress." *Philosophy of Science* 91: pp. 543–60.

Stewart, G.R., Meisner, G.P., and Ku, H.C. 1982. "Specific Heats of the New High T_c Phosphide Superconductors HfRuP and ZrRuP." In *Superconductivity in d- and f-Band*

Metals: Proceedings of the IV. Conference, edited by W. Buckel and W. Weber, pp. 331–35. Karlsruhe: Kernforshungszentrum Karlsruhe.

Stolberg, M. 2003. "A Woman down to Her Bones: The Anatomy of Sexual Difference in the Sixteenth and Early Seventeenth Centuries." *Isis* 94: pp. 274–99.

Strassler, M. April 2, 2012. "OPERA: What Went Wrong" [blog]. https://profmattstrassler.com/articles-and-posts/particle-physics-basics/neutrinos/neutrinos-faster-than-light/opera-what-went-wrong

Student [W.S. Gosset]. 1931. "The Lanarkshire Milk Experiment." *Biometrika* 23: pp. 398–406.

Sudduth, W.M. 1978. "Eighteenth-century Identifications of Electricity with Phlogiston." *Ambix* 25: pp. 131–47.

Suppe, F. 1998. "The Structure of a Scientific Paper." *Philosophy of Science* 65: pp. 381–405.

Sutherland, S. 1992. *Irrationality: Why We Don't Think Straight.* New Brunswick, NJ: Rutgers University Press.

Tal, E. 2016. "Making Time: a Study in the Epistemology of Measurement." *British Journal for the Philosophy of Science* 67: pp. 297–335.

Tang-Martinez, Z. 2020. "The History and Impact of Women in Animal Behaviour and the ABS: a North American Perspective." *Animal Behaviour* 164: pp. 251–60.

Taton, R. 1957. *Reason and Chance in Scientific Discovery.* New York: Philosophical Library.

Taylor, P. 2005. *Unruly Complexity*. University of Chicago Press.

Thackray, A. 1970. *Atoms and Powers.* Cambridge, MA: Harvard University Press.

Thagard, P. 1999. *How Scientists Explain Disease.* Princeton, NJ: Princeton University Press.

Tharp, H. 2022. "Editorial Retraction." *Science* 378: p. 718.

Tharp, H. 2023. "It Matters Who Does Science." *Science* 380: p. 873.

Thorp, H.H., Vinson, V., and Yeston, J. 2023. "Strengthening the Scientific Record." *Science* 380: p. 13.

Time. August 28, 1950. "Science: Weather or Not." http://content.time.com/time/magazine/article/0.9171.813105.00.html

Time. October 19, 1970. "Doubts about Polywater." https://content.time.com/time/magazine/article/0,9171,944149,00.html

Trimble, V. 1996. H_0: The incredible shrinking constant, 1925–1975. *Publications of the Astronomical Society of the Pacific* 108: pp. 1073–82.

Trimble, V. and Mainz, V.V. 2017. "Who Got Moseley's Prize?" In *The Posthumous Nobel Prize in Chemistry: Correcting the Errors and Oversights of the Nobel Prize Committee*, edited by E.T. Strom and V.V. Mainz, Vol. 1, pp. 51–90. Washington, DC: American Chemical Society.

Tröhler, U. 2011. "Adolf Bingel's Blinded, Controlled Comparison of Different Anti-diphtheritic Sera in 1918." *Journal of the Royal Society of Medicine* 104: pp. 302–05.

Tucker, W.H. 1994. *The Science and Politics of Racial Research.* University of Illinois Press.

Tucker, W.H. 2007. *The Funding of Scientific Racism: Wickliffe Draper and the Pioneer Fund.* University of Illinois Press.

Tugend, A. 2011. *Better by Mistake.* Routledge.

Turnbull, D. 1989. *Maps are Territories: Science is an Atlas.* Chicago: University of Chicago Press.

Turner, E.H., Matthews, A.M., Linardatos, E., Tell, R.A., and Rosenthal, R. 2008. "Selective Publication of Antidepressant Trials and its Influence on Apparent Efficacy." *New England Journal of Medicine* 358: pp. 252–60.

Turro, N.J. 1999. "Toward a General Theory of Pathological Science." *21stC.* 3(#4). New York: Columbia University. http://www.columbia.edu/cu/21stC/issue-3.4/turro.html

Underwood, E. 2015. "Measuring Child Abuse's Legacy." *Science* 347: pp. 1408–7.

University of Cambridge. 2014. "Skulls in Print: Scientific Racism in the Transatlantic World." https://www.cam.ac.uk/research/news/skulls-in-print-scientific-racism-in-the-transatlantic-world

Van Fraasen, B. 1980. *The Scientific Image.* New York: Oxford University Press.

van der Linden, S. 2023. *Foolproof.* New York: W.W. Norton.

van Prooijen, J.-W. and van Vugt, M. 2018. "Conspiracy Theories: Evolved Functions and Psychological Mechanisms." *Perspectives on Psychological Science* 13: pp. 770–88.

van Wilgenburg, E. and Elgar, M.A. 2013. "Confirmation Bias in Studies of Nestmate Recognition: A Cautionary Note for Research into the Behaviour of Animals." *PLoS ONE* 8: e53548.

Vavakan, V.V. 2005. "A Physician Looks at the Death of Washington." *Early America Review* 6(1). http://www.earlyamerica.com/review/2005_winter_spring/washingtons_death.htm.

Vazire, S. and Holcombe, A.O. 2022. "What Are the Self-correcting Mechanisms of Science?" *Journal of General Psychology* 26: pp. 212–23.

Vickers, P. 2022. *Future-Proof Science.* Oxford University Press.

von Herrath, M. G, and Nepom, G.T. 2005. "Lost in Translation." *Journal of Experimental Medicine* 202(9): pp. 1159–62.

Wadman, M. 2021. "Journal Retracts Paper Claiming COIVD-19 Vaccines Kill." *Science* 373: p. 347.

Wagner, W. and Steinzor, R. 2006. *Rescuing Science from Politics.* Cambridge University Press.

Walcott, C.D. 1911. *Cambrian Geology and Paleontology II: No.5—Middle Cambrian Annelids.* Washington, DC: Smithsonian Institution.

Wall, J.D. and Kim, S.K. 2007. "Inconsistencies in Neanderthal Genome Sequences." *PLoS Genetics* 3(10), e175.

Waller, J. 2002. *Einstein's Luck: The Truth behind Some of the Greatest Scientific Discoveries.* New York: Oxford University Press.

Waals, J. van der. 1910/1967. "The Equation of State for Gases and Liquids." *Nobel Lectures: Physics* 1901–1921. Amsterdam: Elsevier. https://www.nobelprize.org/uploads/2018/06/waals-lecture.pdf

Walsh, B.D. and Riley, C.V. 1869. "Imitative Butterflies." *American Entymologist* 1: pp. 189–93.

Weber, B.H. 1991. "Glynn and the Conceptual Development of the Chemiosmotic Theory: A Retrospective and Prospective View." *BioScience Reports* 11: pp. 577–617.

Weber, M. 2004. *Philosophy of Experimental Biology.* Cambridge University Press.

Webster, C. 1965. "The Discovery of Boyle's Law, and the Concept of the Elasticity of Air in the Seventeenth Century." *Archive for History of the Exact Science* 2: pp. 441–502.

Wein, H. 2012. "Questions about HDL Cholesterol." NIH Research Matters. https://www.nih.gov/news-events/nih-research-matters/questions-about-hdl-cholesterol

Wein, H. 2016. "When HDL Cholesterol Doesn't Protect Against Heart Disease." NIH Research Matters. https://www.nih.gov/news-events/nih-research-matters/when-hdl-cholesterol-doesnt-protect-against-heart-disease

Weingurt, K.P. 2020. "Vanquishing False Idols, Then and Now." *Science* 376: p. 1312.

Weintraub, D.A. 2007. *Is Pluto A Planet? A Historical Journey Through the Solar System.* Princeton, NJ: Princeton University Press.

Werth, A.J. 2000. "A Kinematic Study of Suction Feeding and Associated Behavior in the Long-finned Pilot Whale, *Globic ephalamelas* (Traill)." *Marine Mammal Science* 16(2): pp. 299–314.

Whitt, L.A. 1992. "Indices of Theory Promise." *Philosophy of Science* 59(4): pp. 612–34.

Widjicks, E.F.M., Varelas, P.N., Gronseth, G.S., and Greer, D.M. 2010. "Evidence-based Guideline Update: Determining Brain Death in Adults: Report of the Quality Standards Subcommittee of the American Academy of Neurology." *Neurology* 74(23): pp. 1911–18. doi: 10.1212/WNL.0b013e3181e242a8

Widom, C.S., Czaja, S.J., and DuMont, K.A. 2015. "Intergenerational Transmission of Child Abuse and Neglect: Real or Detection Bias?" *Science* 347: pp. 1480–84.

Wilkes, J. (Ed.). 1810. *Encylcopedia Londonensis.* London: J. Adlard.

Williams, R. 1961. *Toward the Conquest of Beriberi.* Cambridge, MA: Harvard University Press.

Wilson, E.B., Jr. 1990. *An Introduction to Scientific Research.* New York: Dover.

Wilson, E.O. 2005. "Kin Selection as the Key to Altruism: its Rise and Fall." *Social Research* 72: pp. 159–66.

Wimsatt, W.C. 2007. *Re-engineering Philosophy for Limited Beings.* Cambridge, MA: Harvard University Press.

Winther, R.G. 2020. *When Maps Become the World.* University of Chicago Press.

Wise, J. 2021. "Covid-19: Vaccines Journal Retracts Controversial Paper after Editorial Board Members Quit." *BMJ* 374: n1726.

Witkowski, J.A. and Ingris, J.R. 2008. *Davenport's Dream: 21st Century Reflections on Heredity and Eugenics.* Cold Spring Harbor Laboratory Press.

Wood, R.W. 1904. "The *N*-rays." *Nature* 70: pp. 530–31.

Woodriff, R. 1935. *Allison's Method of Magneto-Optic Analysis* [Master's thesis]. Oregon State Agricultural College.

Woodward, J. 2003. *Making Things Happen: A Theory of Causal Explanation.* Oxford: Oxford University Press.

Wright S. 1968. *Evolution and the Genetics of Populations,* Vol. 1. University of Chicago Press.

Wynne, B. 1976. "C. G. Barkla and the J Phenomenon: A Case Study in the Treatment of Deviance in Physics." *Social Studies of Science* 6: pp. 307–47.

Yoh, J. 1998. "The Discovery of the b Quark at Fermilab in 1977: The Experiment Coordinator's Story." *AIP Conference Proceedings* 424: pp. 29–42.

Youngson, R.M. 1998. *Scientific Blunders.* New York, NY: Carroll & Graf.

Zichichi, A., Sirri, G., and Sioli, M. 2012. "OPERA: An Update." Presented at the LNGS Mini-Workshop on Neutrino Velocity (March 28. 2012). https://agenda.infn.it/getFile.py/access?contribId=9&resId=0&materialId=slides&confId=4876 (November 2, 2015).

Ziman, J. 1968. *Public Knowledge: The Social Dimension of Knowledge.* Cambridge: Cambridge University Press.

Ziman, J. 1978. *Reliable Knowledge.* Cambridge, UK: Cambridge University Press.

Ziman, J. 2000. *Real Science: What It Is, and What It Means.* Cambridge University Press.
Zimring, J. 2019. *What Science Is and How It Really Works.* New York: Cambridge University Press.
Zvereva, E.L. and Kozlov, M.V. 2019. "Biases in Studies of Spatial Patterns in Insect Herbivory." *Ecological Monographs* 89: e01361.

Index

For the benefit of digital users, indexed terms that span two pages (e.g., 52–53) may, on occasion, appear on only one of those pages.

Figures are indicated by an italic *f*.